Advances in Communication, Signal and Image Processing

Nabil Derbel · Quanmin Zhu
Editors

Advances in Communication, Signal and Image Processing

Springer

Editors
Nabil Derbel
Department of Electrical Engineering
University of Sfax
Sfax, Tunisia

Quanmin Zhu
School of Engineering
University of the West of England
Bristol, UK

ISBN 978-981-95-0400-8 ISBN 978-981-95-0401-5 (eBook)
https://doi.org/10.1007/978-981-95-0401-5

This Springer imprint is published by the registered company Springer Nature Singapore Pte Ltd.
The registered company address is: 152 Beach Road, #21-01/04 Gateway East, Singapore 189721, Singapore

Preface

The current book publishes recent developments in the fields of communication, signal and image processing covering a wide range of applications for monitoring, treatment, and assistance. This book provides insights in theory, applications, and perspectives relevant to the field of communication, signal and image processing, as well as the general paradigms and methodologies behind them. Each chapter provides an enrichment of understanding on a research topic along with a balanced treatment of the relevant theories, methods, or applications. It reports on the latest advances achieved in the corresponding field. This book is a good reference for graduate students, researchers, educators, engineers and scientists and includes 15 chapters, structured into four parts as the following.

Part I of this book focuses on communication and networking systems. It comprises five chapters:

- Chapter 1 is dedicated to the use of the artificial intelligence, the Internet of things, and the mixed reality technologies for the development of remote laboratories, allowing students to engage with laboratory experiments that include both virtual and physical units from any location and at any time, using their mobile phones, tablets, or personal computers. The object of this chapter consists on investigating the role that artificial intelligence and internet of things technologies play in the development process of remote laboratories that are based on mixed reality technology.
- Chapter 2. For this, this chapters begins with the historical evolution of SCADA. Then, it discuss about the Communication Protocols and Architectures in SCADA Systems, and it presents the integration of SCADA and HMI in PLC-Driven Industrial Systems. This allows to presents the modern SCADA architectures, the IIoT, and the edge computing.
- Chapter 3 is focused on energy consumption in direct and relayed D2D communications. The object this chapter is the maximization of the network's lifetime. This chapter studies the basic model of D2D communication, in order to compute the energy consumed by different proposed topologies. This study is completed by evaluating the total energy consumed by each topology.

- Chapter 4 concerns the optical communication R&D community that has reached to orbital angular momentum (OAM) of light to fastly transmit information via optical fiber. It deals with a new kind of OAM-fiber known as inverse raised cosine few mode fiber (IRC-FMF). It discusses the design flexibility, the optimization process, and most achievable results in terms of various performances metrics, including the inter-channel separation, the differential groups delays, the dispersion, the nonlinearity effects, the mode coupling under several perturbations, and the tolerance of the designed fiber to bending effects.
- Chapter 5 aims to highlight the design strategies and the recently reported results using specialty fibers. It presents and investigates the IGF key parameters that contribute to support robust OAM modes. This has been realised by designing optimizing four IGFs designs that enhance the intermodal separation between their supported channels. This chapter is intended to offer a comprehensive analysis and a boost aiming to motivate R&D optical fibers community to adopt/conduct further analysis using the designed fibers.

Part II is dedicated to video and signal processing, and comprises three chapters:

- Chapter 6 investigates the effect of the MLE (Maximum likelihood Estimation), Integer order moments, fractional-order moments, and $z \log(z)$ estimators on the detection performance of the GLRT-LTD procedure. A comparison between these approaches has been presented through simulations.
- Chapter 7 proposes an original idea to insert digital video in GSM interfaces without modifying or changing any parameters of the norm. The main task is to make reduction of the bit rate coming from the video encoder using H265 advanced norm by interposing a new design. Moreover, this chapter proposes a new architecture of the receiver used in the application of hand-held video reception.
- Chapter 8 focuses on the identification of MI of left and right hand movements based on EEG signals. New features based on wavelet transform (WT) and Mel-frequency cepstrum coefficients (MFCC) are proposed and are fed to support vector machine (SVM) classifier. Genetic Algorithm (GA) is employed as features selection method. The performance of the classification is evaluated and compared with the previous approaches.

Part III concerns researches on image processing, and comprises three chapters:

- Chapter 9 proposes watermarking approach based on DCT-PMF-DWT-SVD. The proposed algorithm details step by step all instructions that are used and the experimental results determine the efficiency of the proposal. Evaluation metrics used in this algorithm are the parameters PSNR (Peak Signal to Noise Ratio) and NC (Normalized Correlation).
- Chapter 10 proposes a hybrid filtering technique based on the Sigma filter and the non-local mean filter. Pixels used in this method are selected by similarity to the non-local filtering principle. Non-similar pixels in the search area having high patch distances are discarded. To estimate the variability of the pixels, the coefficient variation CV is used which select homogeneous pixels in the search

area. Compared with Lee filter and the improved Sigma filter, results show that the proposed hybrid filter enhances the filtering criteria in terms of speckle reduction and spatial detail preservation.

- Chapter 11 proposes an integer DTT-based image compression scheme. For several standard test images with varying statistical features, a new efficient quantisation matrix can be generated using the energy distribution of the approximated integer coefficients of the DTT. Furthermore, as the majority of integer DTT matrices are not normal, their diagonal matrices have to be merged into the quantisation process.

Part IV details several approaches applied to medical images, and comprises four chapters:

- Chapter 12 presents a meta-heuristic algorithm for the segmentation of MRI images and proposes to apply the technique of ant colonies to estimate the segmentations of brain MRI images from a public MR datasets of 30 MS patients, which are imaged with a 3T MR scanner with conventional sequences in order to optimize their overall rendering and to compare with consensual segmentation. In order to evaluate the performance of the proposed algorithm, it has been optimized by maximizing the intra-class variance criterion MVar.
- Chapter 13 describes a solution for early cancer disease detection. It consists of analyzing miRNAs expression for many cancer types using new machine learning method based on the Support Vector Machine Theory: Twin Support Vector Machine (twin SVM). The achieved AUC outperforms those obtained by methods existing in the state-of-the-art. Further, a comparison with SVM shows that TWSVM outperformed SVM in terms of accuracy, and AUC. Indeed, SVM with linear kernel reaches the best accuracy.
- Chapter 14 introduces the GGD technique as a tool for medical images texture analysis. It presents the GGD analysis technique with theoretical details, and then it describes related works before giving details about the GGD based approach proposed to detect lung nodules, bone sarcoma and breast abnormalities. Results given by using GGD analysis on these several tumors detection have been illustrated and compared to previous techniques.
- Chapter 15 presents a survey of several image processing techniques that have been developed since 2000 to detect and remove pectoral muscle from digital mammograms from the mini-MIAS database. In fact, breast cancer is one of the most deadly cancers that threatens women lives all over the world. Moreover, detecting breast lesions at an early stage is the best way to reduce breast cancer mortality rate as it usually increases recovering chances.

Sfax, Tunisia
Bristol, UK
March 2025

Nabil Derbel
Quanmin Zhu

Contents

Chapter 1
The Future of Remote Laboratories: Enhancing Personalized Learning Through the Internet of Things, Artificial Intelligence, and Mixed Reality

Kasim M. Al-Aubidy

Abstract Accelerated advancements in communication and computer technology have facilitated the creation of e-learning systems to meet the increasing demand for education and challenges caused by crises and wars. Interactive e-learning systems provide students with access to courses at any time and from any location through the internet. Applied science programs faced challenges with incorporating e-learning and blended learning into existing educational environments due to their dependence on laboratories and workshops for practical training. Consequently, efforts have been initiated to utilize modern technologies in the creation of laboratories accessible through the internet. In recent years, there has been an increasing interest in utilizing artificial intelligence, Internet of things, and mixed reality technologies for the development of remote laboratories. These applications allow students to engage with laboratory experiments that include both virtual and physical units from any location and at any time, using their mobile phones, tablets, or personal computers. The purpose of this chapter is to investigate the role that artificial intelligence and internet of things technologies play in the process of developing remote laboratories that are based on mixed reality technology. The laboratory is comprised of a variety of experiments that are designed to improve the practical portion of the specialization. To motivate students to improve their theoretical understanding and develop skills, the experiment is designed to make use of both real components that are present in the laboratory as well as simulated components that are part of an interactive environment. Artificial intelligence algorithms are employed to manage the educational process in the laboratory, assess the student's academic proficiency, and provide them with suitable educational resources for the lab. Additionally, the design of experiments for the mechatronics lab and the renewable energy lab will be discussed. These labs serve as an interactive educational platform that enables students to engage with experimental components, acquire knowledge regarding the operation of each element, carry out tests, and submit technical reports. By incorporating the Internet of things and mixed reality technologies into the design of remote

---

K. M. Al-Aubidy (✉)
Mechatronics Engineering Department, Tishk International University, Erbil, Iraq
e-mail: qasim.obaidi@tiu.edu.iq

N. Derbel and Q. Zhu (eds.), *Advances in Communication, Signal and Image Processing*,
https://doi.org/10.1007/978-981-95-0401-5_1

labs, an interactive environment is created that improves learning quality and skill acquisition. These labs give educational institutions the chance to work together to create innovative, high-caliber, and successful programs that resolve the problems caused by the availability and high cost of laboratory equipment, as well as the lack of technical staff needed by traditional labs.

Keywords Remote laboratory · E-learning · Mixed reality · Virtual reality · Internet of Things · Mechatronics · Laboratory · Renewable energy laboratory

1.1 Introduction

There has been a significant increase in the worldwide university student population over the past twenty years, pushed by demographic expansion and economic development. Fast technological advancements and their applications have increased the value of higher education. The student population is projected to increase from 235 million in 2023 to 470 million by 2034, reflecting an average yearly increase of around 11.93% (Precedence Research 2025). Despite an overall increase in university enrollment, female students in low-income countries accounted for just approximately 30% of undergraduate students in 2014 (UNESCO 2023). A considerable and ongoing demand for higher education can be seen across the world, as evidenced by available statistics. As of 2025, the global market for higher education was estimated to be worth approximately $929.47 billion, and it is anticipated that it will grow to approximately $2,557.93 billion by the year 2034, as given in Fig. 1.1. This growth is

Fig. 1.1 Higher education market size (Precedence Research 2025)

expected to be driven by several factors, including the requirement for policies that guarantee equitable access to opportunities for higher education and the growing demand for online learning.

E-learning is a successful educational strategy that offers a special chance to individuals who have not been able to fulfill their educational objectives in a manner that is suitable for their needs (Badawy et al. 2024; Al-Kashashneh and Al-Aubidy 2019). E-learning has become a key component of modern education systems over the last 10 years, particularly during the COVID-19 epidemic, due to its capacity to offer affordable academic programs at a low cost (Kreiter et al. 2020). The effectiveness of e-learning systems has not been very encouraging in applied disciplines, particularly engineering disciplines. This is mostly because these disciplines rely heavily on practical applications, laboratories, and interactive learning settings. In applied disciplines, the use of virtual classrooms as an alternative to traditional classrooms has a negative impact on the results of learning. Blended learning has therefore developed to integrate e-learning and conventional education under a single framework (Al-Tikriti and Al-Aubidy 2013). Due to the requirement for specialized laboratories that provide students with the opportunity to gain scientific information and skills, applied disciplines faced challenges when attempting to incorporate e-learning into their educational programs (Al-Aubidy et al. 2021). In many applied disciplines, the cost of constructing specialized laboratories is extremely costly. As a result, universities are unable to develop their present academic programs or even investigate the possibility of developing new ones that are in line with the demands of the market. To replace real laboratories, virtual laboratories must be developed using simulated components (Lee et al. 2022). On the other hand, these laboratories do not offer an interactive environment in the same way as conventional laboratories do, which make use of real components.

The Internet of Things (IoT) is a collection of physical objects equipped with sensors and software to facilitate data exchange and connect these objects to other devices and systems connected to the Internet. IoT technology provides a framework that facilitates the interconnection of applications, allowing objects to interconnect via advanced communication methods. IoT technology has been used in a variety of fields, such as manufacturing, agriculture, healthcare, smart cities, transportation, and more. Due to the significant role the IoT plays in gathering information and providing an effective environment for sharing it, it has emerged as an effective technology in modern educational systems, aiming to enhance the quality of education and facilitate student interaction with the learning platform.

With advances in computer and communications technology, virtual reality (VR) technologies have seen significant growth in many applications, using computer software to create an interactive virtual environment. Therefore, numerous VR-based systems have emerged to simulate flight training, microsurgery, and video games. VR can be leveraged in e-learning systems to simulate laboratory experiments, allowing students to interact with experimental components. Smartphones have also contributed to accessing and interacting with VR applications, not only in gaming but also in areas such as learning and training. Virtual reality and mixed reality technologies have significantly enhanced the role of laboratories in modern e-learning and blended learning systems, by facilitating access to actual experimental data using physical devices used in educational institutions (Al-Aubidy 2022). Statistics indicate that the global virtual reality market in education was valued at approximately $2.9 billion in 2022 and is expected to reach $14.2 billion by 2028 (Markets and Markets 2025). Major IT companies have invested heavily in virtual reality (VR) applications in the education sector, aiming to enhance learning experiences and improve outcomes. Meta (formerly Facebook) has established a $150 million fund to support the integration of VR into the education sector. Meta's VR platforms enable virtual field trips, museum visits, and other educational experiences (Kingson 2023). Microsoft has reported that a mixed reality application can reduce training times by approximately 30% and operating costs by 10% (Microsoft 2025). Google has previously introduced the Expeditions app to help students take virtual field trips to different destinations using virtual reality. Google has also developed affordable virtual reality solutions to help schools implement virtual reality experiences using smartphones (Robertson et al. 2015). Apple has developed more than 1,000 mixed reality applications for use in educational institutions (Lin et al. 2024). Lenovo has introduced its "VR Classroom 2" educational platform, which integrates hardware, software, and support to facilitate the use of virtual reality in education.

As a result, there has been a significant shift from the traditional e-learning model to e-learning enhanced by the Internet of Things (IoT), artificial intelligence (AI), and mixed reality (MR) technologies, which provide students with access to information, experiential learning, and an effective learning environment (Badawy et al. 2024). Several research papers (Al-Aubidy et al. 2021; Munir et al. 2022; Saleem and Al-Aubidy 2008) discuss the integration of virtual and mixed reality, as well as IoT, into the design of experiments for remote laboratories in applied disciplines. AI is expected to contribute significantly to the design and management of laboratory experiments, enabling students to obtain knowledge that matches their abilities, enhancing learning outcomes. Integrating AI with IoT and MR technologies provides an effective solution for developing e-learning and blended learning systems in applied disciplines that require laboratories and workshops. These technologies also contribute to providing e-learning systems and establishing remote laboratories at the lowest possible cost and open the way for cooperation between educational institutions to enter new disciplines that meet market requirements.

This chapter will explore the role of AI and IoT technologies in designing remote laboratories based on mixed reality. These laboratories offer an interactive environment that allows students to participate in experiments comprising both real and simulated components. Students can access content customized to their academic level and abilities, aimed at optimizing the learning process and augmenting skills. A remote laboratory consists of physical and simulated units, as well as an IoT system and software for managing and controlling laboratory operations. The chapter reviews the design of two experiments: the first in the mechatronics laboratory and the second in the renewable energy laboratory. This chapter is organized as follows. Section 1.2 explains the impact of AI, IoT, and mixed reality technologies on remote laboratory design. Section 1.3 explains the classification of laboratories and how to access them. Section 1.4 discusses the design of remote laboratories based on mixed reality, while Sect. 1.5 addresses the design of the smart laboratory management unit. Section 1.6 covers the design of experiments in the mechatronics and renewable energy laboratories. Finally, Sect. 1.7 summarizes the main conclusions.

1.2 Development of E-Learning Systems

In general, the development of e-learning systems has been significantly influenced by modern technology, and in particular, the design of remote laboratories has been the subject of major influence. A significant number of opportunities for accessibility, attendance, course management and personalized learning are presented by these innovations. For students and teachers to be able to successfully navigate the new technology that is used in modern e-learning systems, they need to acquire training. A variety of technologies that have significantly contributed to the evolution of e-learning systems will be explored.

1.2.1 Internet of Things Technology

The term "Internet of Things" refers to a network that is comprised of a collection of physical objects that are equipped with sensors and software to facilitate the exchange of data and the connectivity of these objects to other devices and systems that are linked to the internet. In addition to providing a worldwide framework that makes it easier to connect modern applications and systems, the IoT also makes it possible for things to communicate with one another via more sophisticated means. The global IoT market was projected to reach 714.48 billion US dollars in 2024, with an anticipated growth to 4,062.34 billion US dollars by 2032 (Fortune Business Insights 2025). IoT is employed across various sectors, including industry, agriculture, healthcare, smart cities, and transportation. IoT significantly influences information collection and facilitates efficient information sharing, establishing itself as a crucial component in contemporary educational systems.

The IoT profoundly impacts e-learning systems, improving educational quality and student engagement with the learning platform. The introduction of interactive whiteboards and IoT-enabled smart classrooms has significantly enhanced the engagement and effectiveness of learning environments. IoT-enabled educational systems can monitor student engagement in real time to assess attendance, focus, and academic progress. Moreover, they are utilized in adaptive learning systems to deliver knowledge tailored to the requirements of each learner. Furthermore, the integration of virtual reality and mixed reality technologies with IoT facilitates the creation of engaging and interactive courses for students. Gamification and feedback can enhance motivation. The IoT greatly enhances remote access to courses by enabling connection across diverse devices and settings. IoT technology can enable the automation of campus operations, encompassing lighting, temperature regulation, attendance tracking for students and staff, equipment management, and safety issues. Effective resource utilization leads to cost savings and the establishment of more sustainable learning environments.

Internet of Things technologies have impacted the design of e-learning systems, especially in terms of laboratory design and remote access (Al-Obaidi and Derbel 2023). IoT devices facilitate the establishment of virtual and hybrid laboratories. This technology creates a secure platform for testing, overcoming geographical limitations and improving internet accessibility. IoT technology also enhances the design of laboratory investigations through the utilization of mixed reality technologies.

1.2.2 Artificial Intelligence

AI has changed e-learning systems by modeling students and teachers and intelligently managing the educational process. The following are some of the significant advances in remote labs that offer evidence of the impact of AI:

- Personalized learning experiences are made possible by AI algorithms that examine student performance and data. This is done by providing content that matches each learner's needs, improving their understanding and allowing them to explore further education in their field of interest.
- By offering adaptive learning platforms, artificial intelligence (AI) can be used to modify the material level in response to real-time data on student performance. This guarantees that learners are fully challenged and can learn at their own speed, enhancing the learning process.
- To manage the learning process, AI technologies are used to model learners according to their capacity for learning and interaction with content, giving them relevant information depending on their present level of comprehension. Intelligent management of the learning process makes it possible for learners with higher levels of capacity to finish learning modules in a straightforward and expedient manner, whereas learners with lower levels of capability may take additional time and effort to finish each module.
- AI-based modules can be used as an alternative to hardware or software components when designing mixed reality lab experiments. AI modules frequently do jobs with more efficiency and less costs. Additionally, AI modules make it possible to investigate new experimental approaches and procedures that were previously impossible or unattainable. These systems enable flexibility in the design of laboratory experiments and testing arrangements, which is something that standard components might not be able to give.

1.2.3 Virtual and Mixed Reality

The term "virtual reality" refers to a computing system that allows users to enter a digital world that is simulated using software, displays, sensors, and controls. Virtual reality enables users to explore virtual surroundings by utilizing specialized software and equipment such as headsets and 3D glasses. Virtual reality is applied not only in the gaming sector but also in the training, teaching, and learning systems, where it has proven to be successful. The use of smartphones and computers as interfaces has allowed for the creation of virtual environments that have the possibility of providing students and teachers with engaging experiences at a cost that is minimal. Designers are working on VR experiences, such as virtual classrooms, in which users navigate three-dimensional settings and participate in activities that have been scheduled. The use of VR provides students with opportunities to participate in activities within a virtual world that would otherwise be difficult to access (Moro et al. 2021). Virtual reality has a discernible influence on online learning platforms, except for disciplines that necessitate the use of laboratories and workshops. To obtain proficient skills, it is necessary to deal with real instruments and tools in conducting applied disciplines. With technological progress, access to authentic laboratories has been available online. Technology has facilitated the amalgamation of virtual reality with actual reality to establish an interactive environment that enhances e-learning systems in applicable fields. There are a variety of practical applications that can be made possible using mixed reality technology, which involves the interaction of both real and virtual elements (Al-Aubidy 2022). Mixed reality can provide students with a variety of choices for addressing e-learning issues in applied disciplines. A large amount of investment interest from important information technology companies has recently been shown in mixed and virtual reality. It has been estimated that by the year 2025, the number of people who utilize virtual and mixed reality would have reached 443 million (Lin 2022).

In summary, AI, IoT, and mixed reality technologies can enhance remote laboratory work by facilitating student interaction with actual experimental components and enabling the processing and analysis of results, so aiding in the attainment of experimental objectives and the enhancement of acquired abilities.

1.3 Classification of Laboratories

Laboratories have a crucial role in improving students' knowledge of theoretical concepts, strengthening their problem-solving skills, and forming conclusions based on reading and data analysis. As a result, laboratories are an essential component of student education in applied sciences disciplines. Real laboratories are utilized in conventional educational institutions, and students perform hands-on activities with real pieces of equipment that are readily available in the laboratory. Experiments are often carried out by students with only a limited amount of information, with

Table 1.1 Lab classifications according to their components and access

Laboratory	Access	Equipment
Real Lab.	Face-to-face	Physical components
Virtual Lab.	Face-to-face/remotely	Simulated components
Hybrid Lab.	Remotely/face-to-face	Physical and simulated components

greater emphasis focused on the guidance provided by the laboratory supervisor. Such laboratories are costly in certain applied fields and require technicians and supervisors to instruct and assist students during experiments. The advancement of computer and software technologies has enabled the execution of simulated experiments both remotely and in laboratory settings. These simulated laboratories have improved understanding of theoretical concepts, although they have failed to provide students with fundamental experimental abilities. The incorporation of virtual reality, the Internet of Things, and artificial intelligence technologies have expanded laboratory design and management techniques. In general, laboratories can be divided into three categories: real laboratories, virtual laboratories, and hybrid laboratories. The corresponding classifications are presented in Table 1.1.

1.3.1 Real Laboratories

These are the conventional laboratories that are carried out on-site and contain real physical components, instruments, and materials. Experiments can be carried out in the laboratory by students in accordance with a schedule that has been established by the management of the academic program. A laboratory supervisor is necessary in these laboratories since they are responsible for preparing the requirements for the experiment and ensuring that the processes are carried out correctly before the students carry them out. When it comes to building new genuine laboratories in certain areas of expertise, educational institutions may have difficulties since these laboratories are costly and difficult to offer technicians and supervisors to manage the requirements of the lab.

1.3.2 Virtual Laboratories

These facilities emulate real lab components using computer simulations or online platforms. Hybrid laboratories come in several types, including:

- Simulation-based laboratories: These labs simulate experiments using software.
- Remote laboratories: These labs give students the ability to control actual lab equipment using the internet.

- Virtual reality-based labs: These involve the utilization of virtual reality resources, such as headgear and 3D courses, to deliver immersive laboratory experiences.
- Cloud lab platforms: These are online platforms that facilitate virtual experiments.

1.3.3 Hybrid Laboratories

These labs mix genuine components with simulation modules, enabling students to engage with them either on-site or remotely. Real components are often used in hybrid laboratories. The Internet of Things and artificial intelligence are two other technologies that can be utilized for the purpose of developing laboratory tests. The most important features of such laboratories are:

- Mixed reality technology is utilized in the process of experimental design.
- One of the most important aspects of these laboratories is that they offer possibilities to use real equipment in a variety of different labs and places.
- By introducing extra simulated components, typical laboratories can be improved.
- Increasing laboratory efficiency to increase student learning and skill development.
- A flexible schedule for the usage of the laboratory.
- Enhancing interaction chances.
- Guaranteeing safety, especially in laboratories containing dangerous substances or high electrical voltage.
- Effectively analyzing, managing, and disseminating data. - Providing high-quality applied education at a reasonable cost.

In conclusion, the utilization of mixed reality in the design of remote laboratories will make it possible for educational institutions and universities to collaborate on the development of applied specializations, particularly in domains that are scarce and costly.

1.4 Remote Laboratories with Mixed Reality Integration

A mixed reality setting facilitates the development of various experiences by integrating both virtual and real physical components appropriate to the requirements of the experiment. This environment facilitates real-time monitoring of system behavior by observing generated signals and mimicking interactions with real laboratory equipment. However, creating a mixed reality environment presents difficulties, particularly in achieving smooth real-time interaction when incorporating real components into virtual reality. Interface modules are essential for enabling bidirectional information exchange between real and virtual components, as seen in Fig. 1.2. Real components may include electrical, mechanical, hydraulic, and pneumatic devices,

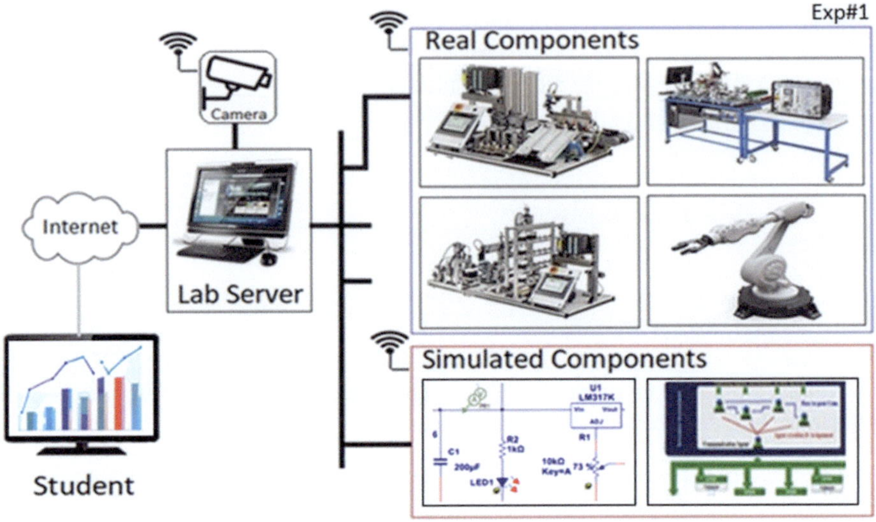

Fig. 1.2 Mixed reality based remote lab

each managing analog or digital signals. On the other hand, virtual components are simulation units that manage digital signals.

It is possible to install both virtual and real components at remote locations with internet access. Students can access online experiments using an intelligent management module. In addition, they can carry out a variety of tasks on virtual workbenches, as well as utilize the performance of real components and the outcomes of virtual experiments while doing analysis and creating reports. Through the utilization of web technology, both real and virtual components can function as web services that are hosted by the laboratory server. A series of tests can be conducted utilizing physical apparatus accessible in the laboratory, alongside simulated components that provide remote access via the internet, as seen in Fig. 1.3. Every experiment possesses a local controller linked to the laboratory server through an internal network or Internet of Things technology. The local controller coordinates the experimental activities and communicates data to the laboratory server. The Experiment page outlines the theoretical foundation and characteristics of each experimental component, encompassing:

- The experiment's objective.
- The components utilized in preparation.
- Experimental procedures.
- Methodology for acquiring data, doing analysis, and submitting reports.

When building mixed reality experiences, designers need to take into consideration the requirements of the curriculum, the needs of the students, and the limits to accomplish the experiment goals. Some of the most important things to consider throughout the design process are as follows:

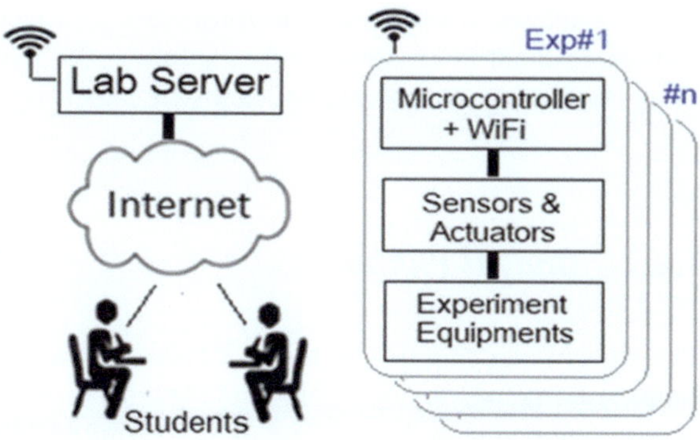

Fig. 1.3 General layout of a mixed reality lab

- Making sure that students conduct experiments within safe parameters and in a controlled environment.
- Putting in place a real-time monitoring system that allows for immediate control and reflection in the real world.
- Delivering real-time video feeds showcasing experimental components, measurement devices, and controls.
- Integrating auditory elements to augment users' spatial awareness within the surroundings.

1.4.1 Lab Platform

The lab server is accountable for the scheduling of experiments, the collection of data from experiments, the delivery of experiment instructions to students, and any other information that may be relevant. It is equipped with the requisite software to offer the following services:

- Lab Manual: It includes the theoretical foundations of the experiments, specifications, and functionalities of each component, as well as complete procedures for experiment handling. These resources are designed to accommodate various student ability levels (Al-Aubidy 2005).
- Student's guide: It offers personalized educational content according to individual ability. If the student has adequate knowledge, the information on the experiment will be short, whereas, if the student lacks sufficient understanding, the laboratory server will provide the student with detailed information.
- Data collection: It involves collecting data from experiments in real time and transferring it to a web server.
- Students can interpret data and make connections between the results of experiments and the objectives that have been specified.

- Lab manager: It is responsible for a variety of tasks, including the management of experiment reservations and scheduling, the selection of experiments, and the supervision of student involvement with experiments.

1.4.2 Student Platform

The student can gain access to the laboratory and interact with the components of each experiment by using a computer (or a smartphone) that is linked to the Internet and is equipped with the software that is required. Several features are available on the student platform, including:

- Make a reservation and choose the experiment that you choose.
- Several steps, including setup, operation, and data gathering, must be followed to have access to the experiment that has been selected.
- Review the results and do data analysis to accomplish the objectives of the experiment.
- Any questions should be submitted through the chat room.
- Watch a video of the activities taking on in the laboratory in real time.

1.5 Smart Laboratory Management

The management of laboratories that use IoT and MR technologies requires the coordination of hardware, software, network infrastructure, user interactions, and data to provide users with the ability to access and control scientific or engineering equipment remotely. The management of remote laboratories can be greatly improved by artificial intelligence, particularly when combined with MR technology. An efficient management of the laboratory is dependent on the utilization of intelligent algorithms to carry out the following tasks:

- Setting up the infrastructure, including computers and peripherals, software, communication networks, cameras, and servers.
- Smart scheduling of resources, the efficient allocation of those resources, and the regulation of access to those resources to guarantee that only authorized users can visit the laboratory.
- Controlling the operation of both real and virtual component's function.
- Real-time monitoring laboratory operations and executing predictive maintenance to avert malfunctions.
- Using artificial intelligence to create models of students and give them material based on their abilities.
- Managing data storage and analysis.

- Utilizing virtual assistants and chatbots to assist users.
- Ensuring the security of data and applying cybersecurity procedures.

This chapter explains how fuzzy logic can be used in student modeling. The student model is utilized in the management of laboratories for a variety of reasons, the most significant of which are (Al-Aubidy 2005):

- Determining whether the learner has a correct comprehension of the laboratory experiment.
- Making sure the student has learned what they need to know.
- Determining the student's current level of knowledge to personalize the instructional materials being used.
- Managing educational resources in a manner that is very effective.
- Performing data analysis on students for the goal of quality control.

The student model is a significant component of the intelligent remote laboratory management system. Depending on the student's current level of knowledge, the model makes use of three different learning paths, as illustrated in Fig. 1.4:

- The Fast Learning Path (FLP) provides a condensed summary of the experiment, which is meant primarily for experienced students who are already familiar with the material.
- Intermediate Learning Path (ILP): This path offers learners with an average level of comprehension standardized educational content to enhance their learning experience.
- Slow Learning Path (SLP): Provides learners who are less familiar with the module content with extensive explanations and assistance to support them in their learning process, hence encouraging a more gradual learning process.

In the context of fuzzy set theory, a linguistic variable may consist of words or phrases as its values, with each term linked to a fuzzy number. A membership function, represented as $\mu(x)$, allocates a value between 0 and 1 to each element x in a fuzzy set, signifying the extent of membership in that set. For instance, let x represent the variable "student's average grade" in an experiment. This variable can

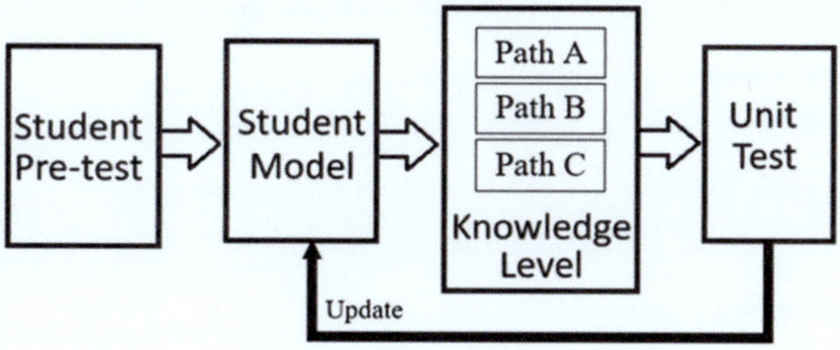

Fig. 1.4 Block diagram of the proposed educational system

be described using the following linguistic terms: excellent (E), very good (VG), good (G), pass (P), fail (F), and unsatisfactory (U), as depicted in Fig. 1.5.

The second input variable for the intelligent management module is the student's learning level, represented by three fuzzy sets: high learning path (HLP), medium learning path (MLP), and slow learning path (SLP). The principal objective of this module is to modify students' learning paths according to 18 fuzzy rules, as illustrated in Table 1.2.

The three learning paths make it possible for students who are performing very well to move through the learning units in a short amount of time, whereas students who want greater support may require additional time and effort to finish each unit. Figure 1.6 is a diagrammatic representation of the educational system that is being suggested, which contains the following components (Al-Aubidy 2005):

Step 1: Give the student a pre-test to see how much they already know before they start the training unit. The system gives each student learning tools that are right for their level based on the results.
Step 2: As illustrated in Fig. 1.7, this step involves evaluating the student's level of comprehension in each educational unit to bring their model up to date and establish the upcoming learning path.

- Students who finish the educational unit with grades of "Good" or "Very Good" can advance from the MLP to the HLP or from the SLP to the MLP in the subsequent level of the educational unit.

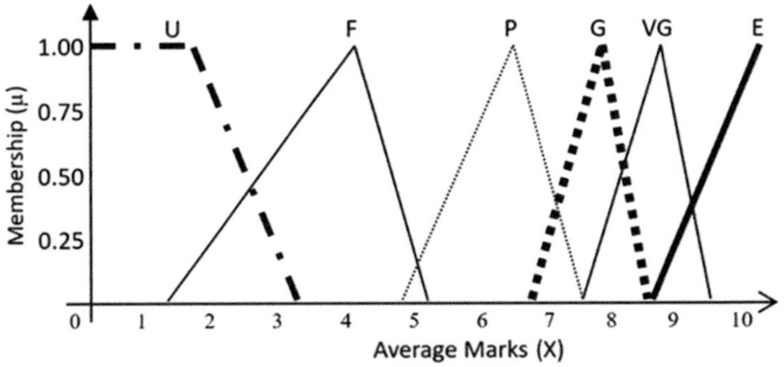

Fig. 1.5 Fuzzy notation of the variable "test result"

Table 1.2 Fuzzy rules for smart management module

		Student's average test score					
		E	VG	G	P	F	U
Current	CHL	NHL	NHL	NHL	NML	NML*	NSL*
Learning	CML	NHL	NHL	NML	NSL	NSL*	NSL*
Path	CSL	NML	NML	NSL	NSL	NSL*	OUT

Note C: Current, N: Next, *: Stay in the same unit

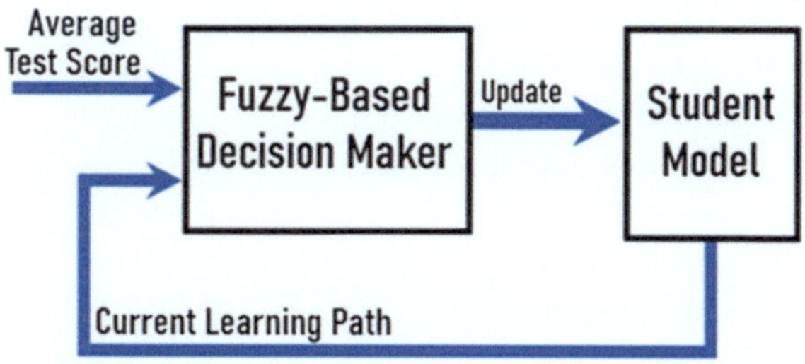

Fig. 1.6 Smart management module

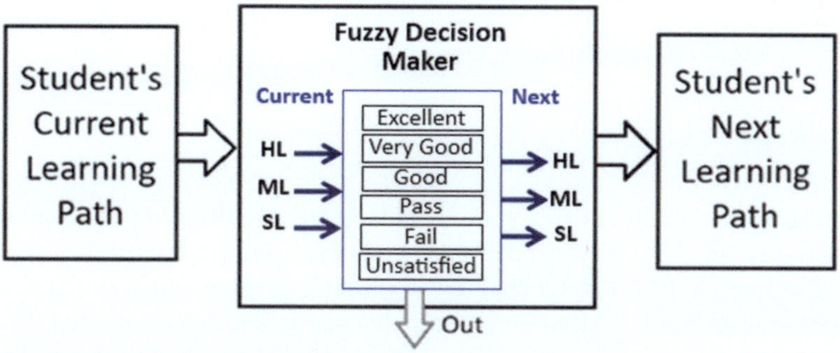

Fig. 1.7 Layout of the decision-making process

- Those students who are given a grade of "Fail" will be moved from the educational track that they are currently on (HLP or MLP) to the track that they are currently on (SLP), and they will continue to be a part of the same educational unit.
- Students who are currently on a different learning route will continue to be placed in the same educational unit and on the same learning path (SLP) if the test result at a particular learning unit is "Unsatisfied". This is regardless of whether they are currently on a higher or lower learning path.
- If a student who is at the knowledge level (SLP) earns a "Fail" on the test, the system will suggest that the learner stop participating in this experiment.

1.6 Experiments Design

The design and implementation of a remote laboratory utilizing mixed reality and Internet of Things technologies represent an important jump in the enhancement of e-learning within practical disciplines. IoT technology allows students to remotely access laboratory experiments at any time and from any location via an internet-enabled computer or smartphone. This section outlines two experiments: one implemented in the mechatronics laboratory and another in the renewable energy laboratory.

Table 1.3 Electrical characteristics of the PV module

PV panel characteristics	Value
Voltage at max. power point (V_{mpp})	18.0 V
Current at max. power point (I_{mpp})	2.78 A
Peak power	50 Wp
Open circuit voltage (Voc)	21.6 V
Short circuit current (Isc)	3.1 A
Serial resistor	0.203 Ω
Parallel resistor	109.3 Ω

1.6.1 Renewable Energy Remote Laboratory

In renewable energy laboratories, equipment is frequently situated at a distance from the laboratory area (Al-Kashashneh and Al-Aubidy 2019; Al-Obaidi and Derbel 2023). The Internet of Things technology facilitates the transmission of sensor signals from equipment to a laboratory computer that is internet-connected. Students can obtain real-time data for renewable energy experiments using the internet. The laboratory photovoltaic (PV) prototype system is illustrated in Fig. 1.8, with nine PV panels organized into three parallel-connected strings to provide direct current electrical energy. Each string contains three PV panels. Table 1.3 enumerates the electrical parameters of the photovoltaic module under standard test conditions (25 °C and 1000 W/m^2).

1.6.1.1 Hardware Design

The photovoltaic module is outfitted with a collection of sensors to detect voltage (V), current (I), irradiance (G), and temperature (T) for each string. As shown in Fig. 1.7, each string has a single-chip microcontroller that scans, measures, and analyzes sensor data. The microcontroller is responsible for receiving data from sensors and

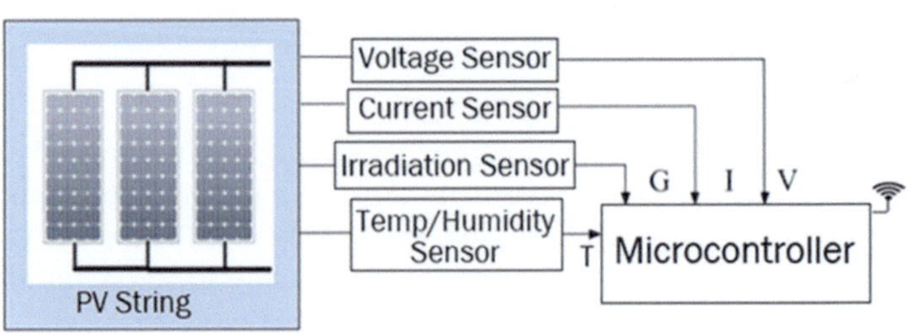

Fig. 1.8 General layout of the PV system

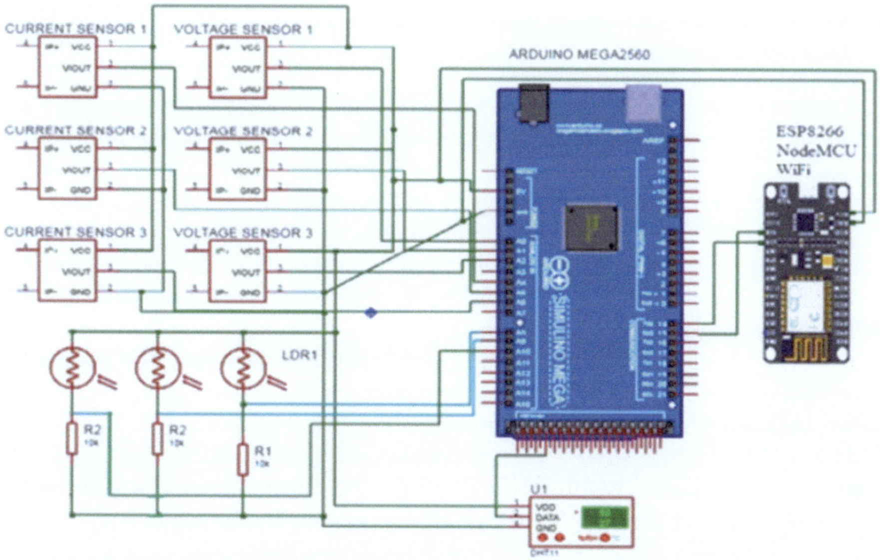

Fig. 1.9 Hardware design of the laboratory prototype

calculating a variety of metrics and indications. Each string is also outfitted with a WiFi microcontroller that is serially coupled to the main microcontroller.

The suggested integrated data acquisition unit for a single string is illustrated in Fig. 1.9. It is possible to facilitate the interchange of data between the embedded microcontrollers and the main computer unit in the laboratory by utilizing IoT technology.

1.6.1.2 Software Design

The software is divided into three main parts: the main program, the data acquisition, the fuzzy-based fault detection, and the fault allocation. Figure 1.10 illustrates the general flowchart of the required software.

1.6.1.3 Experiment Procedures

The laboratory includes several experiments, and the experiment "Measuring the Energy Production Efficiency of Photovoltaic Panels" will be presented. This experiment aims to:

- Measure current and voltage and calculate power at regular intervals under varying atmospheric conditions, including temperature and radiation.

- Measure current and voltage, and calculate the power generated by simulating photovoltaic panels.

Under a variety of environmental conditions, such as changes in temperature and irradiance, students can access the experiment online and utilize the results from the sensors to compute the amount of renewable energy that is generated. Students can also use a PV model to demonstrate and compute the amount of energy that is generated. The influence of solar radiation on the current-voltage ($I - V$) and power-voltage ($P - V$) characteristics of a photovoltaic (PV) module is illustrated in Fig. 1.11. The irradiance levels used in this illustration are 1000 W/m^2 and 750 W/m^2, respectively.

MATLAB/Simulink was also used to simulate the photovoltaic (PV) system to investigate the influence of ambient temperature on the voltage and power curves under standard test settings and different ambient temperatures (30, 25, 20, and 15 °C), as Fig. 1.12 illustrates. The maximum power point (MPP) rises as the surrounding temperature falls. In addition to this, the voltage inside the open circuit rises, whilst the current in the short circuit stays essentially the same.

1.6.2 Mechatronics Remote Laboratory

Mechatronics combines electronic, mechanical, and computer engineering and is one of the disciplines that require the use of virtual reality technology to design remotely accessible experiments. These experiments combine real physical components and simulated elements in the experimental setup (Saleem and Al-Aubidy 2008). The experiment was designed to control a DC motor using computer-implemented control algorithms, as shown in Fig. 1.13. To control the motor speed, three basic components are used: a DC motor, a tacho generator to measure speed, and a motor driver. The goal of the experiment is to enable students to design a closed-loop control system to regulate motor speed. They can also create the required control method using virtual components, such as Matlab/Simulink applications. Students can do many activities through the student platform, which supports:

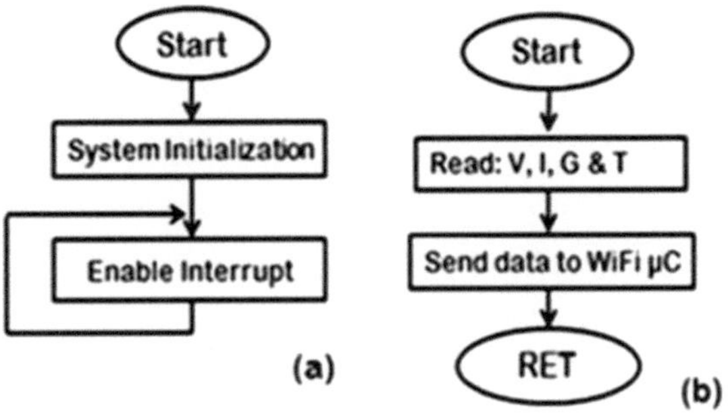

Fig. 1.10 Flowchart of the main program **a**, and data acquisition **b**

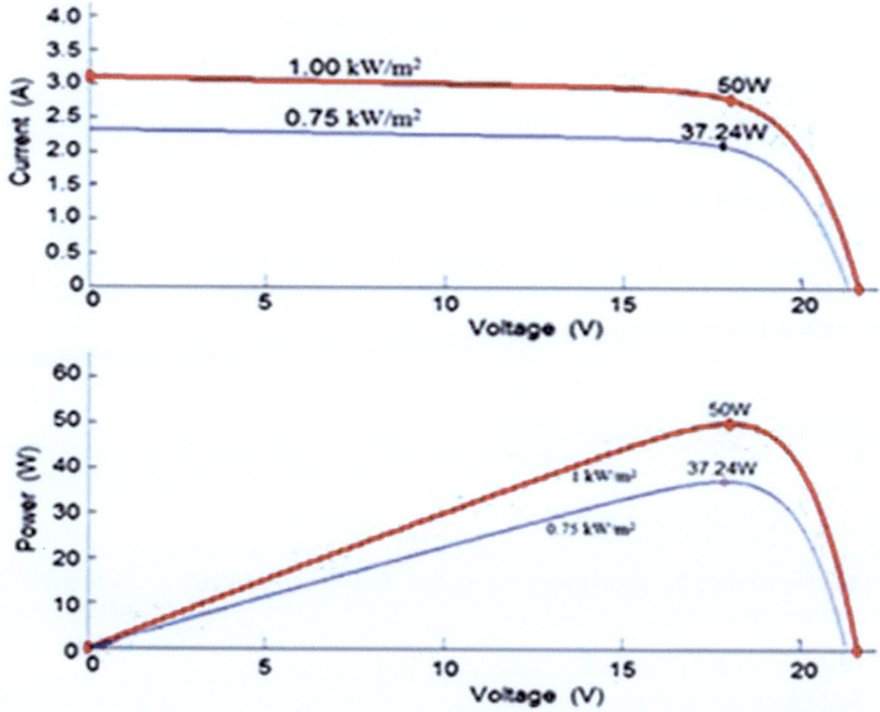

Fig. 1.11 $I - V$ and $P - V$ curves of the real PV module

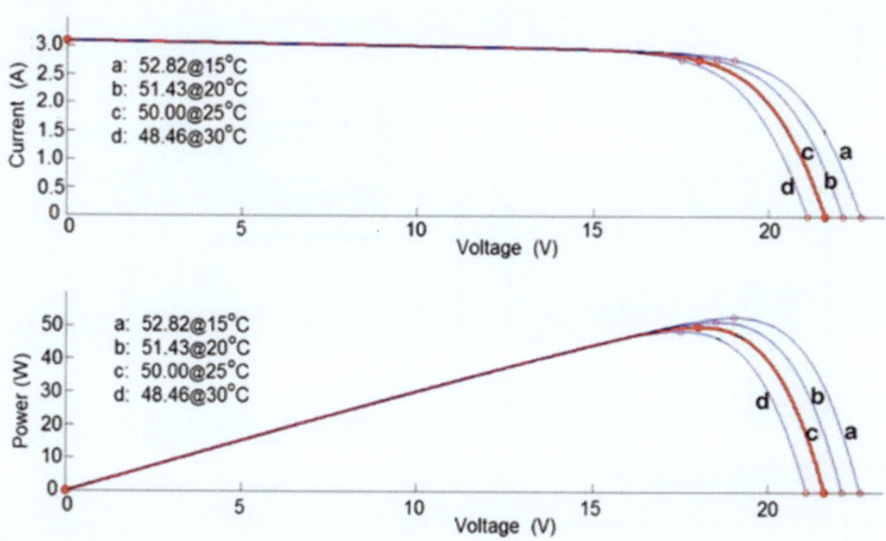

Fig. 1.12 Impact of temperature on simulated I-V and P-V curves of the simulated PV module

- Setting up the experiment using actual and virtual components.
- Carrying out the experiment in accordance with the given procedures.
- Changing the parameters of the control algorithm.

1.6.2.1 Hardware Design

The ATmega328P microcontroller generates two control signals for the motor drive unit. Motor speed is sensed by the tacho generator, which produces an analog signal proportional to the speed. In this experiment, the main microcontroller can be serially connected to another microcontroller type (ESP8266) for wireless communication with the laboratory server. IoT technology can also facilitate signal transfer between experiment components.

1.6.2.2 Experiment Procedures

Upon completing the experiment setup, the student is required to follow several procedures, including:

- Designing a digital controller, such as a PID controller.
- Testing the speed control response.
- Adjusting the PID controller parameters to observe their impact on the motor time response, (see Fig. 1.14).

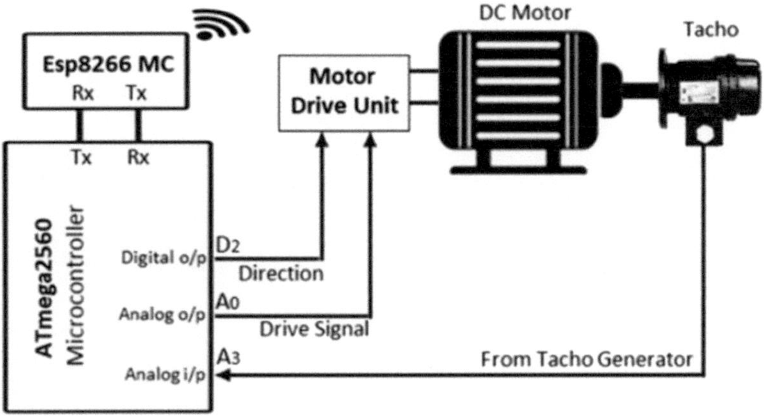

Fig. 1.13 Layout of the motor speed control experiment

- Assessing the synchronization between the physical environment (represented by the DC motor and its drive unit) and the virtual environment (represented by the PID control algorithm).

Following the selection of the real and virtual components that are necessary for the design of the experiment, the student is expected to complete a set of procedures, which includes the following:

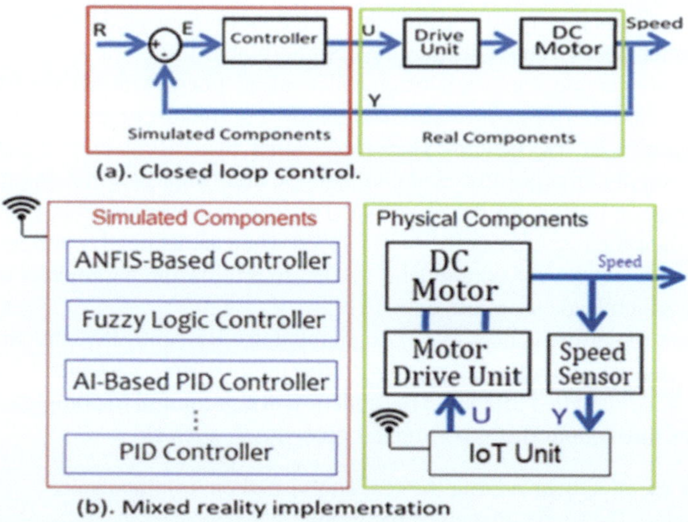

Fig. 1.14 Experiment setup for DC motor control

- The development of a digital controller to control motor speed.
- Selecting controller parameters to improve control system performance; testing the motor speed control system's time response.
- The evaluation of the synchronization between the physical environment, which is represented by the DC motor and its drive unit, and the virtual environment, which is represented by the control algorithm.

Figure 1.15 demonstrates the capability of the PID controller to track speed commands during acceleration to 40 and 60 rpm, as well as during deceleration to 20 rpm. A major impact on the speed performance of the motor is caused by the fact that the PID controller needs to have its parameters fine-tuned. Using the Ziegler-Nichols's algorithm, for example, can result in controller values that provide satisfactory performance; but additional adjustments are frequently required to compensate for changes in the dynamics of the system. Therefore, rather than using human tuning, it is desirable to apply an algorithm that can tune the controller parameters online. This optimization can be accomplished online. This can be observed in Fig. 1.16, which illustrates the time response of the motor speed when an artificial intelligence-based tuner, which is a software component, is used to modify the parameters of the PID controller, which leads to in less overshoot.

To assess the effectiveness of the implemented PID controller, the student will determine several indicators pertaining to the motor's time response based on the results, as illustrated in Table 1.4. The student will be required to complete a number of activities throughout the experimental procedures, such as:

- The process of synchronizing the real and virtual environments.
- Determining the time constant of the motor.
- Choosing a suitable sampling time.
- Using the processor clock for calculating the time required to implement the PID control algorithm.

Another experiment prepared at the same laboratory is focused on the design and implementation of an AI-based online tuner for the PID controller that makes use of fuzzy logic, neural networks, or genetic algorithms. Additionally, students might be challenged with developing an intelligent controller for the purpose of regulating

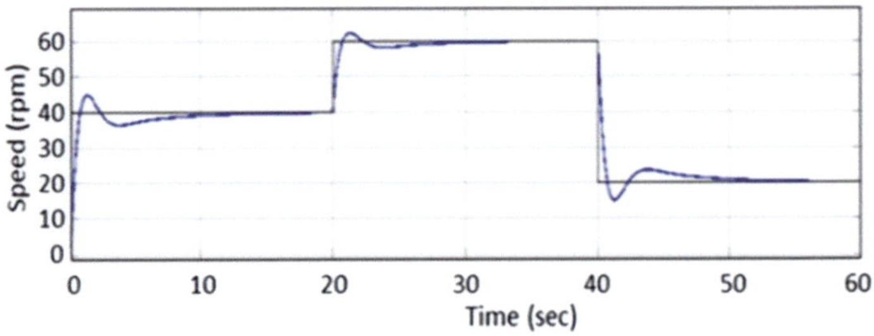

Fig. 1.15 Transient response of motor speed with manually tuned PID controller

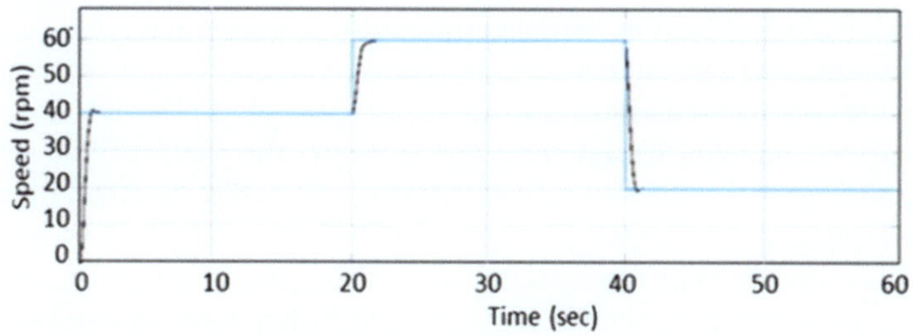

Fig. 1.16 Transient response of motor speed with intelligently tuned PID controller

Table 1.4 Calculation of time response parameters

Test No.	Settling time (s)	Rising time (s)	Max overshoot
Test 1 (Fig. 1.15)	1.025	0.058	13%
Test 2 (Fig. 1.16)	0.028	0.017	0.1%

the speed of a motor by making use of AI-based controllers. It is possible to conduct additional experiments within the laboratory by substituting the speed sensor with a position sensor. This will allow for designing other experiments.

1.7 Conclusion

As an outcome of the rapid advancements in technology and the digital revolution, the adoption of e-learning and blended learning strategies will be needed. This is because the number of students enrolling in higher education is growing simultaneously. Specialized laboratories and workshops are incorporated into the curriculum of applied disciplines in general, and engineering disciplines in particular, with the purpose of developing and strengthening students' abilities to cope with the requirements of the market. Therefore, to incorporate e-learning into applied disciplines, it is necessary to both develop laboratories that already exist and expand laboratories that do not yet exist. Research in university research centers and IT companies is increasingly focusing on the design of remote interactive laboratories that incorporate mixed reality, IoT, and AI technologies. Mixed reality laboratories give students the opportunity to acquire essential skills by enabling them to access experiments remotely using smartphones or laptops connected to the internet. This allows students to acquire skills without being limited by time or location.

The purpose of this chapter is to investigate the role that artificial intelligence and emerging technologies, such as the internet of things and mixed reality, play in the process of enabling remote interactive laboratories that integrate real and virtual components. These laboratories provide several benefits, including the following:

- The purpose of mixed reality-based purpose of mixeexperiment design is to combine both real components that are available on-site and virtual components into laboratory studies. This allows for economically feasible and scalable laboratories. It is possible to add new experiments by modifying the configuration settings.
- These laboratories are open seven days a week, which has the effect of increasing student availability and improving their capabilities in specialized sectors.
- Students can interact with both real and virtual components remotely, giving them the opportunity to experience the experiment as if they were physically present in the laboratory.
- Because of the sophisticated systems that control lab operations and communicate with students, mixed reality-based laboratories require less direct supervision than traditional laboratories do. This reduced dependence on supervisors is a significant benefit.
- The educational content is adapted to the learning ability of each individual student, which enables the high learning student to understand the material more quickly and complete the laboratory tasks in a more efficient manner.
- Remote labs encourage collaboration between educational institutions, which enables shared use of resources to serve a larger student population. This results in enhanced collaboration within the educational community.
- Mixed reality-based laboratories make a substantial contribution to the development of applied curricula and open the way for the introduction of new specializations that are less expensive, which in turn supports higher enrollment in universities.

Acknowledgements This research was supported by Tishk International University, Erbil, Iraq. The researcher is grateful for the support and encouragement provided by the Department of Mechatronics Engineering and the College of Engineering.

References

Al-Aubidy, K.M.: Applying Fuzzy logic for learner modeling and decision support in online learning systems. i-Manag. J. Educ. Technol. **2**(3), 76–85 (2005)

Al-Aubidy, K.M.: IoT-based mixed reality laboratories in applied disciplines. ARID Int. J. Sci. Technol. (AIJST) **5**(10) (2022). https://doi.org/10.36772/arid.aijst.2022.5102

Al-Aubidy, K.M., Al-Mutairi, A.W., Al-Kashashneh, H.Z.: IoT based remote laboratory for solar energy experiments: design and implementation. In: 2021 18th International Multi-Conference on Systems, Signals & Devices (SSD21), Tunisia, 22–25 March (2021)

Al-Kashashneh, H.Z., Al-Aubidy, K.M.: Wireless sensor network based real-time monitoring and fault detection for photovoltaic systems. In: 16th IEEE International Multi-Conference on Systems, Signals & Devices (SSD19), Istanbul, Turkey, 21–24 March (2019)

Al-Obaidi, M.Q., Derbel, N.: Design of IoT based remote renewable energy laboratory. Int. J. Emerg. Technol. Learn. (iJET) **18**(12), 75–87 (2023). https://doi.org/10.3991/ijet.v18i12.38659.

Al-Tikriti, M.N., Al-Aubidy, K.M.: Embedding mixed-reality laboratories into e-learning systems for engineering education. i-Manag. J. Educ. Technol. **9**(4), 25–35 (2013)

Badawy, I., Bassiuny, A.M., Darwish, R., Tolba, A.S.: A digital twin of a remote real-time accessible labs. In: Auer, M.E., Cukierman, U.R., Vendrell Vidal, E., Tovar Caro, E. (eds.) Towards a Hybrid, Flexible and Socially Engaged Higher Education. ICL 2023. Lecture Notes in Networks and Systems, vol. 900. Springer, Cham (2024). https://doi.org/10.1007/978-3-031-52667-1_21

Fortune Business Insights: Internet of Things (IoT) market size, share and industry analysis (2025). https://www.fortunebusinessinsights.com/industry-reports/internet-of-things-iot-market-100307/?utm_source=chatgpt.com

Kingson, J.A.: Virtual reality takes a seat in college classrooms. Axios (2023). https://www.axios.com/2023/09/12/meta-virtual-reality-classrooms?utm_source=chatgpt.com

Kreiter, C., Klinger, T.: Development of a low-cost remote lab concept for electronic engineering education based on NI myDAQ and NI ELVIS. Int. J. Online Biomed. Eng. (iJOE), **16**(14), 18–29 (2020)

Lee, J., Natarrajan, R., Rodriguez, S.S., Panda, P., Ofek, E.: Remotelab: a VR remote study toolkit. In: The 35th Annual ACM Symposium on User Interface Software and Technology (UIST '22), October 29–November 2 (2022)

Lin, B.: Apple pitches its $3,500 vision pro headset to businesses. Will It Catch On? Wall Street J. (2024). https://www.wsj.com/articles/apple-pitches-its-3-500-vision-pro-headset-to-businesses-will-it-catch-on-0e20f9fc?utm_source=chatgpt.com

Lin, Y.: 10 Virtual Reality Statistics Every Marketer Should Know in 2022. Published 12 February 2022 (2022). Retrieved from https://www.oberlo.com/blog/virtual-reality-statistics

Markets and Markets: Augmented and virtual reality in education market size, share analysis. Mark. Mark. (2025). https://www.marketsandmarkets.com/Market-Reports/virtual-classroom-market-203811025.html

Microsoft: Microsoft XR Enterprise Success Stories: How Enterprises are Scaling with Microsoft XR. Published: February 19 (2025). https://www.xrtoday.com/mixed-reality/microsoft-xr-enterprise-success-stories-how-enterprises-are-scaling-with-microsoft-xr/

Moro, C., Phelps, C., Redmond, P., Stromberga, Z.: HoloLens and mobile augmented reality in medical and health science education: a randomized controlled trial. Br. J. Educ. Technol. **52**(2), 680–694 (2021)

Munir, T., Akbar, M.S., Ahmed, S., Sarfraz, A. , Sarfraz, Z., Sarfraz, M., Felix, M., Cherrez-Ojeda, I.: A systematic review of internet of things in clinical laboratories: opportunities, advantages, and challenges. Sensors **22**(20), 8051 (2022). https://doi.org/10.3390/s22208051

Precedence Research: Higher Education Market Size, Share, and Trends 2025–2034 (2025). Available online: www.precedenceresearch.com/higher-education-market?utm_source=chatgpt.com

Robertson, A.: Google has a new cardboard headset, and it supports iPhones. The Verge (2015). Retrieved from https://www.theverge.com/2015/5/28/8662873/google-cardboard-virtual-reality-io-2015

Saleem, A., Al-Aubidy, K.M.: Mixed reality environment for web-based laboratory interactive learning. Int. J. Online Biomedi. Eng. (iJOE) **4**(1), 40–45 (2008)

UNESCO: UNESCO paper shows governments not keeping pace with growing demand for higher education (2022). www.unesco.org/en/articles/unesco-paper-shows-governments-not-keeping-pace-growing-demand-higher-education?utm_source=chatgpt.com

Chapter 2
The Evolution and Integration of SCADA and HMI Systems in Industrial Automation

Ezideen A. Hasso

Abstract In the modern industrial landscape, the convergence of SCADA (Supervisory Control and Data Acquisition), advanced HMI (Human-Machine Interface) systems, and cutting-edge communication technologies have revolutionized how production facilities are managed and optimized. These technologies collectively serve as the digital backbone of intelligent industrial operations, providing real-time visibility into processes, assets, and resource flows. By enabling the instantaneous collection, analysis, and visualization of statistical data, SCADA and HMI platforms empower management teams to make informed decisions rooted in accurate, actionable knowledge rather than; random guessing and assumptions; or historical approximations. The immediacy with which data is now available, facilitated by Ethernet, industrial wireless protocols, cloud integration, and mobile HMI access, has dramatically shortened the feedback loop between the shop floor and the decision-making hierarchy. Managers can monitor key performance indicators (KPIs), detect deviations from operational baselines, and respond to anomalies in real time, thus, minimizing downtime and maximizing throughput. Decisions are now data-driven and timely, enhancing the agility and competitiveness of manufacturing operations. Moreover, this data-centric visibility has extended into the supply chain domain. Real-time inventory tracking, consumption trends, and predictive usage models, fueled by SCADA data analytics, allow for more accurate and just-in-time procurement. This results in optimized sourcing strategies, reduced inventory carrying costs, and minimized material shortages. At the same time, predictive maintenance capabilities have emerged as a critical advantage. By leveraging continuous equipment health monitoring and pattern recognition, SCADA-integrated systems can preemptively schedule maintenance activities, thereby avoiding unexpected breakdowns and ensuring asset longevity. Collectively, these advancements have ushered in a new era of industrial efficiency. The seamless integration of real-time data acquisition, intelligent human interfaces, and robust communication protocols has not only improved productivity and reduced unplanned downtime but also transformed the very nature of industrial decision-making; from reactive to predictive, from manual to autonomous.

E. A. Hasso (✉)
Mechatronics Engineering Department, Tishk International University, Erbil, Iraq
e-mail: ezideen.hasso@tiu.edu.iq

N. Derbel and Q. Zhu (eds.), *Advances in Communication, Signal and Image Processing*,
https://doi.org/10.1007/978-981-95-0401-5_2

As industries continue to evolve, the strategic use of SCADA, HMI, and communication technologies will remain a cornerstone of sustainable, data-informed production systems.

Keywords PLC · Supervisory control and data acquisition · HMI · Idustrial applications · Integrated systems

2.1 Historical Evolution of SCADA: A Global Perspective (Ingram and Green 2023)

2.1.1 Introduction to Early Automation Concepts (Pre-1950s)

Before the advent of SCADA systems, industrial control was largely manual or electromechanical, relying on local instrumentation panels, analogue meters, and relay-based logic systems. Feedback mechanisms were either operator-led or basic PID control loops using pneumatic instruments. These systems lacked centralized visibility and scalability, especially across distributed processes (Boyer 2010; Stallings and Miller 2017).

2.1.2 1950s–1960s: The Birth of Remote Monitoring and Control

SCADA systems began to take form in the 1950s, driven by the need to monitor and control large utility infrastructures such as electrical grids, oil pipelines (Rashad et al. 2022), and water distribution networks. These early systems utilized:

- Telemetry systems with dedicated analogue telephone lines or radio frequency links to transmit data.
- Central control stations using electromechanical relays and hardwired logic.
- Mimic boards to visually represent the physical layout of the monitored system.

Military radar and NASA telemetry systems were pivotal in advancing the idea of long-distance monitoring. Some of the first implementations occurred in municipal water treatment plants in the United States and Europe.

2.1.3 1970s: Centralization and Mainframe Integration

As mainframe computers became more accessible, SCADA systems transitioned to centralized digital control. Key developments included:

- Introduction of RTUs (Remote Terminal Units) to collect data from field devices.
- Use of proprietary protocols by companies like GE, Westinghouse, and ABB.
- Integration with central computing units for rudimentary data logging and alarm management.

Regional deployments in the United States' power sector and France's national water management efforts marked this era. These systems, however, were expensive, difficult to scale, and vendor locked as they were purpose built to a specific task.

2.1.4 1980s: The Microprocessor Revolution

The rise of microprocessors enabled decentralization and cost reduction. Major milestones:

- Distributed Control Systems (DCS) emerged in oil refineries and chemical plants.
- Standard bus architectures (e.g., RS-232, RS-485) improved connectivity.
- Birth of Modbus protocol (1979 by Modicon) based on request-response protocol embracing a master-slave technique, which remains widely used.
- SCADA adoption spread to Asia, especially in Japan's industrial sector and India's railway electrification.

These systems supported larger networks and basic SCADA-HMI interactions.

2.1.5 1990s: SCADA and HMI Merge on Windows Platforms

The 1990s witnessed an explosion of PC-based control systems. Graphical HMIs replaced static interfaces, and Windows NT and Windows 95 platforms supported open SCADA architectures. Developments included:

- WYSIWYG HMI editors with tag browsing, (what you see is what you get).
- OPC standardization (Open Platform Communications) (Object Linking and Embedding (OLE) for Process Control) to unify disparate data sources.
- Integration of SCADA with LANs and early Ethernet networks.
- Global vendors like Siemens, Schneider, and Rockwell released integrated HMI-SCADA solutions.

This decade also marked the rise of PLCs as affordable, rugged control units tightly coupled with SCADA/HMI software.

2.1.6 2000s–2010s: Web-Based SCADA and Global Standards

- Introduction of web-based HMI panels and thin-client architectures.
- Evolution of OPC into OPC UA (Unified Architecture), providing cross-platform and encrypted communication.
- Protocols like IEC 61850, DNP3 Secure Authentication, and MQTT gained traction.
- Cloud technologies began integrating with SCADA for remote asset management.
- Siemens introduced SIMATIC WinCC flexible and WinCC OA, allowing modular and scalable deployments.

2.1.7 2010s–Present: Smart SCADA and Industry 4

Modern SCADA systems emphasize real-time analytics, cybersecurity, and mobility. Innovations include:

- Cloud-hosted SCADA with REST APIs and MQTT brokers.
- Edge computing devices interfacing PLCs with cloud dashboards.
- Use of AI/ML for predictive maintenance and anomaly detection.
- SCADA platforms integrated with Industrial IoT (IIoT) ecosystems like Siemens MindSphere, enabling enterprise-level visibility (Mekki et al. 2020; Zhang et al. 2023).

2.1.8 Comparative Global Trends

North America focuses on grid modernization and NERC-CIP cybersecurity compliance.

- Europe prioritizes interoperability through standards like IEC 60870 and ISO 27001.
- Asia-Pacific leads in scale—smart cities in China and India deploy SCADA in transportation, utilities, and environmental monitoring (Monticelli and Annakkage 2024; Hossain et al. 2024; Li et al. 2023).

2.2 Communication Protocols and Architectures in SCADA Systems

2.2.1 Overview of SCADA Communication Layers

Communication in SCADA systems can be visualized using the OSI model. The layers span from:

- Physical Layer: Cables, RF, and fiber optics
- Data Link Layer: Ethernet, RS-485, CAN
- Network Layer: TCP/IP, UDP
- Application Layer: Modbus, OPC, DNP3, IEC 60870, MQTT (VanDoren et al. 2017).

2.2.2 Legacy Protocols and Their Limitations

Early SCADA systems relied on manufacturer-specific protocols, leading to compatibility issues. Examples include:

- Modbus RTU/ASCII: Simple serial communication; lacks encryption
- DNP3: Designed for power systems; supports time-stamped events
- IEC 60870-5-101/104: Widely used in Europe for telemetry.

2.2.3 Modern Protocols and Interoperability Standards

- OPC UA: Platform-independent, secure, and extensible
- MQTT: Lightweight, publish/subscribe messaging for IIoT
- BACnet and LonWorks: Used in building automation
- IEC 61850: Rich object models for substations (Saber et al. 2023).

2.2.4 Communication Media and Topologies

Depending on geography and criticality:

- Wired (Ethernet, fiber) for reliability
- Wireless (Wi-Fi, 4G/5G, LoRaWAN) for remote areas
- Hybrid architecture; combining both wired as well as wireless.

2.2.5 Cybersecurity Considerations

With IT-OT convergence, SCADA networks face cybersecurity challenges:

- Firewalls, DMZs, VLANs for segmentation
- Deep Packet Inspection (DPI) for anomaly detection
- Zero Trust Architecture models are emerging
- Protocols like OPC UA incorporate encryption, authentication, and redundancy (Beigh and Rather 2021).

2.2.6 Role of SCADA Gateways and Edge Devices

Modern SCADA deployments utilize:

- Edge gateways to normalize and preprocess field data.
- Protocol converters to integrate legacy systems.
- IIoT platforms like Siemens Industrial Edge to bridge field and cloud (Fig. 2.1).

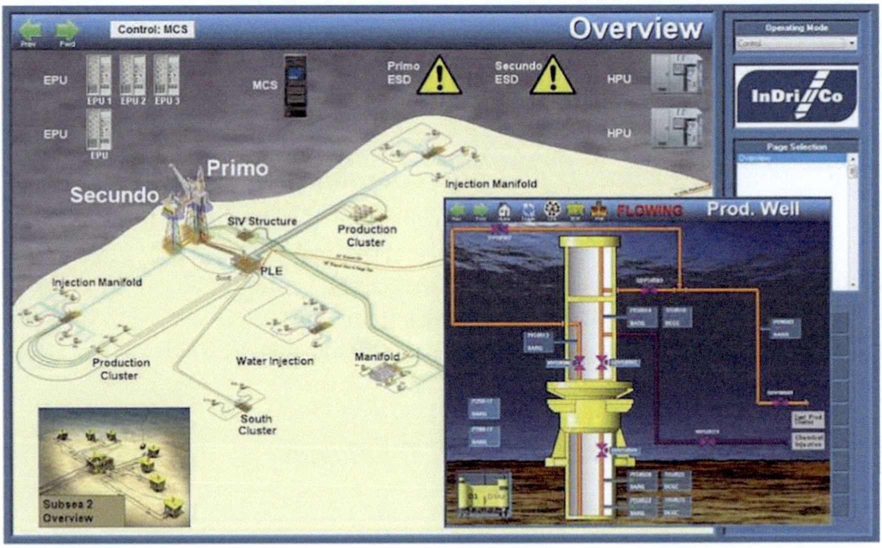

Fig. 2.1 Integrating HMI with SCADA under PLC control (Trihedral 2024)

2.2.7 *Integration of HMI and SCADA in Communication Context*

Graphical HMIs now serve as real-time interfaces driven by SCADA backends. Notable trends include:

- Tag-based architectures in tools like Siemens SIMATIC WinCC.
- Bidirectional OPC UA communication for diagnostics and control.
- Remote HMIs via VPNs or cloud-based HMIs.

2.2.8 *Real-Time Communication and Determinism*

Industrial automation demands deterministic response:

- Time-Sensitive Networking (TSN) ensures bounded latency.
- PROFINET IRT and EtherCAT for motion-critical applications.
- SCADA systems with event-driven architectures to prioritize alarms.

2.2.9 *A Complete SCADA Installation Generally Delivers Four Core Functions*

1. Collecting data from the process.
2. Transmitting that data across a communications network.
3. Displaying the information to operators in a meaningful way.
4. Issuing control actions back to the field.

To accomplish these tasks, a SCADA architecture relies on four principal component groups:

1. Field sensors and control relays (digital or analog)—Devices that make direct, physical contact with the equipment or process being monitored and actuated.
2. Remote Telemetry Units (RTUs)—Compact, rugged computers installed on-site that aggregate sensor readings and pass control commands to the relays.
3. Master stations/Human–Machine Interfaces (HMIs)—Centralized computer systems that provide the operator interface, run supervisory logic, and automatically adjust the process based on incoming data.
4. The communications backbone—The network medium (wired, wireless, or mixed) that links the master station with all distributed RTUs in the field (Fig. 2.2).

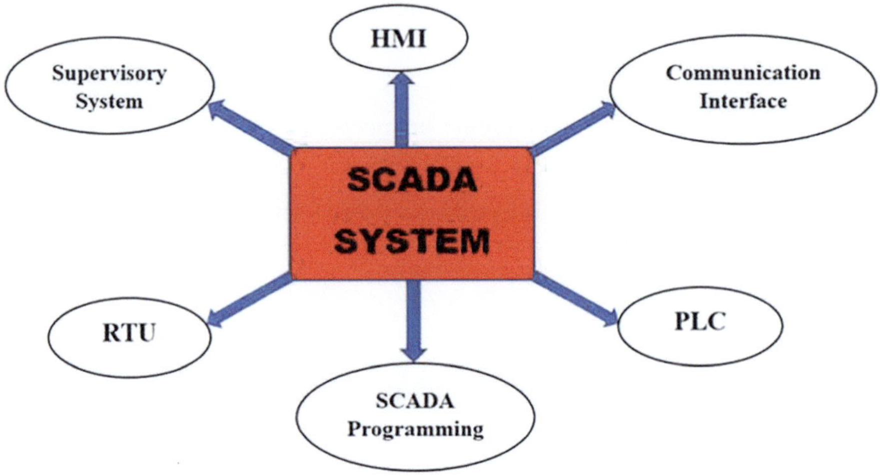

Fig. 2.2 Schematic diagram of SCADA based PLC-driven system

2.3 Integration of SCADA and HMI in PLC-Driven Industrial Systems

2.3.1 The Central Role of PLCs in Automation

Programmable Logic Controllers (PLCs) are foundational components in industrial automation. They act as the intermediary between physical field devices and supervisory control systems. Their deterministic behavior, rugged hardware, and real-time I/O processing make them ideal for critical control tasks. With the evolution of SCADA and HMI, PLCs have increasingly become integrated into holistic control architectures.

2.3.2 Architectures of Integration

Integration models vary across industries, but common topologies include:

- PLC–HMI–SCADA Stack: PLCs handle control, HMIs handle local visualization, and SCADA systems provide enterprise-level monitoring.
- Integrated Platforms: Solutions like Siemens SIMATIC WinCC embed SCADA and HMI directly into the PLC programming environment (e.g., TIA Portal), reducing complexity.

2.3.3 Communication Standards Enabling Integration

- PROFINET and PROFIBUS: Siemens-led protocols enabling fast, deterministic communication.
- EtherNet/IP and Modbus TCP: Common in North American industries.
- OPC UA and MQTT: Layered on top for cross-vendor and cloud connectivity.

2.3.4 Siemens SIMATIC WinCC Case Study

Siemens' WinCC family offers a scalable SCADA-HMI ecosystem that integrates deeply with its PLCs:

- WinCC Professional and WinCC Unified allow flexible deployment.
- WinCC OA (Open Architecture) supports complex, large-scale systems.
- Seamless integration with TIA Portal ensures consistent tag databases, version control, and deployment.
- Integration with Siemens S7-1500 PLCs enables native support for diagnostic functions, trending, and alarms.

2.3.5 Data Flow and Tag Management

- Tag mapping ensures consistency between PLC addresses, HMI elements, and SCADA logging.
- Symbolic addressing in tools like TIA Portal reduces errors and improves scalability.
- Real-time data is displayed on HMI panels while historical data is archived in SCADA databases.

2.3.6 Examples of Human-Centered Design in HMI

- Emphasis on alarm rationalization and color standards (ISA-101, ISO 11064).
- Use of gesture-based HMI, multi-touch panels, and mobile clients.
- Integration of voice-assisted interfaces and Augmented Reality (AR) in maintenance.
- Water Industries.
- Car Manufacturing Industries.
- Food and Packaging Industries.
- Warehouse Automations.
- Smart oil well drilling heads communication. (Namuq et al. 2024; Shastry 2023; Vicente and Lee 2024; Parasuraman et al. 2024; Sanderson et al. 2023; Al-Nuaimi et al. 2023).

2.3.7 Troubleshooting and Diagnostics

Modern integrated systems allow for:

- Live diagnostics from PLCs to SCADA dashboards.
- Scripted alerts in WinCC when PLC faults occur.
- Remote maintenance using VPN-secured engineering stations.

2.3.8 Benefits of Tight Integration

- Reduced engineering time and lower TCO.
- Higher system reliability due to coherent fault management.
- Future readiness with cloud and AI integrations.

2.4 Modern SCADA Architectures, IIoT, and Edge Computing

2.4.1 Convergence of SCADA and IIoT

Industry 4.0 envisions factories with intelligent assets capable of self-optimization. In this landscape, SCADA evolves beyond traditional supervisory roles into being a central orchestrator within IIoT ecosystems. Key enablers include:

- Smart sensors and actuators with native IP stacks
- Plug-and-play connectivity via protocols like OPC UA and MQTT
- Cloud SCADA dashboards accessible from any device.

2.4.2 Edge Computing for Real-Time Processing

Instead of sending raw data to the cloud, edge devices perform preprocessing, filtering, and analytics locally. Benefits include:

- Reduced latency and bandwidth usage
- Improved reliability during connectivity outages
- Support for AI-based models (e.g., anomaly detection, quality prediction).

Edge devices are typically deployed as:

- Rugged industrial PCs or dedicated edge gateways
- Equipped with multi-protocol support and Docker containers for modularity.

2.4.3 *SCADA in the Cloud: Benefits and Risks*

Cloud-based SCADA offers scalability, centralized access, and easier maintenance. Common use cases:

- Remote monitoring of distributed assets (e.g., solar farms, pump stations)
- Multi-site aggregation and reporting
- Disaster recovery and backup services.

Risks include:

- Data sovereignty and compliance concerns
- Connectivity dependence
- Attack surface expansion.

2.4.4 *Digital Twins and Virtual Commissioning*

Modern SCADA systems can link with digital twins—real-time virtual models of physical assets:

- Simulation of control strategies prior to physical deployment.
- Predictive diagnostics based on historical and real-time data.
- Integration with CAD and PLM platforms for design feedback.

2.4.5 *AI and Machine Learning Integration*

SCADA platforms are beginning to support ML modules:

- Anomaly detection using unsupervised learning
- Predictive maintenance leveraging regression models
- Energy optimization via reinforcement learning agents.

Vendors such as Siemens, AVEVA, and Honeywell offer platforms with integrated AI toolkits.

2.4.6 *Case Study: Smart Water Management with SCADA-IIoT Integration*

In Singapore's PUB water agency:

- Thousands of smart sensors feed data into a centralized SCADA platform
- Edge devices at pump stations preprocess flow and pressure data
- AI models predict pipe bursts and optimize chlorine dosing
- Operators access real-time dashboards via cloud-based HMIs (Kumar and Garg 2021; Seno and Wibowo 2022; Sharma and Bansal 2022).

2.4.7 *Future Trends*

- 5G and private cellular networks enabling ultra-low latency SCADA.
- Blockchain for secure event logging and audit trails.
- AR/VR for immersive control room interfaces.
- Self-healing networks with AI-driven reconfiguration.

2.4.8 *Conclusion*

Modern SCADA is no longer just about supervision—it is the digital nervous system of smart factories, cities, and infrastructures. With IIoT and edge computing, SCADA becomes adaptive, intelligent, and scalable.

2.5 Cybersecurity in SCADA and IIoT Environments

2.5.1 *The Unique Nature of SCADA Cybersecurity*

Unlike traditional IT networks, SCADA systems operate in real-time, manage critical infrastructure, and often use legacy hardware that cannot be easily patched. Security in these environments must balance protection with system availability, making traditional IT cybersecurity strategies insufficient without adaptation.

2.5.2 *Threat Landscape and Common Attack Vectors*

SCADA systems are increasingly targeted by both state-sponsored and criminal actors. Common attack vectors include:

- Spear phishing and social engineering to gain initial access.
- Exploiting insecure protocols (e.g., Modbus without encryption).
- Insider threats from disgruntled employees or third-party contractors.
- Remote access exploits through poorly configured VPNs or cloud portals.

2.5.3 High-Profile Cyber Attacks on SCADA Systems

Several attacks have highlighted SCADA vulnerabilities:

- In 2010 a highly sophisticated worm targeting Siemens PLCs, it disrupted Iran's uranium enrichment program. Stuxnet worm "targeted high-value Iranian assets", Published by BBC 23 September 2010.
- Cyber-Attack Against Ukrainian Critical Infrastructure in 2015 disrupting Ukraine's power grid, highlighting SCADA's role in national infrastructure (as reported by major news channels).
- Oldsmar Water Plant of Florida, USA in 2021 a hacker, remotely, attempted to poison a Florida water supply using SCADA access (reported by BBC 9 February 2021).

2.5.4 Defense-In-Depth for SCADA Systems

A multi-layered security approach is essential. Key measures include:

- Network segmentation: Separating IT and OT environments using firewalls, VLANs, and DMZs.
- Least privilege access: Ensuring users only access necessary systems/functions.
- Patch management: Controlled, tested updates especially on field devices.
- Whitelisting and application control: Preventing unauthorized software from executing. (ISA 2023; Stouffer et al. 2024a, b; Radvanovsky and Bongard 2022).

2.5.5 Security Standards and Compliance Frameworks

Organizations must adhere to cybersecurity standards tailored to industrial systems (Stouffer et al. 2024a, b; Knowles and Nodari 2023):

- NERC CIP (Critical Infrastructure Protection) for North American energy systems.
- IEC 62443 for industrial control system cybersecurity.
- NIST SP 800-82: Guide to Industrial Control System Security.
- ISO/IEC 27019: Information security for energy automation.

2.5.6 Secure Communication Protocols and Practices

Modern SCADA networks adopt secure protocols and practices:

- OPC UA with encryption and authentication.
- TLS/SSL for secure web interfaces.
- Use of VPNs with two-factor authentication for remote access.
- DNP3 Secure Authentication for power systems (Al-Anzi and Medoff 2020).

2.5.7 Intrusion Detection and Monitoring in OT Networks

Specialized intrusion detection systems (IDS) for OT environments monitor traffic for anomalies:

- Deep Packet Inspection (DPI) tools tailored for ICS protocols.
- Passive network monitoring to avoid impacting performance.
- SIEM integration with IT security dashboards. (Ahmed and Kim 2020; Lemay et al. 2018).

2.5.8 AI-Driven Threat Detection in SCADA Environments

AI and ML are increasingly used to enhance SCADA security (Alcaraz and Lopez 2024; Koay et al. 2023):

- Behavioral baselining to detect deviations in operator or device behavior.
- Anomaly detection using time-series models.
- Threat intelligence fusion from multiple sources for proactive alerts.

2.5.9 Supply Chain and Third-Party Risks

Vulnerabilities may be introduced through vendors, integrators, or component suppliers:

- Secure procurement policies and vetting of OEMs.
- Firmware signing and validation to ensure authenticity.
- Access control audits for third-party service providers.

2.5.10 *Future Directions in SCADA Cybersecurity*

As SCADA becomes more connected, the future will see:

- Zero Trust Architectures in OT environments
- Blockchain for tamper-proof logging and authentication
- Post-quantum cryptography for long-term resilience
- Cyber ranges and digital twins for simulated security training and testing (Al-Anzi and Medoff 2020; Mitra and Sikdar 2021).

2.6 Conclusion

SCADA systems have undergone a profound transformation over the past seven decades–from the rudimentary days of analogue telemetry and relay-based control panels to today's advanced, intelligent platforms powered by AI, edge computing, and the Industrial Internet of Things (IIoT). This evolution has not only increased the functional capabilities of SCADA but has also redefined the way industries monitor, control, and optimize their processes. The once isolated and hardware-centric SCADA infrastructures have become interconnected, software-driven ecosystems enabling holistic, real-time operational visibility across geographically distributed assets.

As industries worldwide continue to embrace digital transformation, the integration of SCADA with advanced Human-Machine Interfaces (HMIs), robust networking protocols, and secure communication frameworks has transitioned from being a competitive advantage to a business-critical necessity. HMIs, once limited to static, hardwired control panels, now deliver rich, interactive visualizations with data-driven dashboards and mobile accessibility. Together with SCADA, they form the backbone of intelligent control systems capable of supporting data-centric decision-making, anomaly detection, and continuous optimization.

The evolution of Siemens SIMATIC WinCC stands as a prominent example of this shift. Initially conceived as a traditional desktop-based HMI/SCADA solution, WinCC has progressively adapted to the demands of modern industrial environments. The platform has expanded into web-enabled, modular, and scalable architectures, such as WinCC Unified and WinCC Open Architecture (OA). These iterations are not only compatible with cloud-based services but also support containerized deployment, flexible user management, and seamless integration with industrial data lakes and MES/ERP systems. Such advancements underscore a global trend toward openness, interoperability, and cybersecurity by design.

Today's SCADA solutions are expected to go beyond basic data acquisition and control. They must deliver high availability, real-time responsiveness, and compliance with international cybersecurity standards such as IEC 62443, NIST SP 800-82, and OPC UA Security Profiles. As the threat landscape continues to grow in complexity, modern SCADA must include capabilities for role-based access control, encrypted data exchange, secure remote access, and continuous monitoring for threat detection. Failure to secure SCADA infrastructures can have devastating consequences–not only in terms of operational downtime but also in public safety and national security.

Looking forward, SCADA will continue to serve as the nerve center of industrial automation. However, its role is evolving from reactive control to predictive intelligence and even autonomous decision-making. By leveraging AI and machine learning, future SCADA systems will enable condition-based maintenance, intelligent alarming, and self-optimizing control loops, thereby improving operational efficiency and reducing costs. Moreover, as IT (Information Technology) and OT (Operational Technology) converge, SCADA will become a key enabler of enterprise-wide data orchestration–bridging control rooms with business systems to align operational insights with strategic goals.

The integration of emerging technologies such as digital twins, blockchain-based audit trails, edge analytics, and 5G connectivity will further elevate SCADA's capabilities. These innovations promise to make industrial systems more adaptive, transparent, and resilient in the face of dynamic global challenges such as energy transition, sustainability targets, and workforce shortages.

In summary, the future of SCADA is one of intelligent convergence–where systems are not only automated but also context-aware, secure, and adaptive to change. As technology continues to evolve, so too must our strategies for designing, deploying, and safeguarding SCADA infrastructures to ensure they remain robust, agile, and aligned with the needs of tomorrow's industries.

References

Ahmed, C.M., Kim, H.: Behavior rule specification-based anomaly detection for smart grid SCADA systems. IEEE Trans. Smart Grid **11**(3), 2394–2405 (2020). https://doi.org/10.1109/TSG.2019.2941692 SCADA environments

Al-Anzi, T., Medoff, M.: Implementing IEC 62443: A Pragmatic Approach to Cyber-Security. ISA, 3rd edn (2023). IEC 62443-3-3 secure-by-design requirements Hands-on guidelines for segmenting SCADA networks, hardening protocol endpoints, and auditing secure channels

Alcaraz, C., Lopez, J.: Digital Twin-assisted anomaly detection for industrial scenarios. Int. J. Crit. Infrastruct. Protect. **47**, 100721 (2024)

Al-Nuaimi, A.M., Omer, A.M., Aziz, H.A.: A cloud enabled SCADA–IIoT architecture for real time leak detection and energy optimization in municipal water networks. J. Water Res. Plan. Manage. **149**(3), 04023005 (2023). https://doi.org/10.1061/(ASCE)WR.19435452.0001632

Beigh, B.M., Rather, G.M.: Trends and challenges in the security of SCADA systems. Comput. Secur. **105**, 102233 (2021). Explores evolving threats to SCADA networks with mitigation strategies involving AI-based intrusion detection and blockchain technologies

Boyer, S.A.: SCADA: Supervisory Control and Data Acquisition, 4th edn. ISA—The International Society of Automation (2010)

Boyer, A.: SCADA & MES: Past, Present, and Future, 5th edn. ISA—International Society of Automation, Research Triangle Park, NC (2022). A major update of Boyer's classic text. The first three chapters provide a detailed timeline from 1950s utility telemetry to edge-and-cloud SCADA, with new material on OPC UA, MQTT, cybersecurity, and digital twins

Choo, W.T., Lim, T.S.: Digitalising a nation's water supply: Singapore's smart water grid. In: Nero, J.C.D., King, J. (eds.) Smart Water Utilities: Complexity Made Simple, 2nd edn, pp. 141–170. IWA Publishing, London (2022). ISBN 978-1-78065-800-4

Falliere, N., Murchu, L.O., Chien, E.: W32.Stuxnet Dossier. Symantec Security Response (2011). https://www.symantec.com/content/en/us/enterprise/media/security_response/whitepapers/w32_stuxnet_dossier.pdf

GIFA IPC Blog Post: The Evolution of Industrial Controllers (2023). gifaipc.com

Hossain, E., et al.: A Secure, IEC 61850-Compliant SCADA Architecture for Distributed Renewable Energy Management. IEEE Trans. Smart Grid **15**(2), 1551–1564 (2024). Describes a field deployment that fuses solar-farm RTUs, microgrid PLCs, and a WinCC OA-based HMI; details role-based access and TLS-encrypted Goose messaging

Ingram, T., Green, D.: Control, Supervisory, Acquisition, Data: A Comprehensive Guide to Design, Security, and Digital Transformation. IET Press, London (2023). 978-1-83953-514-8, Chapter 2 (A Brief History of SCADA) walks through milestones from relay logic and RTUs to modern WinCC Unified and cloud dashboards. The book then ties each historical phase to current best practices in secure, converged IT/OT designs

ISA International Society of Automation: (ISA) ISA/IEC 62443-3-3:2023—System Security Requirements and Security Levels. Research Triangle Park, NC: ISA, 3rd edn (2023). Defines the multi-layer security model used worldwide for SCADA/ICS. Details zoning & conduits, least-privilege access, secure remote maintenance, and patch-management–core pillars of defense-in-depth

ISO 11064-1 to -7: Ergonomic Design of Control Centres. 3rd edn, International Organization for Standardization (2023), 2023 update. The newest revision of the global standard on control-room layout, display design, and human-centered evaluation techniques. Widely cited in HMI projects

Knowles, W., Nodari, R. (eds.): Securing the Industrial Internet of Things: SCADA, Edge, and Cloud Defenses. Springer, Cham (2023). Multi-author volume devoted to practical defenses: encrypted OPC UA PubSub, secure MQTT brokers, anomaly-detection ML for RTUs, and incident reports on Stuxnet-type attacks. Case studies span oil & gas pipelines, water utilities, and smart-grid substations–ideal context for your "Cybersecurity in SCADA and IIoT Environments" subsection

Koay, A.M.Y., Ko, R.K., Hettema, L.H., Radke, K.: Machine learning in industrial control system (ICS) security: current landscape, opportunities and challenges. J. Intell. Syst. **60**, 377–405 (2023)

Kumar, R., Garg, V.: Cloud-based SCADA systems for water distribution networks. J. Water Res. Plan. Manage. **147**(6) (2021). Focuses on SCADA implementations in smart cities, particularly for water quality monitoring and automated leak detection

Lemay, A., Calvet, J., Menet, F., Fernandez, J.M.: Survey of public datasets for intrusion detection in SCADA systems. In: Proceedings of the 3rd ACM Workshop on Cyber-Physical System Security (2018). https://doi.org/10.1145/3198458.3198464

Li, Z., Wang, Y., Zhang, Q.: An integrated SCADA-HMI platform for real-time utility management in the shanghai smart-city pilot. IEEE Access **11**, 107 365–107 382 (2023). https://doi.org/10.1109/ACCESS.2023.3289475

Mekki, K., et al.: A survey of SCADA systems communication protocols: evolution, vulnerabilities and challenges. IEEE Commun. Surv. Tutor. **22**(3), 1839–1878 (2020). Covers the full range of SCADA communication protocols such as Modbus, DNP3, IEC 60870-5-104, and OPC UA with an emphasis on security gaps and interoperability

Mitra, P., Sikdar, B.: Blockchain-enabled secure communication for industrial IoT and SCADA systems. IEEE Internet of Things J. **8**(4), 2344–2354 (2021). Analyzes how distributed ledger technologies can protect data integrity and access control in SCADA-enabled IIoT environments

Monticelli, A., Annakkage, U.: Power-System SCADA and Energy Management Systems: Foundations and Strategies for the Smart Grid. 3rd edn. Wiley-IEEE Press, Hoboken, NJ (2024). ISBN 978-1-119-83217-6. Comprehensive EMS/SCADA text updated for IEC 61850, CIM,

Synchrophasor (PMU) integration, and advanced HMI dashboards used in modern control centers

Namuq, M.A., Hasso, E.A., Jamal, M.A., Namuq, K.A., Yu, Y.: Simulation and modeling of data transmission process in boreholes using intelligent drill pipe for a laboratory experiment. MDPI Modell. **5**(4), 1961–1979 (2024). https://doi.org/10.3390/modelling5040102

Parasuraman, R., et al.: Human-centered automation in industry 4.0: design guidelines for adaptive HMIs. IEEE Trans. Hum.-Mach. Syst. **54**(1), 15–29 (2024). Empirical study on adaptive interfaces that change complexity based on operator workload; provides measurable HCD metrics

Radvanovsky, R., Bongard, J.: Handbook of SCADA/Control Systems Security, 2nd edn. CRC Press, Boca Raton, FL (2022). Part II (Multi-Layered Protections) provides case studies from power, water, and oil-and-gas sectors illustrating practical deployment of layered firewalls, intrusion-detection sensors, and secure PLC programming practices

Rashad, O., Attallah, O., Morsi, I.: A PLC-SCADA pipeline for managing oil refineries. In: Proceedings of the 5th International Conference on Computing and Informatics (ICCI) (2022). This paper presents a PLC/HMI system for tracking oil products in refineries, utilizing Add-On Instruction (AOI) programming to enhance processor capabilities and streamline ladder logic programs

Saber, M.F., Shareef, A.A., Ali, S.K., Hamad, O.S., Ahmed, M.G.: Innovative SCADA-based oil refinery control with Arduino integration using LabVIEW. Int. J. Multidiscip. Res. Anal. **6**(10) (2023). This study explores the implementation of a SCADA system integrated with Arduino and LabVIEW to monitor and control oil levels in a refinery tank, offering a cost-effective alternative to traditional DCS systems

Sanderson, P.M., et al.: From big data to big insights: human-centered dashboard design for smart manufacturing. J. Manuf. Syst. **68**, 467–480 (2023). Presents a usability study comparing traditional SCADA screens to modern HCD dashboards; offers practical design heuristics and UX metrics

Seno, T.H., Wibowo, Y.: Developing human-machine interfaces (HMI) for real-time monitoring of industrial processes using SCADA and Python. J. Ind. Eng. Res. **8**(2), 45–58 (2022). Demonstrates custom HMI development using open-source tools like Python and Node-RED, showcasing flexibility for small-to-medium industries

Sharma, A., Bansal, R.C.: Integration of IoT and SCADA for real-time industrial automation: a review. Int. J. Electr. Power Energy Syst. **135**, 107510 (2022). This paper reviews how IoT technologies enhance SCADA performance across sectors such as manufacturing, agriculture, and logistics

Shastry, C.: Design and implementation of an oil and gas surface processing plant using digital SCADA. Int. J. Res. Appl. Sci. Eng. Technol. (IJRASET) **11**(4) (2023). The paper details the interfacing of PLC, SCADA, and HMI systems in an oil and gas processing plant, emphasizing the use of OPC KEPServer for communication between devices

Stallings, J., Miller, W.: Distributed Control and Cyber Security Issues in Process Automation Systems (2017). Chapter 2 offers a brief history of control architectures, emphasizing how plant wide visibility was absent prior to DCS/SCADA. Momentum Press

Stouffer, K.A., Torres, C.K., Scarfone, K.A.: Guide to Industrial Control System Security, Rev. 3. NIST Special Publication 800-82r3. National Institute of Standards and Technology, Gaithersburg, MD (2024a). Chapter 6 (Defense-in-Depth Strategies) offers step-by-step implementation for SCADA: network segmentation, DMZ design, layered authentication, and coordinated incident response

Stouffer, K.A., Torres, C.K., Scarfone, K.A.: Guide to Industrial Control System (ICS) Security, Revision 3. NIST Special Publication 800-82r3. National Institute of Standards and Technology, Gaithersburg, MD (2024b). The newest NIST guide (February 2024) adds a dedicated chapter on IIoT/edge nodes in SCADA networks, covers zero-trust zoning, secure-by-design PLCs, and maps controls to IEC 62443-3-3. Highly authoritative for best-practice frameworks and threat modeling

Trihedral Engineering Limited: VTScada Software (2024). https://www.vtscada.com/oil-and-gas-scada/

VanDoren, Vance and ISA Staff: A Guide to the Automation Body of Knowledge, ISA, 3rd ed. Chapter on control system history covers local instrumentation panels, hard-wired relay logic, and the emergence of PLC, DCS, and SCADA (2017)

Vicente, K.J., Lee, J.D.: Engineering for Humans: Cognitive Systems Design in Industrial HMI, 2nd ed. CRC Press. A fresh edition of a cognitive-engineering classic, expanded with case studies on IIoT dashboards, alarm design, and augmented-reality HMIs (2024)

Zhang, Y., Wang, H., Xu, Y.: A digital twin approach for smart manufacturing based on SCADA and edge computing. J. Manuf. Syst. **66**, 421–433. Discusses integration of SCADA and digital twins in Industry 4.0, with case studies from automotive and electronics manufacturing

Chapter 3
Direct or Relay D2D Communication for Network Lifetime Enhancement

Anouar Ben Abdennour, Taoufik Ben Ahmed, Mohamed Ouwais Kabaou, and Belgacem Chibani Rhaimi

Abstract Device-to-device (D2D) communication stands out as a promising technology for future wireless networks. It enables direct communication between devices, improving latency, reliability, throughput, and power consumption. These reduce also network load and increases cellular coverage. In addition, a device can act as a relay between the base station and another device located outside. Therefore, D2D communication can help meet the requirements of future 5G (5th Generation) wireless networks (Pirinen in 1st International conference on 5G for ubiquitous connectivity, pp. 17–22, 2014). In this chapter, we will focus on energy consumption in direct and relayed D2D communications. We will study the basic model of D2D communication. This will be used for calculating the energy consumed by different proposed topologies. The study will be completed by evaluating the total energy consumed by each topology.

Keywords D2D · Communication · Energy · 5G · Relay

3.1 Introduction

Telecommunications have undergone a significant evolution since their former appearance. The progress has firstly referred to analog models that are completely different compared to actual numerical ones. The accomplished effort has introduced many models that were built number of limitations. Various generations have been proposed but each time an enhancement must be introduced. Among these limitations, one can name e.g. only simple voice communication via an analog network, as proposed by the first-generation. However, researchers have deployed many efforts such that a revolution was induced thanks to the discrete processing's technology.

A. Ben Abdennour (✉) · T. Ben Ahmed · M. O. Kabaou · B. Chibani Rhaimi
National School of Engineering of Gabes, University of Gabes, Gabes, Tunisia
e-mail: ben.abdennour.anouar@gmail.com

B. Chibani Rhaimi
e-mail: abouahmed17@gmail.com

N. Derbel and Q. Zhu (eds.), *Advances in Communication, Signal and Image Processing*,
https://doi.org/10.1007/978-981-95-0401-5_3

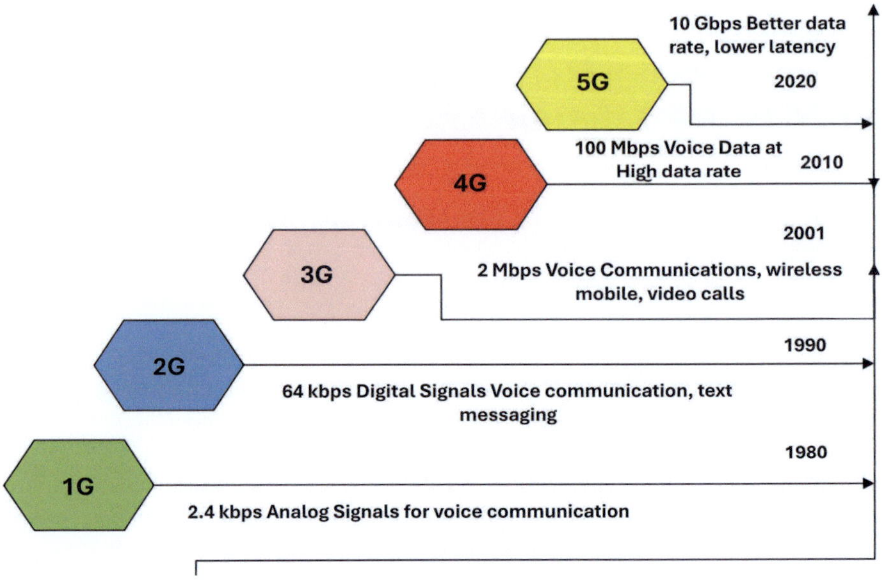

Fig. 3.1 Wireless communications technology evolution

Services have becoming plenty diversified. Added to voice's exchanging, customers have possibility to choose other multiple and various services as offered based on digital networks. The transmission of data (text messages, etc.) has becoming possible. Currently, with the latest generation of mobile technologies, communication could be established between mobile customers with respect of high definition video requirements. Figure 3.1 illustrates the significant and continuous advancement of communication technologies throughout the past several decades (Reka et al. 2019; Agyapong et al. 2014; Hossain et al. 2014).

This is conducted by research activities at the different layers of the protocol stack (Gandotraa et al. 2017). Nowadays, an extension is observed imposing to Future systems being able to offer an interconnection between objects. This has becoming possible thanks to wide world web i.e. internet. This wide network has offered the possibility for objects to send or receive needed information. This defines what technically known Internet of Things (IoT) is. So, a new communication mode named device to device (D2D) has appeared. The mobile stations, in a D2D communication are able to detect each other and communicate via a direct link. Resource management is done, with the aim of not only allowing D2D communication but also supporting future system applications. Nevertheless, the rapid evolution of terminals, their variety (smartphone, tablet,) induces an increased traffic can cause a high energy consumption (Abolfazli et al. 2014; Zhang and Li 2016; Kulkarni et al. 2016).

In this chapter, we are interested to the problem of 5G radio resource management in order to optimize energy consumption in a D2D communication. One will be able to satisfy the needs of all users and therefore could decide implementation of services with good quality. A comparison between direct and relay-based communications will be conducted to reduce the energy consumption of devices involved in D2D communication.

3.2 System Model and Problem Formulation

This section concentrates first on the fundamental model of direct D2D communications, followed by an analysis of relay-assisted D2D communications. We will explain our approach in terms of calculating distances communications. Thereafter, we will determine the different models for calculating the energy consumed. Finally, we will try to explain the optimal model that consumes the least energy in D2D communication.

3.2.1 D2D Communications Model

This part addresses various device-to-device communication topologies. According to the literature, the authors of Tehrani et al. (2014) categorize D2D communications into four distinct types, referred to as "topologies" (Tehrani et al. 2014; Ben Abdennour et al. 2021), as illustrated in Fig. 3.2:

- DR-OC (Device Relaying with Operator Controlled Link Establishment): In this D2D communication mode, a device located at the cell edge or in an area with weak coverage and poor signal quality can communicate with the base station by relaying its data through nearby peripheral devices. A control link is established, allowing the base station to partially or fully manage resource allocation (Olaobaju 2018; Choi et al. 2014; Chevillon et al. 2017; Tang et al. 2007; Iqbal et al. 2019).
- DC-OC (Direct D2D Communication with Operator Controlled Link Establishment): In this mode, the source and destination devices engage in direct D2D communication, transmitting data without passing through the base station. Nevertheless, the base station is responsible for establishing control links to manage radio resources (Olaobaju 2018; Choi et al. 2014; Chevillon et al. 2017; Tang et al. 2007; Iqbal et al. 2019).
- DR-DC (Device Relaying with Device Controlled Link Establishment): This mode is similar to DR-OC; however, unlike DR-OC, no control link is required from the base station. Instead, the source and destination devices coordinate directly and utilize relay devices to enable data transmission (Olaobaju 2018; Choi et al. 2014; Chevillon et al. 2017; Tang et al. 2007; Iqbal et al. 2019).

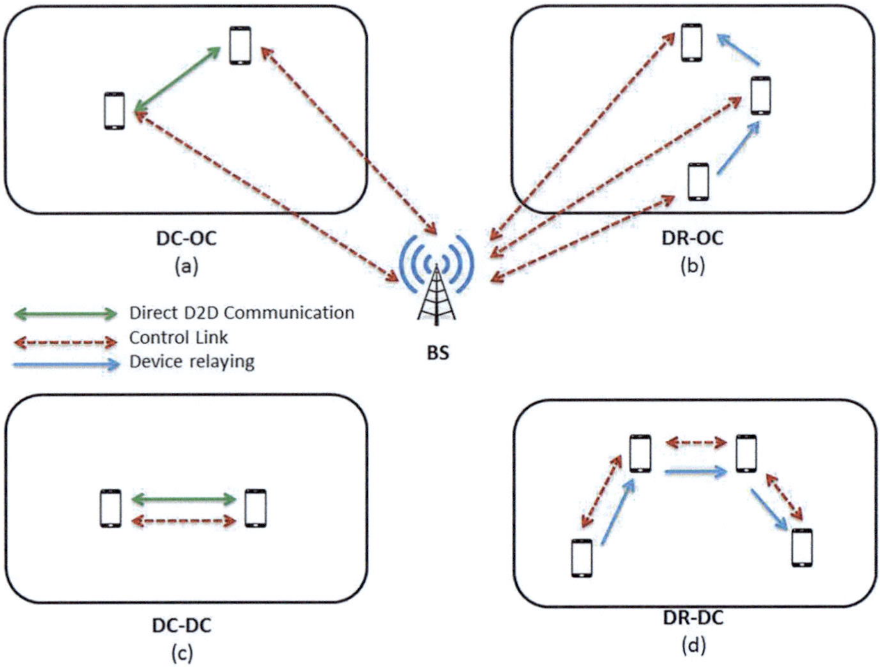

Fig. 3.2 D2D communications topologies

Table 3.1 Used notations

Symbol	Description
UEA	User equipment A
UER	User equipment B
UER	User equipment R (Relais)
d_{AB}	Distance between UEA and UEB
d_{AR}	Distance between UEA and UEA
d_{BR}	Distance between UAB and UAR
D_{UEA}	Data transmitted by UEA
D_{UEB}	Data transmitted by UEB
D_{UER}	Data transmitted by UER

- DC-DC (Direct D2D Communication with Device Controlled Link Establishment): In this D2D communication mode, the source and destination devices communicate directly without any involvement from the base station in establishing control links (Olaobaju 2018; Choi et al. 2014; Chevillon et al. 2017; Tang et al. 2007; Iqbal et al. 2019).

The various notations employed throughout this work are summarized in Table 3.1.

In this study, two topologies are considered: DC-DC and DR-DC. The primary objective is to determine whether there could be an overall power gain in a network consisting of three devices, namely UE_A, UE_B et UE_R.

- UE_A is the source device transmitting its own data.
- UE_B is the destination device, receiving data from UE_A.
- UE_R is the relay device, relaying data from UE_A to UE_B.

In DC-DC topology, UE_A directly transmits its data D_{UEA} to UE_B. This topology involves two transmissions:

1. UE_B sends a discovery signal to UE_A, and
2. UE_A sends its own data D_{EUA} to UE_B.

In DR-DC topology, UE_A transmits its data D_{EUA} to UE_B via the relay node UE_R. Thus, four transmissions occur within the network, as follows:

1. UE_B sends a discovery signal to UE_R
2. UE_R sends a discovery signal to UE_A
3. UE_A sends its data D_{EUA} to UE_R, and
4. UE_R sends data from UE_A (D_{EUA}) to UE_B.

3.2.2 *Modeling Distances Between Devices*

In this part, we will represent the Cartesian coordinates for the devices UE_A, UE_B and UE_R. We move UE_R along a line parallel to the line (UE_A–UE_B) see Fig. 3.3. The coordinates of UE_A, UE_B and UE_R are, respectively, (0, 0), (d_{AR}, 0) and (x_R, d). Thus, we have the following distances:

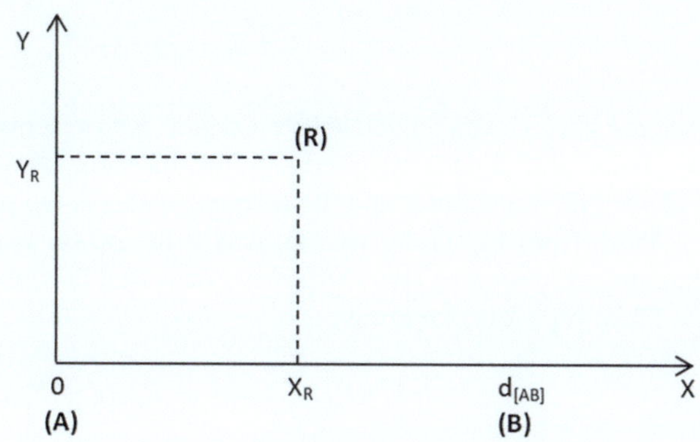

Fig. 3.3 Cartesian representation of the position of UE_R with respect to the positions of UE_A and UE_B

$$d_{AR} = \sqrt{(x_A - x_R)^2 + (y_A - y_R)^2}$$
$$d_{AR} = \sqrt{x_R^2 + d^2}$$
$$d_{BR} = \sqrt{(x_B - x_R)^2 + (y_B - y_R)^2}$$
$$d_{BR} = \sqrt{(d_{AB} - x_R)^2 + d^2}$$

3.2.3 Models for Calculating Energy Consumed

In this section, we will give an idea of calculating the power consumed by each device, as well as the total power in the DC-DC and DR-DC topologies.

3.2.3.1 Transmit Powers

First, it is assumed that the bandwidth used for transmission remains constant over time for all transmitted data. To assess the potential gain in energy efficiency within a D2D network, we will analyze the overall energy consumption of the entire network (Cui et al. 2014).

Initially, we apply Shannon's capacity formula, as referenced by the authors of Choi et al. (2014). This formula provides the relationship between the data rate transmitted by the user equipment i (D_i) and the power transmitted by the user equipment i for the transmission of the data rate D_i (P_i) as follows:

$$C = W \log_2 \left(1 + \frac{S}{N}\right) \quad (3.1)$$

where W, the width of the bandwidth, S the signal strength and N is the power of Additive White Gaussian Noise (AWGN)

$$S = P_i \, |h|^2 \, kr^{-\alpha} \quad (3.2)$$

where k is a constant related to antenna gain, h represents channel fading at transmitter and receiver (fading is assumed to follow the same law at both ends), r is the distance between the transmitter and the receiver, and α is the path loss exponent.

Thus, from Eq. (3.1), we can derive the power consumed by the transmission equipment (assuming that the power consumed is equivalent to the power emitted).

From Eq. (3.1), we know:

$$P_T = P_{UEA} + P_{UEB} + P_{UER} \quad (3.3)$$

The power can be derived by expanding Shannon's capacity formula as follows:

$$\frac{C}{W} = \log_2\left(1 + \frac{S}{N}\right) \tag{3.4}$$

$$P_i = \frac{N}{|h|^2\, k} d_r^{\alpha} \Phi(C_i) \tag{3.5}$$

$$P_{UEA} = \frac{N}{|h|^2\, k} d_{AR}^{\alpha} \Phi(C_A) \tag{3.6}$$

$$P_{UEB} = \frac{N}{|h|^2\, k} d_{BR}^{\alpha} \Phi(C_B) \tag{3.7}$$

$$P_{UER} = \frac{N}{|h|^2\, k} d_{AR}^{\alpha} \Phi(C_R)$$

Using Eqs. (3.6)–(3.8), we can derive Eq. (3.3):

$$P_T = \frac{N}{|h|^2\, k}[\Phi(C_A)\ (d_{AR}^{\alpha} + d_{BR}^{\alpha}) + \Phi(C_B)\ (d_{AR}^{\alpha} + d_{BR}^{\alpha})] \tag{3.8}$$

$$P_i = \frac{N}{|h|^2\, k.r^{-\alpha}}[\exp(\ln 2\ C_i) - 1] = \frac{N}{|h|^2\, k.r^{-\alpha}} \Phi(C_i) \tag{3.9}$$

where $C_i = \frac{D_i}{W}$ (in bit/s/Hz) and $\Phi(C_i) = \exp(\ln 2\ C_i) - 1$.

Our goal is to determine the overall gain in energy efficiency for the entire network. Thus, we define the total power consumption of the network as follows:

$$P_T = P_{UEA} + P_{UEB} + P_{UER} \tag{3.10}$$

where P_{UEA} is the total power consumed by UE_A, P_{UEB} is the total power consumed by UE_B and P_{UER} is the total power consumed by UE_R.

For the DC-DC, DR-DC topologies, we have P_T^D, P_T^R the total power, respectively (Ben Abdennour et al. 2023a, b).

3.2.3.2 DC-DC Topology

As we saw in the previous section in DC-DC topology, the data transmitted by UE_A is not relayed by UE_R. Thus, we can calculate the total energy consumption from Eq. (3.8) as.

As observed in the previous section, in the DC-DC topology, the data transmitted by UE_A is not relayed by UE_R. Therefore, we can calculate the total energy consumption from Eq. (3.3) as follows:

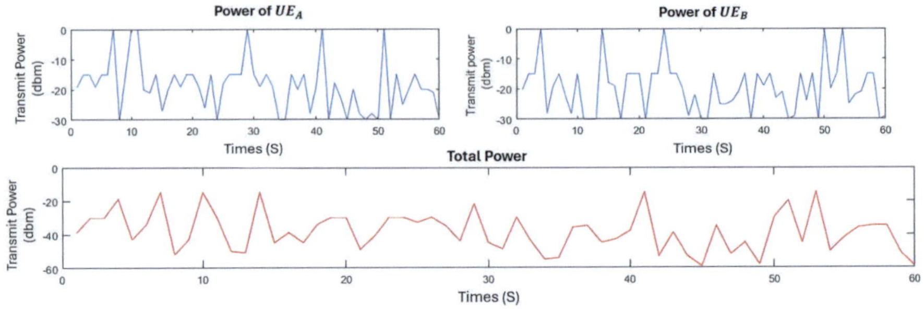

Fig. 3.4 DC-DC transmission power

$$P_T^D = P_{UEA} + P_{UEB} \tag{3.11}$$

Using Eqs. (3.6) and (3.7), we can then determine the total power consumed by the network:

$$P_T^D = \frac{N}{|h|^2 k}(d_{AB}^{\alpha}) \left(\Phi(C_A) + \Phi(C_{beacon})\right), \tag{3.12}$$

where $C_A = \dfrac{D_A}{W}$, $C_{beacon} = \dfrac{D_{beacon}}{W}$ and d_{AB} is the distance between UE_A and UE_B.

Figure 3.4 illustrates the DC-DC transmission power.

3.2.3.3 DR-DC Topology

In this topology, communication links are initiated between UE_A, UE_B and UE_R. As we saw in the previous section, UE_B and UE_R must transmit a discovery signal so that the data sent by UE_A is relayed by UE_R to be sent to UE_B. As a result, we can derive Eqs. (3.5) and (3.6) the P_T^R value as Chevillon et al. (2017):

$$P_T^R = \frac{N_0}{|h|^2 k} \Upsilon(d_{AR}, d_{BR}) \left(\Phi(C_A) + \Phi(C_{beacon})\right) \tag{3.13}$$

where $\Upsilon(x, y) = x^{\alpha} + y^{\alpha}$, and d_{AR} and d_{BR} are the distance between UE_A and UE_R and the distance between UE_B and UE_R, respectively.

Figure 3.5 represents the DR-DC transmission power.

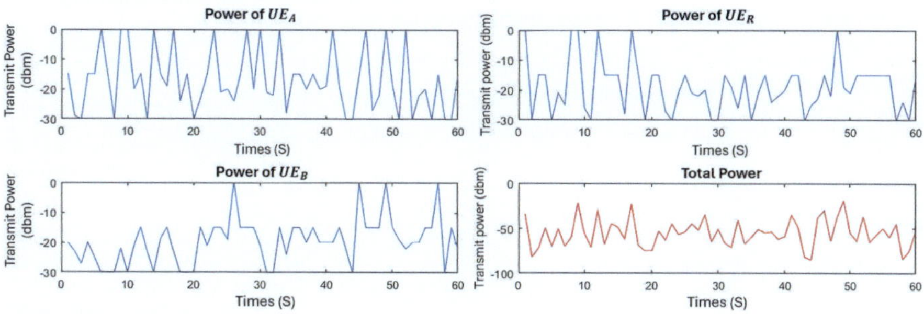

Fig. 3.5 DR-DC transmission power

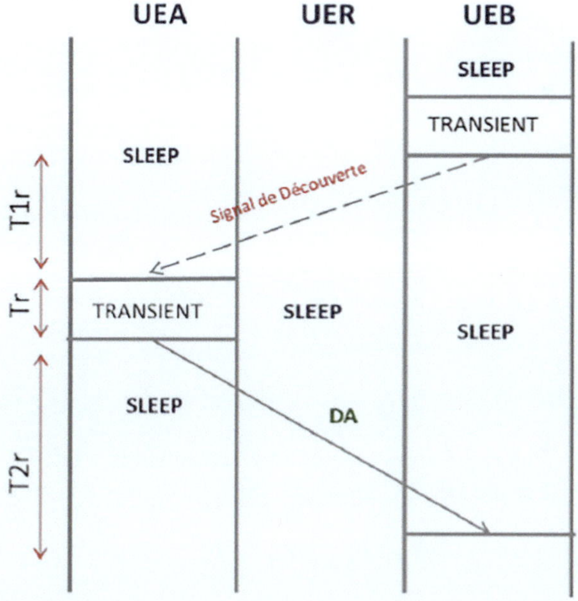

Fig. 3.6 Time sequence process of transmissions in DC-DC topology

3.2.3.4 Time Sequence Process

Let us now look at the temporal processes that are taken into account for all topologies. The following processes are inspired by the works of Tang et al. (2007). Figures 3.6 and 3.7 graphically show the time processes for DC-DC, DR-DC topologies.

3.2.3.5 Total Energy for Used Topologies

From temporal processes, we can deduce the total energy consumed for all topologies. Thus, for the DC-DC topology, the total energy consumed by the network is expressed as following (Tang et al. 2007):

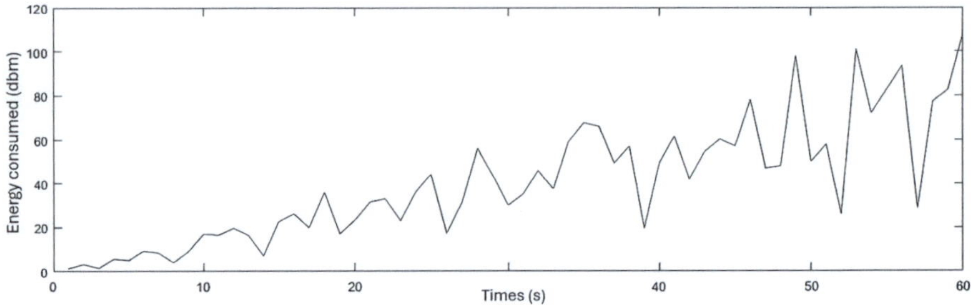

Fig. 3.7 Time sequence process of transmissions in DR-DC topology

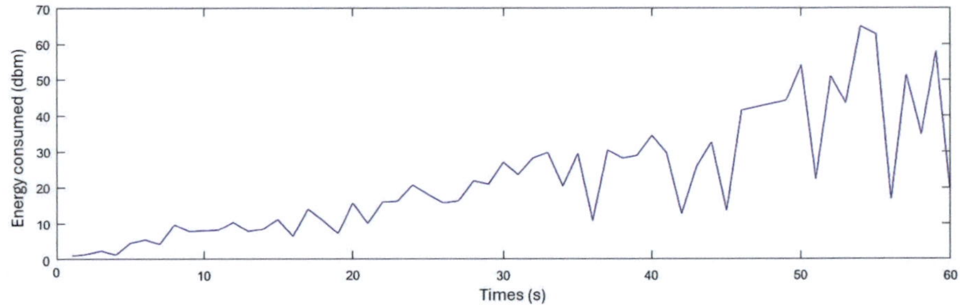

Fig. 3.8 Energy consumed in a DC-DC communication

$$E_T^D = P_A * T_{1D} + P_B * T_{2D}$$

Figure 3.8 shows the energy consumed in a DC-DC communication.

For the DR-DC topology, the total energy consumed by the network is given by:

$$E_T^R = P_{A\to B}^x.T_{2R} + P_{B\to R}^x.T_{1R} + P_{R\to A}^x.T_{1R} + P_{R\to B}^x.T_{3R}$$

Figure 3.9 illustrates the energy consumed in a DR-DC communication.

3.3 Discussion

In this section, we compare the two D2D communication topologies (DC-DC and DR-DC) in terms of total power consumption.

As shown in Figs. 3.8 and 3.9, the energy consumed in a DR-DC communication is significantly lower than that consumed by the DC-DC mode. This is because, in DR-DC, relays with low power levels, and those situated at a greater distance from the transmitter or receiver, often operate in sleep mode and only transition to active mode when required. This approach effectively reduces energy consumption during idle periods.

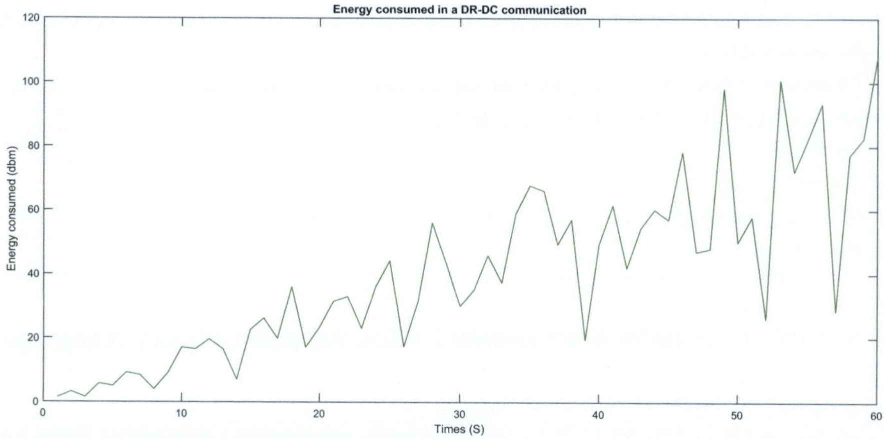

Fig. 3.9 Energy consumed in a DR-DC communication

Figures 3.8 and 3.9 also show that the DR-DC mode extends network lifetime, even when the network density fluctuates. This is a crucial benefit since, as the nodes switch between detection and relay modes, they dynamically adjust their energy consumption according to their role. This mode shift enables a more efficient use of available resources, especially in scenarios where maintaining continuous communication would otherwise deplete the energy reserves of individual nodes.

In addition, the distance between devices plays a vital role in selecting the optimal relay node. By considering the residual energy of each node, the system ensures that energy is conserved. Periodically, when a node needs to transmit data, it selects the relay node with the highest residual energy, minimizing unnecessary energy expenditure. This periodic energy-efficient relay selection mechanism helps in reducing energy waste, ensuring a balanced energy load across the entire network, which further contributes to the longevity and stability of the network.

Thus, implementing intelligent relay selection strategies, based on distance and energy status, can lead to significant improvements in the overall energy efficiency of the network, particularly in environments with fluctuating device density and varying distances between communicating nodes. This makes DR-DC a more sustainable choice for large-scale deployments, where energy efficiency is a major concern.

3.4 Conclusion

Device-to-device communications represents a promising strategy in modern mobile networks, particularly for the upcoming 5G systems. This approach offers significant benefits, making it a highly attractive option. Such communication models could allow for a greater number of subscribers to communicate simultaneously, optimizing network efficiency.

However, while this could lead to higher network loads, it does not necessarily imply a reduction in communication quality.

In this paper, we have conducted a comparative study of different topologies, including DC-DC and DR-DC. We have analyzed the total energy consumption in various scenarios, providing insights into how different configurations impact overall power efficiency.

The main objective of this work was to maximize the network's lifetime by minimizing energy consumption.

To validate our findings, we presented simulation results obtained using Matlab. These simulations demonstrated that low power consumption can be achieved with careful architecture selection, such as the DR-DC mode. This result holds true for most common scenarios. We can confidently recommend this architecture for extended network lifetime.

Further improvements could be made by exploring adaptive management strategies. This adaptive approach could potentially yield even better results in terms of energy efficiency and network longevity.

Additionally, we suggest that a well-coordinated cooperation plan between communications network operators and the national electric power company could facilitate the implementation of more energy-efficient policies. Such collaboration could lead to further enhancements in network performance and sustainability.

References

Abolfazli, S., Sanaei, Z., Ahmed, E., Gani, A., Buyya, R.: Cloud-based augmentation for mobile devices: motivation, taxonomies, and open challenges. IEEE Commun. Surv. Tutor. **16**(1), 337–368 (2014)

Agyapong, P.K., Iwamura, M., Staehle, D., Kiess, W., Benjebbour, A.: Design considerations for a 5G network architecture. IEEE Commun. Mag. **52**(11), 65–75 (2014)

Ben Abdennour, A., Ouwais Kabaou, M., Flah, A., Chibani Rhaimi, B.: Optimizing energy efficiency in multidevice communication with artificial intelligence assistance. Int. Trans. Electr. Energy Syst. (2023a)

Ben Abdennour, A., Ouwais Kabaou, M., Chibani Rhaimi, B.: Novel approach to energy optimization in relay-assisted device-to-device communication. In: IEEE 3rd International Maghreb Meeting of the Conference on Sciences and Techniques of Automatic Control and Computer Engineering (MI-STA) (2023b)

Ben Abdennour, A., Ouwais Kabaou, M., Chibani Rhaimi, B.: Energy optimization in a D2D communication based on a fuzzy system. In: IEEE 2nd International Conference on Signal, Control and Communication (SCC) (2021)

Chevillon, R., Andrieux, G., Diouris, J.-F.: Energy optimization of D2D communications using relay devices and data entropy. In: 28 IEEE International Symposium on Personal, Indoor and Mobile Radio Communications (PIMRC), Montreal, QC, Canada (2017)

Choi, C., Park, S., Cho, D.-H.: User-cooperation scheme based on clustering for energy efficiency in cellular networks with D2D communication. In: 25 IEEE International Symposium on Personal, Indoor and Mobile Radio Communications, PIMRC, September, pp. 1365–1369 (2014)

Cui, S., Goldsmith, A.J., Bahai, A.: Energy-constrained modulation optimization. IEEE Trans. Wirel. Commun. **4**(5), 2349–2360 (2014)
Gandotraa, P., Kumar Jha, R., Jain, S.: A survey on device-to-device (D2D) communication: architecture and security issues. J. Netw. Comput. Appl. **78**, 9–29 (2017)
Hossain, E., Rasti, M., Tabassum, H., Abdelnasser, A.: Evolution toward 5G multi-tier cellular wireless networks: an interference management perspective. IEEE Wirel. Commun. **21**(3), 118–127 (2014)
Iqbal, J., Iqbal, M., Ahmad, A., Khan, M., Qamar, A., Han, K.: Comparison of spectral efficiency techniques in device-to-device communication for 5G. IEEE Access **7** (2019)
Kulkarni, P., Khanai, R., Bindagi, G.: Security frameworks for mobile cloud computing: a survey. In: International Conference on Electrical, Electronics, and Optimization Techniques (ICEEOT), pp. 2507–2511 (2016)
Olaobaju, A.: Device-to-Device Communication in 5G Cellular Networks, University of Vaasa (2018)
Pirinen, P.: A brief overview of 5g research activities. In: 1st International Conference on 5G for Ubiquitous Connectivity, pp. 17–22 (2014)
Reka, S.S., Dragičević, T., Siano, P., Prabaharan, S.R.S.: Future generation 5G wireless networks for smart grid. A comprehensive review. Energies **12**(11), 2140 (2019)
Tang, Z., Glover, I.A., Evans, A.N., He, J.: Energy efficient transmission protocol for distributed source coding in sensor networks. In: ICC '07. IEEE International Conference on Communications, June, pp. 3870–3875 (2007)
Tehrani, M.N., Uysal, M., Yanikomeroglu, H.: Device-to-device communication in 5g cellular networks: challenges, solutions, and future directions. IEEE Commun. Mag. **52**(5), 86–92 (2014)
Tehrani, M.N., Uysal, M., Yanikomeroglu, H.: Device-to-device communication in 5G cellular networks: challenges, solutions, and future directions. IEEE Commun. Mag. **52**(5), 86–92 (2014)
Zhang, Z., Li, S.: A survey of computational off loading in mobile cloud computing. In: 4th IEEE International Conference on Mobile Cloud Computing, Services, and Engineering (MobileCloud), pp. 81–82 (2016)

Chapter 4
Towards Next-Generation OAM Fibers: Inverse Raised Cosine Few-Mode Fiber: Design and Performance Evaluation

Alaaeddine Rjeb, Hussein Seleem, Habib Fathallah, and Mohsen Machhout

Abstract With the growing demand for high-speed online services and data-driven applications, the optical communication research community has turned its attention to utilizing the orbital angular momentum (OAM) of light for high-capacity data transmission through optical fibers. Among the promising advancements in this area are specialized OAM fibers, which offer a viable pathway for supporting stable and high-throughput OAM channels. This chapter introduces a novel type of OAM-supporting fiber: the inverse raised cosine few-mode fiber (IRCFMF). We explore its design methodology, optimization techniques, and performance across key parameters such as mode separation, differential group delay, chromatic dispersion, nonlinear behavior, resistance to mode coupling under perturbations, and bending tolerance. This chapter consolidates all our IRC-FMF analysis findings, aiming to provide the research community with a solid foundation for future developments and innovations involving this new class of OAM fibers.

Keywords Space division multiplexing · Mode division multiplexing · Orbital angular momentum (OAM) · Specialty fibers design OAM mode stability · Low crosstalk fiber · Optical communication · Fiber bending tolerance · Dispersion engineering · Mode coupling analysis

---

A. Rjeb (✉) · M. Machhout
Laboratory of Electronics and Microelectronics, Physics Department, Faculty of Sciences of Monastir, Monastir, Tunisia
e-mail: alaaeddine.rjeb@gmail.com

H. Seleem
Department of Electronics and Electrical Communications, Faculty of Engineering, Tanta University, Tanta, Egypt
e-mail: eltaibee@f-eng.tanta.edu.eg; hussein.eltaibee@pua.edu.eg

Electical Engineeing Department, Faculty of Engineering, Pharos University in Alexandria, Alexandria, Egypt

H. Fathallah
Laboratory of Artificial Intelligence and Data Engineering Applications, Computer Department, Faculty of Sciences of Bizerte, Bizerte, Tunisia

N. Derbel and Q. Zhu (eds.), *Advances in Communication, Signal and Image Processing*,
https://doi.org/10.1007/978-981-95-0401-5_4

4.1 Introduction

During the last decade, the demand for online data services including video streaming, video conferencing, online gaming, social networking, online banking and online store has been raised exponentially (Cisco VNI Global 2018). Intuitively, these services and data technologies are likely to further boost demand for more and more capacity transmission and spectral efficiency. Hence, strong government and industry efforts have been conducted in optical communication and networking realm aiming to cope with the increase of these services users, where various innovative optical communication technologies have been developed and explored (Chralyvy 2009; Essiambre and Tkach 2012).

Recently, space division multiplexing (SDM) has seen tremendous interest due to its capability to enhance the capacity of deployed communication systems and unleash the spectral efficiency (Richardson et al. 2013). One scheme of SDM is based on carrying data on spatial modes and transmitting them through an optical fiber. This direction is known as mode division multiplexing (MDM) (Berdague and Facq 1982). MDM consists of establishing different data path (i.e. modes/channels) over a transmission medium with the requirement that these modes are independents. Hence, the overall capacity is multiplied by the number of available channels, hence multiplied by the number of co-propagating modes sharing the same fiber (Saridis et al. 2015). MDM uses multimode fibers (MMFs) deployed over short haul interconnect transmission or few mode fibers (FMFs) as transmission medium over long haul transmission link (Yaman et al. 2010). This is caused by the fact that MMFs enable high mode coupling issues (i.e. channels crosstalk) due to the huge number of considered modes (i.e. hundreds). A multiple input multiple output digital signal processing MIMO DSP is required at the receiver side aiming to compensate the crosstalk effects among channels (Bae et al. 2019). MIMO DSP is considered as an extreme solution to decipher channels since it is heavy and complex. Its complexity is came from the direct proportionality between the transmission distance, the number of modes from one side and the number of required electrical filters from other side. This tend to allow the MIMO to be impractical in real time and threats the scalability of MDM in next generation optical communication system (Ryf et al. 2012a). Moreover, since each mode has its own velocity, a high differential mode delay (DMD) will occur especially if we consider a long transmission link (e.g. accumulative delays). In order to cope with the crucial issues that tend to degrade the link performances and the scalability of MDM over MMFs, optical communication R&D community had oriented their effort towards the FMFs (Sillard et al. 2011). An FMF supports few number of spatial modes with controlled/moderated DMD and very low interactions among them (i.e. reduced crosstalk). Several optical fiber modal basis have been utilized for FMF-MDM showing its effectiveness to scale up the capacity transmission from (Terabit to Petabit) and unleash the spectral efficiency (from dozen to hundred (bit/s/Hz)) over optical fiber (Ryf et al. 2012b).

In recent years, there has been a growing interest in exploring the angular momentum (AM) of light, which quantifies the rotational characteristics inherent in an electromagnetic wave (Ramachandran and Kristensen 2013). AM consists of two distinct components: Spin Angular Momentum (SAM), which is linked to the polarization state of light (e.g., left- or right-handed circular polarization), and Orbital Angular Momentum (OAM), which arises from the helical phase structure of the beam, typically represented as $exp(jl\phi)$, where l denotes the topological charge and ϕ is the azimuthal coordinate (Trichili et al. 2019). OAM, in particular, has emerged as a highly attractive solution for expanding the capacity of optical communication systems, both in free-space links and guided wave platforms like optical fibers (Wang et al. 2012; Bozinovic et al. 2013). Its ability to provide an unbounded set of mutually orthogonal modes makes it ideal for mode multiplexing schemes. By encoding separate data streams onto different OAM states, OAM-based mode division multiplexing (OAM-MDM) significantly enhances transmission capacity and spectral efficiency. However, the practical deployment of OAM-MDM in few-mode fibers (FMFs) introduces a number of engineering obstacles. Among the most pressing are mode coupling, differential group delay (DGD), chromatic dispersion, and intermodal crosstalk–all of which degrade mode integrity during propagation. Additional factors such as modal purity, nonlinear interactions, limited effective mode area, and sensitivity to structural imperfections in the fiber also pose challenges. Addressing these limitations is essential to fully unlocking the potential of OAM multiplexing in next-generation optical networks (Zhang et al. 2019; Rusch et al. 2018; Rjeb et al. 2021b).

Since the incorporating of MIMO DSP is considered as an extreme solution to decipher channels, another direction of research has seen tremendous interest, which is the optical fiber engineering realm. The fiber engineering is related to the design, the optimization and the improvement of fiber key parameters including the refractive index profile (shape, materials refractive indexes), and the fiber layers sizes. Through deliberate engineering of key fiber design parameters–such as the refractive index profile, core geometry, and material composition–it becomes feasible to regulate modal interactions and optimize performance characteristics. This strategic control facilitates a more efficient and seamless transition from conceptual design to practical fabrication. Over the past few years, a range of specially engineered fibers tailored for orbital angular momentum (OAM) transmission have been introduced, demonstrating strong potential for integration into high-capacity, next-generation OAM-based mode division multiplexing (OAM-MDM) systems capable of operating at terabit-scale data rates. Vortex fibers (Ramachandran et al. 2009), air core fibers (Brunet et al. 2014), ring core fibers (Brunet et al. 2015), graded index ring core fibers (Zhu et al., 2017), hyperbolic tangent fibers (Rjeb et al. 2020c, 2021a), inverse parabolic graded index fibers (Ung et al. 2014) and so on, are all examples of OAM fibers that have been proved and demonstrated. Recently, we proposed a novel OAM few mode fiber that we refer to as Inverse raised cosine few mode fiber (IRC-FMF) (Rjeb et al. 2019). The function raised cosine has been borrowed from digital filters realm (Proakis and Salehi 2008) and incorporated as a refractive index profile in optical fiber scenario. Therefore, this chapter addresses the IRC-FMF. We touches various aspects of the proposed fiber from design process till performance metrics assessment. This chapter acts as a survey or a scan of all demonstrated results approved by the designed IRC-FMF.

The structure of this chapter is organized as follows. Section 4.2 introduces the fundamental parameters and the design methodology of the proposed Inverse Raised Cosine Few-Mode Fiber (IRC-FMF). In Sect. 4.3, we examine the supported spatial modes by analyzing the modal distribution and the effective refractive index spacing. Section 4.4 focuses on the supported orbital angular momentum (OAM) modes, where we assess their quality in terms of mode purity and inherent crosstalk. Section 4.5 extends the investigation to encompass the performance of OAM modes across the C and L communication bands, including an evaluation of intermodal spacing, chromatic dispersion, differential group delay, effective mode area, and fiber nonlinearity. Section 4.6 explores the sensitivity of mode coupling in the presence of structural perturbations, specifically ellipticity and birefringence. Finally, Sect. 4.7 evaluates the robustness of the IRC-FMF under real-world conditions by analyzing the impact of bending on mode count, mode separation, walk-off behavior, and bending-induced losses.

4.2 The Inverse-Raised-Cosine Fiber Design

This section highlights the IRC refractive index profile. We discuss the origin of RC function and its incorporation in optical fiber scenario.

4.2.1 *Raised-Cosine Function*

The Raised-Cosine Function (RCF) is commonly employed in digital modulation systems, particularly for pulse shaping in digital filtering. It derives its name from the shape of its frequency spectrum, specifically, in its basic form (when the rolloff factor $\beta = 1$), the non-zero spectral region resembles a cosine wave that is "raised" to sit atop the frequency axis. In the frequency domain, the transfer function of a raised-cosine filter is mathematically expressed as follows (Proakis and Salehi 2008):

$$H(f) = \left| \begin{array}{lll} 1 & \text{if} & |f| \leqslant \frac{1-\beta}{T} \\ \frac{1}{2}\left\{1+\cos\left(\frac{\pi T}{\beta}\left[|f| - \frac{1-\beta}{T}\right]\right)\right\} & \text{if} & \frac{1-\beta}{T} \leqslant |f| \leqslant \frac{1+\beta}{T} \\ 0 & \text{otherwise} & \end{array} \right. \tag{4.1}$$

where T is the symbol-period of the communications system (i.e. the reciprocal of the symbol-rate), and $0 \leq \beta \leq 1$ is the roll-off factor, which describes the excess bandwidth of the filter. Figure 4.1 shows the frequency response of a raised-cosine filter with various roll-off factors β (i.e. smooth parameter). In standard RCF, this

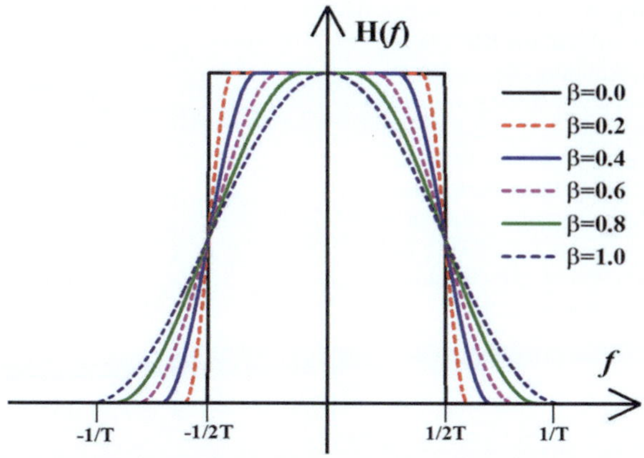

Fig. 4.1 The frequency response of raised-cosine filter with various roll-off factors (β)

smooth behavior (supervised by the parameter β) is a key characteristic that gives the unique filtering properties of RCF in digital signal processing. In practical applications, when the RCF is applied to a wideband signal, it effectively suppresses unwanted high-frequency components outside the intended bandwidth, resulting in a significantly cleaner and more confined spectral shape. One of the key advantages of using the RCF in signal processing lies in its ability to minimize intersymbol interference (ISI), which can otherwise degrade data integrity. This makes the RCF an essential tool in digital communication systems, where preserving signal clarity and maintaining channel efficiency are critical objectives (Proakis and Salehi 2008).

4.2.2 *Inverse-Raised-Cosine Profile*

The first incorporation of the RCF as a refractive index profile for optical fibers was with chebaane and co-workers in Chebaane et al. (2016, 2018). The aim was to design a refractive index profile that minimizes the DMD in FMFs. Following OAM fiber design guidelines, we first inverse this RCF in order to obtain a depressed core (i.e. to obtain a doughnut-shaped core = ring) and adapt it to the context of a fiber profile by tail truncation. The latter enables to limit the extent of the RCF to the core region. The resulted function is expressed as follows (Rjeb et al. 2019):

$$n(r) = \left| \begin{array}{ll} n_2 & \text{if} \quad 0 \leqslant r \leqslant a\frac{1-\alpha}{2} \\ n_1 - \frac{1}{2(n_2-n_3)} \left\{1 + \cos\left(\frac{\pi}{a\alpha}\left[r - a\frac{1-\alpha}{2}\right]\right)\right\} & \text{if} \quad a\frac{1-\alpha}{2} \leqslant r \leqslant a\frac{1+\alpha}{2} \\ n_2 & \text{if} \quad a\frac{1+\alpha}{2} \leqslant r \leqslant a \\ n_3 & \text{if} \quad r > a \end{array} \right. \tag{4.2}$$

where a, n_1, n_2, and n_3 are the core radius, the maximum refractive index at the core, the minimum refractive index at the core, and the refractive index of the cladding

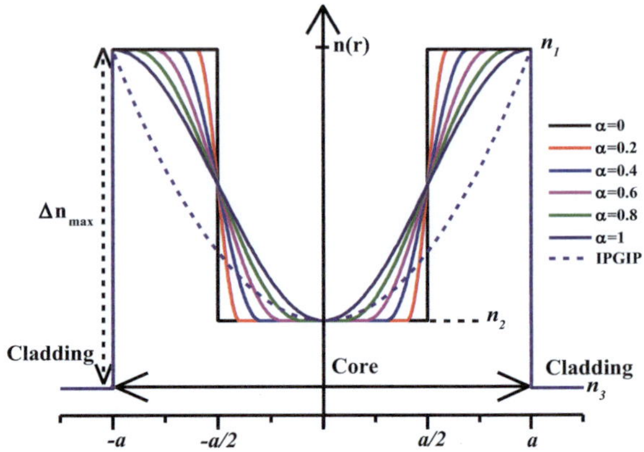

Fig. 4.2 Refractive index profiles of the IRC fiber (solid lines), for different values of shape parameter α ($\alpha = [0, 1]$) and the refractive index profile of IPGI fiber (dashed line). Both fibers have the same values of a, n_1, n_2, and n_3. $\Delta n_{\max}$ is the maximum refractive index contrast defined as $n_1 - n_3$

($r > a$), respectively. The exponent α is the profile shape that controls the shape behavior of refractive index from core center ($r = 0$) till the interface between the core and the cladding ($r = a$). In order to assess performances of the proposed fiber, the inverse-parabolic graded index (IPGI) fiber (Ung et al. 2014) has been used as a benchmark (i.e. for comparison). The refractive index of IPGI fiber is given as follows:

$$n(r) = \left| \begin{array}{ll} n_2 \sqrt{1 - 2N\Delta\left(\dfrac{r^2}{a^2}\right)} & \text{if } \; 0 \leqslant r \leqslant a \\ \quad n_3 & \text{if } \quad r > a \end{array} \right. \tag{4.3}$$

where the parameters a, n_1, n_2, and n_3 are the same as those of the designed IRC fiber. Figure 4.2 depicts both refractive indexes profiles: IRC fiber (solid lines) for different values of shape parameter α, and IPGI fiber (shown with dashed line). Both profiles share the same values of a, n_1, n_2, and n_3. From Fig. 4.2, the IRC profile when α approaching 1, is more similar to the IPGI profile. In contrast, IRC-FMF become a step ring-core fiber when the smoothness α get close to 0. Figure 4.2 shows also that the IRC profile is practically thinner (or more concentrated around the fiber axis) than the IPGI profile. In addition, IRC profile becomes much smoother (i.e. slow variation) when reaching the interface between the core and the cladding ($r = a$).

4.3 Fiber's Modes Analysis

In this section, we performed numerical simulations in order to perform a modal analysis using finite element method (FEM) packaged in Comsol multiphysics software. Our analysis include the number of supported modes, the separation among them and the impact of each fiber parameter on them.

4.3.1 Eigenmodes Number: Modal Map

Fiber eigenmodes are the true solutions of propagation equation over an optical fiber. These modes are $HE_{l,m}^{even/odd}$, $EH_{l,m}^{even/odd}$, $TE_{0,m}$ and $TM_{0,m}$. l and m subscript are the azimuthal and radial indices of modes. The investigation of eigenmodes number versus the shape parameter α for different values of core radius a are depicted in Fig. 4.3. This Figure shows that the number of eigenmodes increases as the core radius (a) increases. Moreover, a slight increase in the shape parameter (α) raises the number of supported eigenmodes. In addition to the IRC profile, we show the increase of the supported eigenmodes in IPGI fiber as his core radius increases (for purpose of comparison). Particular attention in the modal map is given to the few-mode zone (when the number of supported modes are low). At this zone, especially when $\alpha = 1$, both fibers (IRC-FMF and IPGI-FMF) share quite close number of supported eigenmodes but with different mode properties/features. The latter is among the key motivations of the adaptation of the IRC function in fiber context because it provides insightful understanding of the impact of smoothing fiber index profile using our proposed RCF (i.e. the effects of the shape parameter α).

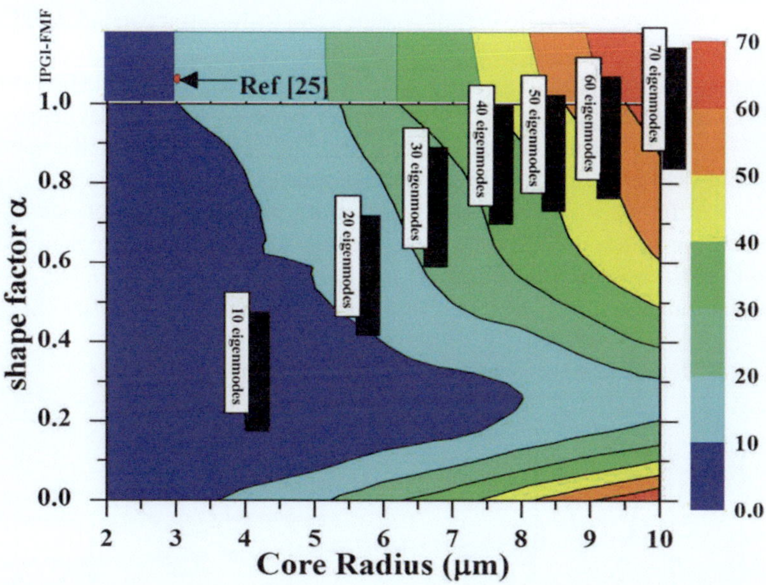

Fig. 4.3 Eigenmodes number versus the shape parameter α and the core radius for IRC and IPGI profiles. Other parameters are kept as $n_1 = 1.494$, $n_2 = 1.454$, and $n_3 = 1.444$

4.3.2 *Modal Separation: Numerical Investigation of IRC Parameters*

In this subsection, we investigate the effect of various IRC-FMF parameters on the intermodal separation. According to the first order perturbation (Rjeb et al. 2020c; Ung et al. 2014), the enhancement of intermodal separation Δn_{eff} requires a high profile gradient magnitude. Figure 4.4a depicts the refractive profile of IRC-FMF ($\alpha = 1$), linked with the normalized E-field intensity of $TE_{0,m}$ mode, while their corresponding gradients are depicted in Fig. 4.4b. From Fig. 4.4b, one can see that the maxima of both gradients are concurred at the core-cladding boundary ($r = a$), and have the same sign (i.e. negative), these features confirm that a wide intermodal separation could be achievable with IRC profile. Implementing the refractive index expression of IRC-FMF into the expression of the gradient (Ung et al. 2014), and evaluating the final expression at $r = a$, we can clearly find the following proportionality:

$$\Delta n_{eff}(IRCF) \propto \frac{1}{\alpha a \Delta n} \tag{4.4}$$

The above relation shows that the intermodal separation is inversely proportional to the smoothness parameter α, to the core radius a and to the refractive index difference Δn, where $\Delta n = n_1 - n_2$. In order to validate and confirm that proportionality relation, a numerical investigation was performed using finite element method (FEM).

In Fig. 4.5a, we plot the minimum of Δn_{eff} between eigenmodes inside LP_{11} group, (TE_{01}, HE_{21}, and TM_{01}) as a function of the parameter α and the maximum contrast between core and cladding refractive indices. The calculation of Δn_{eff} is done for 3 different values of $\Delta n_{\max}$ [0.04, 0.05, and 0.06]. The choice of LP_{11} is based on the fact that TE_{01} and TM_{01} are parasitic and its superposition do not provide OAM mode. Moreover, the higher order modes HE_{lm} or EH_{lm} with the same radial mode number (m) will share the same magnitude of Δn_{eff} (i.e. HE_{31}

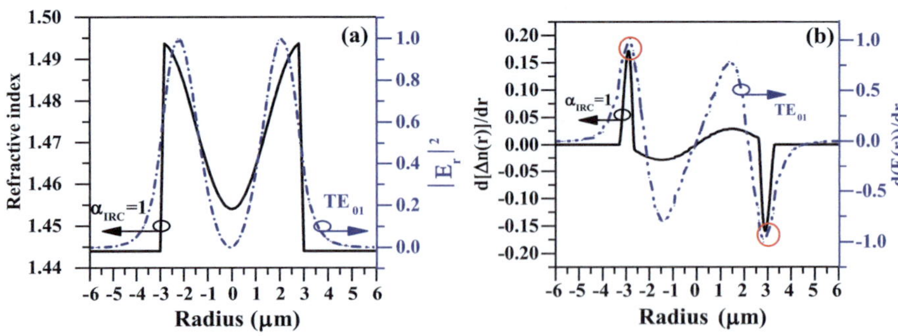

Fig. 4.4 **a** IRC-FMF refractive index profile ($\alpha = 1$) and the simulated normalized E-field intensity of the $TE_{0,m}$ mode at $\lambda = 1.55$ μm, and **b** the gradients of the refractive index distribution (dark solid) and E-field amplitude (blue dashed)

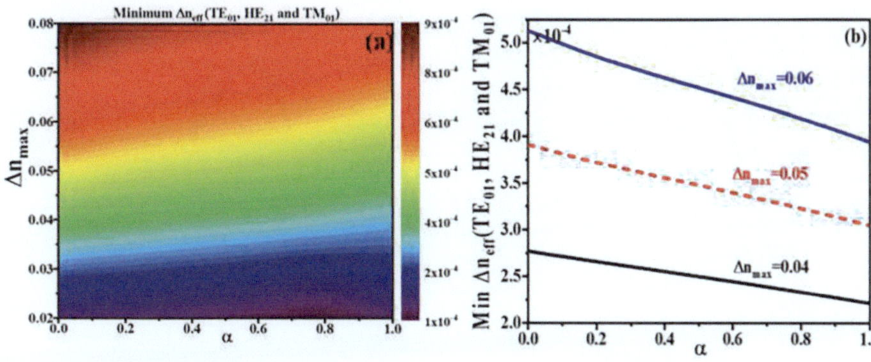

Fig. 4.5 Minimum effective index separation inside LP_{11} group in IRCF-MF **a** as a function of the maximum refractive index contrast **b** as a function of selected $\Delta n_{\max}$= [0.04; 0.05; 0.06], and depending on α

and EH_{11}, HE_{41} and EH_{21}, and so on). From Fig. 4.5, it is clearly seen that Δn_{eff} is enhanced with high refractive index contrast ($\Delta n_{\max}$) while it inversely increases with the shape parameter α. Thus, confirm the well separation between vector modes at high contrast values. An ultra large vector mode separation (8×10^{-4}–9×10^{-4}) are achievable with IRC-FMF since the high values of $\Delta n_{\max}$ are accomplishable with modified chemical vapor deposition (MCVD) process with Germania doped silica or other schemes.

We calculated the minimum separation inside LP_{11} versus the actual difference Δn and the shape parameter α. In Fig. 4.6a, b, and at fixed value of α, it is clear that Δn_{eff} is inversely proportional to Δn, which is in total agreement with expression (4.4). As a last investigation, we calculate the minimum Δn_{eff} among LP_{11} group

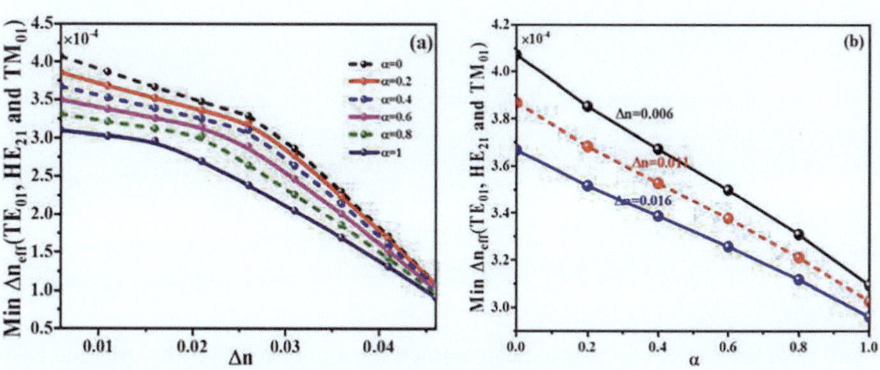

Fig. 4.6 Minimum effective index separation inside LP_{11} group in IRCF ($a = 3\,\mu$m, n2 = 1,444 and $\lambda = 1.55\,\mu$m) **a** as a function of refractive index difference (Δn) and for various shape parameter α, **b** Isolines along selected Δn = [0.006, 0.011, and 0.016]

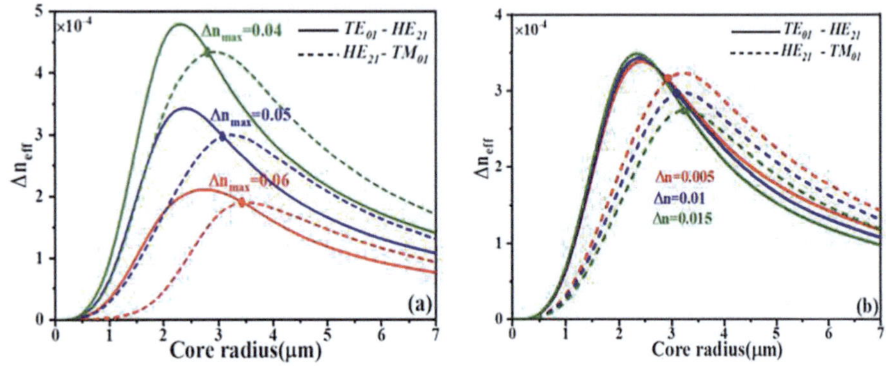

Fig. 4.7 Modal separation in an IRCF ($\alpha = 1$, $n_3 = 1.444$ and $\lambda = 1.55\,\mu$m) as a function of the core radius a and for various values of **a** the maximum refractive index contrast $\Delta n_{\max}$ and **b** the difference Δn

(TE_{01}, HE_{21}, and TM_{01}) as a function of the core radius (a) and the shape parameter (α). The fiber parameters are kept at $\Delta n_{\max} = 0.05$, $n_{eff} = 1.444$, $\Delta n = 0.01$, and $\lambda = 1.55\,\mu$m. The results of investigation are shown in Fig. 4.7a, b. These figures show and confirm the inverse proportionality between Δn_{eff} and the core radius a. Actually, Δn_{eff} increases with a till it reaches a maximum value Δn_{eff}, at an optimum value of a ($a \leq a_{opt}$). After that value ($a \geq a_{opt}$), Δn_{eff} decreases with the core radius, here a_{opt} is located around 3 μm. More detailed results show that in the first zone where $a \leq a_{opt}$, the minimum Δn_{eff} takes place between HE_{21} and TM_{01}. On the other hand when $a \geq a_{opt}$, the minimum Δn_{eff} occurs between TE_{01} and HE_{21} modes.

4.4 OAM Mode Analysis

In this section, we generate the OAM modes supported in the designed IRC-FMF. We assessed the quality of supported modes. This include the OAM mode components, OAM purities, and the intrinsic crosstalk.

4.4.1 OAM Modes in IRC-FMF

OAM modes in optical fibers can be generated by appropriately combining the even and odd components of each HE_{lm} and EH_{lm} vector mode with a relative phase difference of $\pm(\frac{1}{2}\pi)$ (Chen et al. 2017; Chen and Wang 2017). When the spin characteristics of light (i.e., its circular polarization) are included, these modes are expressed as $OAM^{L/R}_{\pm lm}$, where the superscript L/R indicates the spin angular momentum cor-

Table 4.1 Construction formula of OAM modes from fiber eigenmodes (reproduced from Rjeb et al. (2020a))

Combination of cylindrical vector modes (i.e. eigenmodes)	OAM modes
$HE_{1m}^{even} \pm HE_{1m}^{odd}$	$OAM_{0m}^{L/R}$
$TM_{0,m} \pm iTE_{0,m}$	$OAM_{\pm 1,m}^{R/L}$
$HE_{l+1,m}^{even} \pm iHE_{l+1,m}^{odd}$	$OAM_{\pm l,m}^{L/R}$
$EH_{l-1,m}^{even} \pm iEH_{l-1,m}^{odd}$	$OAM_{\pm l,m}^{R/L}$

responding to left or right circular polarization. The subscript indices l and m refer to the azimuthal and radial mode numbers, respectively. Specifically, l represents the topological charge associated with the number of phase helices or twists in the beam profile, while m specifies the number of radial intensity nulls (or rings). The spin angular momentum (SAM) carries $a \pm s\hbar$, with $s = +1$ for left-handed and $s = -1$ for right-handed circular polarization. The orbital angular momentum (OAM) contributes $\pm l\hbar$, making the total angular momentum (AM) the sum of both components, with a combined magnitude of $(\pm l \pm s)\hbar$. A summary of these relationships is provided in the following Table 4.1.

Considering the cylindrical coordinates (r, ϕ, z), the transverse fields of OAM modes are interpolated in a novel basis vector $(\widehat{L}, \widehat{R}, \widehat{Z})$ where are $\widehat{L}$ and $\widehat{R}$ (Left/Right) circular polarization basis vectors, expressed as a function of cylindrical basis vectors $(\widehat{L}, \widehat{R}, \widehat{Z})$ as Rjeb et al. (2020b):

$$\widehat{L} = \frac{\left(\widehat{r} + j\widehat{\phi}\right)\exp(j\phi)}{\sqrt{2}}; \quad \widehat{R} = \frac{\left(\widehat{r} - j\widehat{\phi}\right)\exp(-j\phi)}{\sqrt{2}}; \quad \widehat{Z} = \widehat{z} \tag{4.5}$$

Hence, the transverse fields of OAM modes are written as Rjeb et al. (2020b):

$$\begin{cases} E^{+}_{+l,m}(r,\phi) = a^{+}_{+l,m}(r)\exp(+jl\phi)\widehat{L} + b^{+}_{+l,m}(r)\exp[+j(l+2)\phi]\widehat{R} \\ E^{-}_{-l,m}(r,\phi) = a^{-}_{-l,m}(r)\exp(-jl\phi)\widehat{R} + b^{-}_{-l,m}(r)\exp[-j(l+2)\phi]\widehat{L} \\ E^{-}_{+l,m}(r,\phi) = a^{-}_{+l,m}(r)\exp(+jl\phi)\widehat{R} + b^{-}_{+l,m}(r)\exp[+j(l-2)\phi]\widehat{L} \\ E^{+}_{-l,m}(r,\phi) = a^{+}_{-l,m}(r)\exp(-jl\phi)\widehat{L} + b^{+}_{-l,m}(r)\exp[-j(l-2)\phi]\widehat{R} \end{cases} \tag{4.6}$$

In these expressions, $a^{\pm}_{\pm l,m}(r)$ and $b^{\pm}_{\pm l,m}(r)$ represent real-valued radial functions. They reveal that every OAM mode inherently comprises two components: one exhibits left-handed circular polarization $(\widehat{L})$ and the other right-handed circular polarization $(\widehat{R})$. Additionally, one component carries a topological charge l–referred to as the dominant term–while the second exhibits a charge of either $l+2$ or $l-2$, considered the secondary term. Under the weakly guiding approximation, where

Table 4.2 Components of OAM, TE and TM modes supported by IRC-FMF. The number presents the order l of the generated OAM light beam. Bold l are the dominant components. Normal l are the secondary components

Modes	$\widehat{L}$	$\widehat{R}$	$\widehat{Z}$	Modes	$\widehat{L}$	$\widehat{R}$	$\widehat{Z}$
$OAM^L_{0,1}$	**0**	2	1	$OAM^L_{-2,1}$	**−2**	0	−1
$OAM^R_{0,1}$	−2	**0**	−1	$OAM^L_{2,1}$	**2**	4	1
$OAM^L_{1,1}$	**1**	3	2	$OAM^R_{-2,1}$	−4	**−2**	−3
$OAM^R_{-1,1}$	−3	**−1**	−2	$TE_{0,1}$	1	−1	–
$OAM^R_{2,1}$	0	**2**	1	$TM_{0,1}$	1	−1	0

the refractive indices satisfy $n_1 \approx n_3$, the influence of $b^{\pm}_{\pm l,m}(r)(r)$ becomes minimal, allowing the fiber to support predominantly a single, pure OAM component. However, in a high-index-contrast fiber (i.e., $n_1 > n_3$), the mode energy is split between both components, causing two beams with different topological charges (l and $l \pm 2$) to co-propagate. This overlapping propagation results in intrinsic crosstalk between OAM channels (Guerra et al. 2019). Employing the optimized parameters detailed earlier, the refractive indices and structural parameters are chosen as: $n_1 = 1.494$, $n_2 = 1.454$, $n_3 = 1.444$, core radius $a = 3\,\mu m$, and operating wavelength $\lambda = 1.55\,\mu m$. Using a finite element method (FEM) via COMSOL Multiphysics, the IRC-FMF is found to support the following vector modes: HE_{11}, TE_{01}, HE_{21}, TM_{01}, EH_{11} and HE_{31}. Each HE_{lm} and EH_{lm} vector mode contributes even and odd field distributions, resulting in 10 distinct eigenmodes supported by the fiber. The minimum separation in effective refractive index between any two consecutive vector modes is 1.27×10^{-4} (specifically between EH_{11} and HE_{31}), which helps prevent modal degeneracy and unwanted coupling. By applying the synthetic relations in Table 4.1, eight OAM modes are derived and categorized into three mode groups corresponding to $l = 0, \pm 1$, and ± 2. Due to the large mismatch in effective indices, the TE_{01} and TM_{01} modes do not combine to form valid OAM states and instead pose a risk of interference with the OAM channels generated from the HE/EH mode combinations. Table 4.2 details the order (l) of each OAM component and the TE/TM modes supported in the IRC-FMF, where boldface l identifies the dominant term and standard font l indicates the secondary. Figure 4.8 presents selected spatial phase maps of the dominant OAM components' intensity profiles.

4.4.2 OAM Purity and Intrinsic Crosstalk

The impact of the shape parameter α on the purity of OAM modes and on the induced intrinsic crosstalk are highlighted in Fig. 4.9a–b. The purity increases as α increases. Thus, as the profile is much smoother (i.e. graded), as the purity of OAM is enhanced.

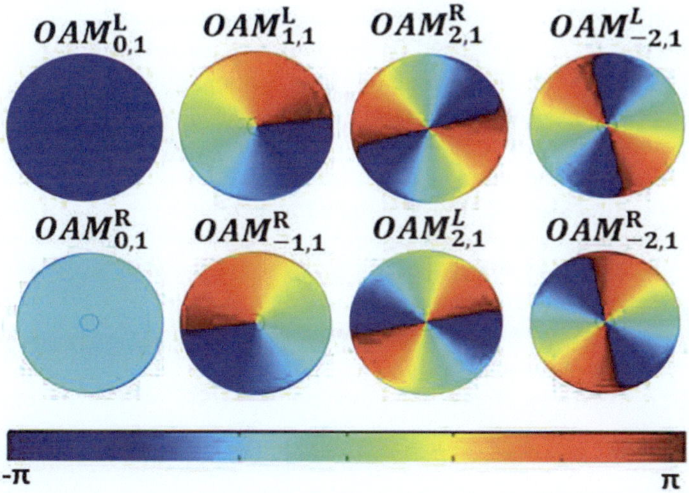

Fig. 4.8 Part of spatial phase distributions of the intensity profiles of the major OAM components in IRC fiber at 1550 nm

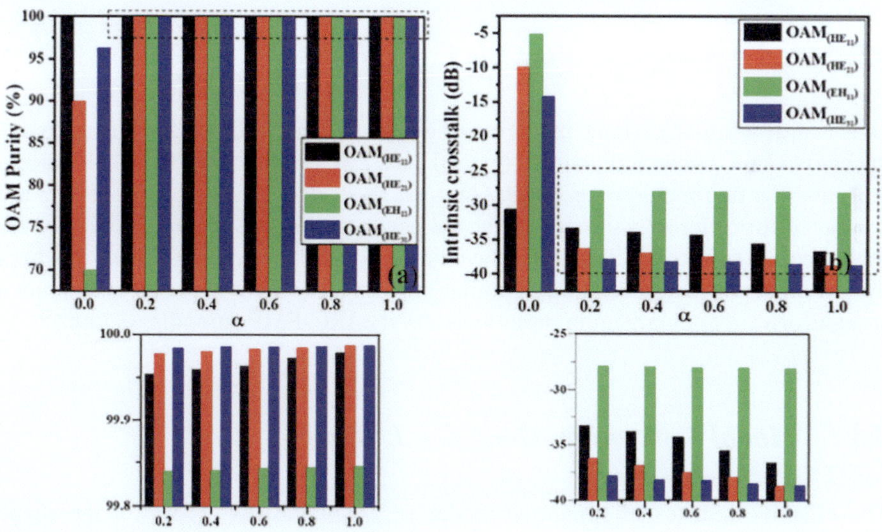

Fig. 4.9 **a** The purity of synthetized OAM modes versus the smoothness parameter α. **b** The equivalent intrinsic crosstalk versus the smoothness parameter α

This has been proved in previous analysis using the conventional graded index fiber. Where the study was concentrated on the graded shape between the cladding and the core (i.e. outer ring). Concerning our investigation, the demonstration of inner smooth variation impact on OAM purity while keeping abrupt outer variation has been performed. This was one of the motivations and targets of our analysis using our proposed IRC-FMF. The worst OAM purities is found at $\alpha = 0$ which corresponds to the ring core structure (RCF) where the refractive index presents two discontinuous areas with large refractive index contrast and abrupt variations (i.e. jump). The worst purity is for OAM synthesized by EH_{11} valuing 69.907% (equivalent to an intrinsic crosstalk of −5 dB). Moreover, at $\alpha = 1$, the OAM purity synthetized by modes HE_{11}, HE_{21}, EH_{11} and HE_{31} are 99.97%, 99.98%, 99.84% and 99.98%, respectively. Their intrinsic crosstalk are −36.65 dB, −38.75 dB, −28.12 dB and −38.75 dB, respectively. Hence, the proposed IRC-FMF with $\alpha = 1$ enables high OAM purity. The lowest purity is possessed by the OAM mode synthetized by EH_{11}. This is attributed to the high the effect of anti-aligned spin-orbit coupling. Moreover, the IRC profile still observes an abrupt index variation ($r = a$) which presenting a discontinuity that tend to degrades the purity of OAM modes (Rjeb et al. 2019).

4.5 Modal Analysis Along the C + L Bands

Over a large wavelength band that is ranging from 1525 to 1625 nm and which covers the entire C + L bandwidth defined by the ITU-T, we evaluate the IRC-FMF modes. Our investigation include Δn_{eff}, DGD, CD, A_{eff}, and γ. Some of the results are computed for different values of the shape parameter α. We conduct a comparison with reported results of IPGI fiber. The comparison are limited to the case in which the two fibers have approximately similar profiles ($\alpha = 1$). This helps to conduct a straightforward comparison in order to assess the IRC-FMF smoothness impact.

4.5.1 Modal Separation Along C + L Bands

The effective indexes of supported modes (n_{eff}) of the IRC-FMF, for the shape parameter $\alpha = 1$ and versus C + L bandwidth, are highlighted in Fig. 4.10a. From that figure, the supported modes are divided into three groups (i.e. a group contains modes with approximately similar n_{eff}): G_1, is formed by $OAM^{L/R}_{\pm 01}$; G_2, contained TE_{01}, $OAM^{L/R}_{\pm 11}$, and TM_{01}; G_3, comprises $OAM^{R/L}_{\pm 21}$ and $OAM^{L/R}_{\pm 21}$. Intermodal separations ($\Delta n_{eff}s$) are evaluated where are divided into inter- and intra-group separation. The results are depicted in Fig. 4.10b versus wavelength. Considering inter-groups separation, the minimum separations are of 6.70×10^{-3} between G_1 and G_2, and of 1.25×10^{-3} between G_2 and G_3. On the other side, the minimum separation of modes belonging to the same group are much lower with minimum

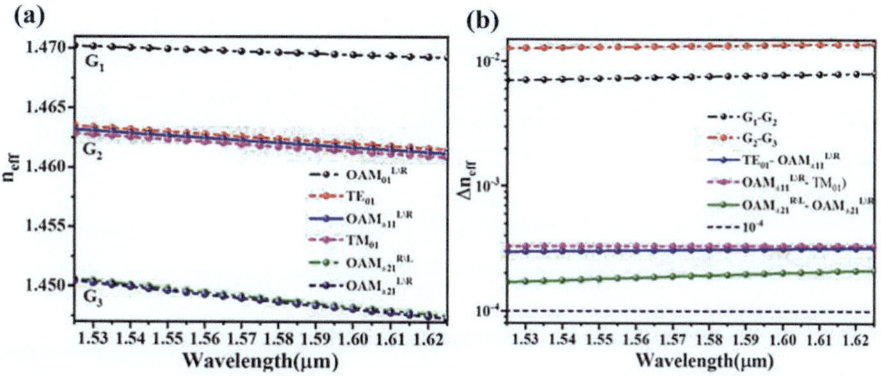

Fig. 4.10 **a** Effective indexes across wavelength, **b** Δn_{eff} between successive modes supported in IRC-FMF versus wavelength

Table 4.3 Modal separation between couple of modes and couple of groups at $\lambda = 1.55$ μm for the considered IRC and IPGI fibers (Ung et al. 2014)

	IRC	IPGI
$OAM_{\pm 01}^{L/R} - TE_{01}$	6.96×10^{-3}	6.61×10^{-3}
$TE_{01} - OAM_{\pm 11}^{L/R}$	3.31×10^{-4}	2.44×10^{-4}
$OAM_{\pm 11}^{L/R} - TM_{01}$	3.04×10^{-4}	2.43×10^{-4}
$TM_{01} - OAM_{\pm 21}^{R/L}$	1.25×10^{-2}	1.07×10^{-2}
$OAM_{\pm 21}^{R/L} - OAM_{\pm 21}^{L/R}$	1.78×10^{-4}	1.27×10^{-4}

intra-modal of 1.78×10^{-4} reached between the $OAM_{\pm 21}^{R/L}$ and the $OAM_{\pm 21}^{L/R}$. Even though this value reaches the target modal separation of 10^{-4} and guarantees the independence of all supported modes. The intermodal separation between successive modes of IRC-FMF are compared, at $\lambda = 1.55$ μm, with those in IPGI-FMF in Table 4.3. The comparison shows that the modal separation achieved by the IRC fiber is always larger, which may reduce the overall coupling.

4.5.2 *Differential Group Delay and Chromatic Dispersion*

The performance of the data-carrying and transverse modes is analyzed in terms of Differential Group Delay (DGD) and Chromatic Dispersion (CD). Both metrics are evaluated across the C-band for multiple values of the shape parameter α. The DGD results are illustrated in Fig. 4.11, while the CD trends are presented in Fig. 4.12. For

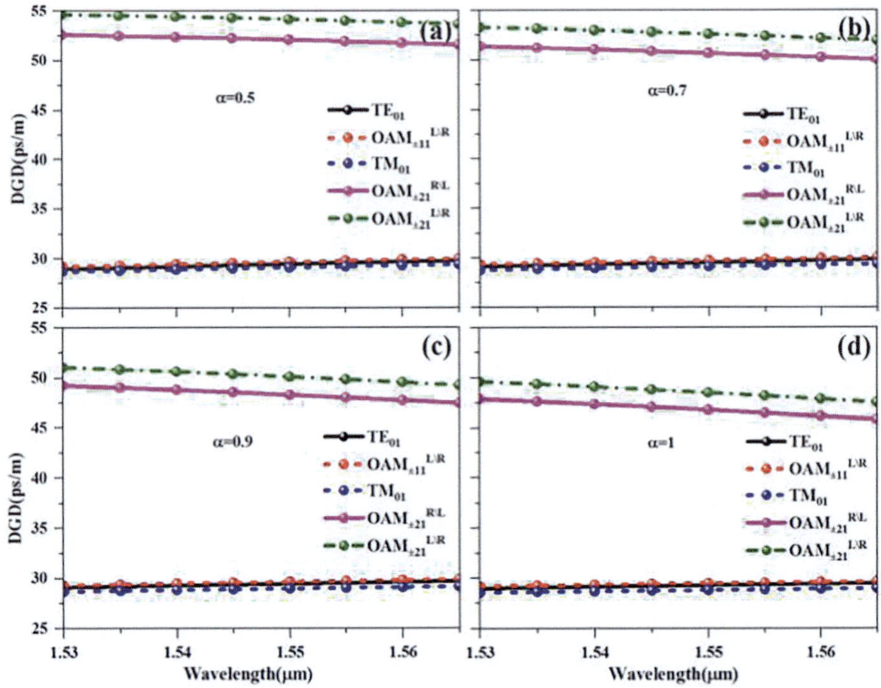

Fig. 4.11 DGD versus the C band and for different values of shape parameter α: **a** $\alpha = 0.5$, **b** $\alpha = 0.7$ **c** $\alpha = 0.9$ and **d** $\alpha = 1$

the DGD analysis, modes within group G_2 exhibit values around 30 ps/m, whereas modes in group G_3 show a larger DGD of approximately 50 ps/m. This difference can be attributed to two main factors: the proximity of G_3's effective refractive index to the cut-off threshold, and the wider separation between G_3 and G_1 (used as the reference group for DGD computation). Notably, the DGD curves remain nearly flat over the entire wavelength range and show minimal sensitivity to α, which aligns with the relatively stable intermodal spacing across different shape parameters. In terms of CD, group G_1 modes present values near -20, ps/(km nm), modes in G_2 are nearly dispersionless (close to 0 ps/(km nm)), while group G3 modes experience significantly negative dispersion, ranging between -50 and -80, ps/(km nm). Except for the G_3 group, all CD curves remain relatively flat with respect to wavelength. In G_3, however, the dispersion increases with wavelength due to the closeness of the effective index to the cladding index, i.e., approaching the cut-off. Like DGD, the CD values remain largely invariant with changing α. For benchmarking purposes, Table 4.4 summarizes the DGD and CD values of all supported modes at $\alpha = 1$ and $\lambda = 1.55\,\mu$m, and compares them with those observed in the IPGI-FMF. It is evident

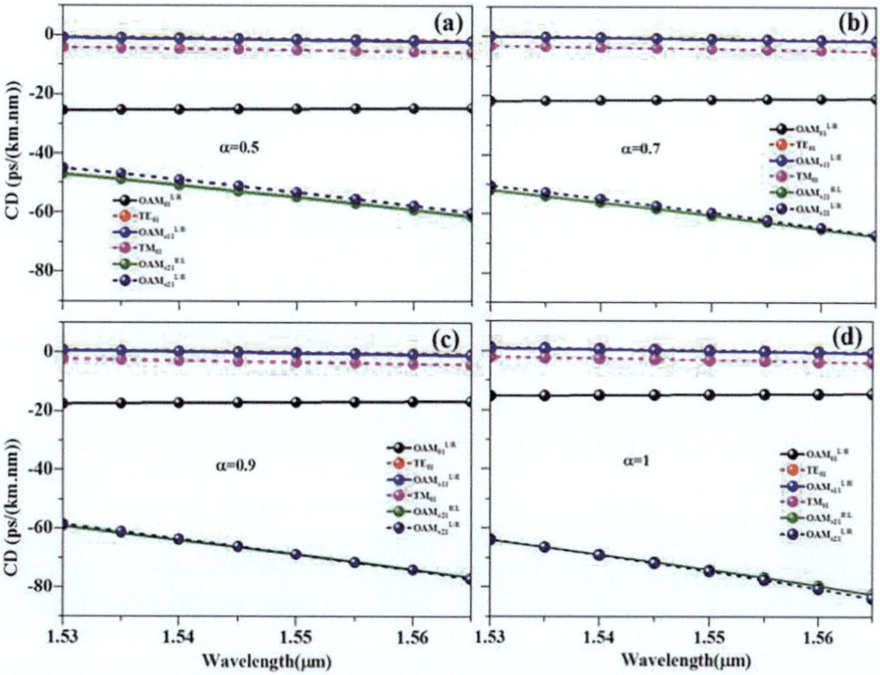

Fig. 4.12 CD versus the C band and for different values of shape parameter α: **a** $\alpha = 0.5$, **b** $\alpha = 0.7$ **c** $\alpha = 0.9$ and **d** $\alpha = 1$

Table 4.4 DGD in [ps/m] and CD in [ps/km nm] at $\lambda = 1.55\,\mu$m for IRC-FMF and IPGI-FMF. DGD of IPGI (Zhu et al. 2017). CD of IPGI (Zhu et al. 2017)

	DGD of IRC	DGD of IPGI	CD of IRC	CD of IPGI
$OAM_{\pm01}^{L/R}$	–	–	–14.86	4.00
TE_{01}	29.33	25.3	0.25	12.00
$OAM_{\pm11}^{L/R}$	29.54	26.75	–0.08	11.83
TM_{01}	28.82	26.4	–3.17	8.56
$OAM_{\pm21}^{R/L}$	46.46	27.32	−74.09	−136.16
$OAM_{\pm21}^{L/R}$	48.18	28.10	−74.68	−138.20

from the table that the IPGI-FMF structure features lower DGD values, while the IRC-FMF demonstrates superior CD performance for most modes, except the OAM01 mode. The markedly higher CD in IPGI-FMF for G_3 modes is again attributed to the group indices near the cut-off condition (Rjeb et al. 2019).

4.5.3 Effective Mode Area and Nonlinearity Coefficient

In this section, we evaluate the effective mode area (A_{eff}) of all modes supported in the designed IRC-FMF. In addition, we evaluate the corresponding nonlinearity coefficient γ. The results are shown in Fig. 4.13a, b for wavelengths ranging from 1.530 to 1.565 μm, (C band). It can be seen that A_{eff} values of all modes are quite limited and not more than 45 μm^2. The maximum A_{eff} value across the C band is reserved for the third group (OAM_{21}); on the other hand, the minimum A_{eff} value is for the fundamental mode OAM_{01} (around 33 μm^2 at $\lambda = 1.53$ μm). We compute γ versus wavelength and find out that the nonlinearity coefficient is around 3 (/dB/km^2) at 1.530 μm for the fundamental mode and around 2.25 (/dB/km^2) at 1.565 μm for $OAM_{\pm 21}$ (generated from EH_{11} and HE_{31}). These results indicate two trends. With the above-cited parameters (from IPGI fiber (Zhu et al. 2017)), IRC-FMF cylindrical vector modes possess low A_{eff}, since the core radius is small, which is detrimental to signal transmission inside the fiber due to nonlinear effects. This can be a significant threat to the system performance, especially to the fiber. On the other hand, low A_{eff} is useful in the case of Raman gain (less pump power required). Furthermore, large A_{eff} causes micro bending loss (Mercy Kingsta and Shantha Selvakumari 2019). On the other hand, A_{eff} and γ values of modes are almost equal; the difference in A_{eff} is no more than 10 μm^2. This property confirms the similarity of modes (OAM) fields propagating inside the core fiber. This similarity (distribution and/or intensity) helps to deal with most modes in the same manner (i.e., facilitate the amplification with the same power for all channels). Hence, the designed IRC-FMF can find applications in diverse areas, such as multicore fiber transmission (Lataoui et al. 2021).

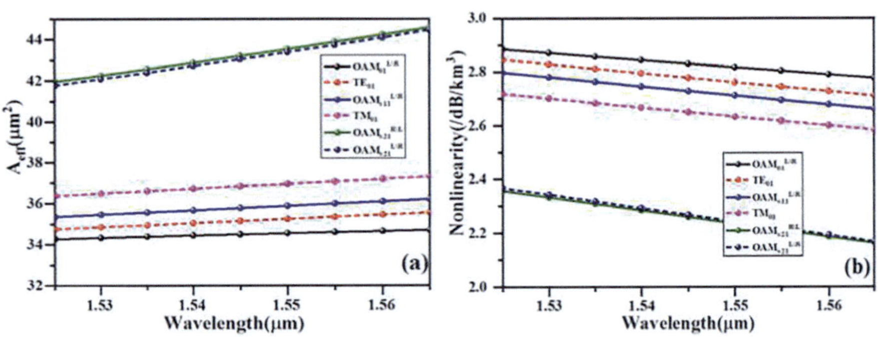

Fig. 4.13 **a** Effective mode area versus wavelength, **b** nonlinearity coefficient versus wavelength, in the proposed IRC-FMF

4.6 OAM Mode Coupling

This section presents the mode-coupling framework as applied to the IRC-FMF structure. In OAM-based space-division multiplexing (SDM) systems, mode coupling that arises during light propagation is a critical limitation, contributing significantly to signal degradation and data loss. Ideally, the guided modes within the fiber are orthogonal, ensuring minimal interaction. However, this ideal condition is often disrupted by physical imperfections or environmental disturbances, such as fiber twisting, intrinsic birefringence, non-circular core geometries (e.g., ellipticity), and mechanical bending (Lonardi et al. 2017). The strength and type of mode coupling are governed by the magnitude and spatial characteristics of these perturbations, as well as the effective separation between neighboring modes or mode groups. A widely adopted theoretical model allows for analyzing mode coupling in multimode and multicore fibers. This model is based on the coupling coefficient, which is structured as an $N \times N$ matrix, quantifying interactions among N guided modes. The coupling between two specific modes, say μ and ϑ, is defined by Palmieri (2014):

$$K_{\mu\vartheta} = \frac{\omega}{4P} \int_{0}^{\infty} \int_{0}^{2\pi} r \; E_{\mu}^{*} \, \widehat{\varepsilon} \, E_{\vartheta} \, d\phi \, dr \tag{4.7}$$

Here, r and ϕ denote the radial and angular coordinates in cylindrical geometry. E_{μ} and E_{ϑ} represent the full vectorial electric field distributions (including both transverse and longitudinal components) of the interacting modes. $\widehat{\varepsilon}$ stands for the dielectric perturbation tensor, and P is a power normalization factor. The dielectric perturbation tensor $\widehat{\varepsilon}(r, \phi)$ is often expressed as the product of a radial function and an angular function. Specifically, it can be written as Rjeb et al. (2020b):

$$\widehat{\varepsilon_{i,j}}(r, \phi) = \varepsilon_{i,j}(r) \exp(\pm jk\phi) \tag{4.8}$$

where $k = 0, \pm 1, \pm 2, \ldots$ denotes the azimuthal order of the perturbation. Since mode profiles typically have an angular dependence of the form $\exp(j\phi)$, meaningful coupling between two modes will occur only if the angular overlap integral is nonzero. This leads to a resonance-like condition. For two modes μ and ϑ with azimuthal indices l_{μ} and l_{ϑ}, coupling is permitted only when the following criterion is met (Rjeb et al. 2020b):

$$l_{\mu} - l_{\vartheta} = \pm k \tag{4.9}$$

It is noteworthy that satisfying this angular condition alone does not guarantee coupling; the radial integral must also yield a nonzero result. Therefore, Eq. (4.9) provides a necessary but not always sufficient condition for coupling. Moreover, the specific structure of the dielectric perturbation tensor $\widehat{\varepsilon}(r, \phi)$ determines which field components (e.g., transverse or longitudinal) are prone to interact. In what follows, we delve into two representative types of perturbations–core ellipticity and material birefringence-and analyze their distinct impacts on mode coupling behavior.

4.6.1 Ellipticity

In the presence of slight core ellipticity, the resulting dielectric perturbation can be described by the following expression (Rjeb et al. 2020b):

$$\widehat{\varepsilon_{i,j}}(r,\phi) = \varepsilon_0 \sin 2\phi \sum_{k-1}^{R-1} \gamma_k (n_k^2 - n_{k+1}^2)\delta(r - r_k) \tag{4.10}$$

Here, r_k denotes the nominal outer radius of the k-th ring, γ_k represents the radial deviation caused by ellipticity, and n_k is the refractive index associated with that ring. The expression introduces a scalar modification to the dielectric tensor, which is directly linked to the structural asymmetry in the core geometry. Such an ellipticity-induced perturbation leads to coupling predominantly between modes of the same polarization. Specifically, coupling is allowed between transverse components sharing the same handedness (i.e., $\widehat{R} \leftrightarrow \widehat{R}$ or $\widehat{L} \leftrightarrow \widehat{L}$), as well as between longitudinal components (i.e., $\widehat{Z} \leftrightarrow \widehat{Z}$). This restriction arises from the symmetry properties of the perturbation and the selection rules governing angular momentum conservation.

4.6.2 Birefringence

In the presence of material birefringence, the dielectric perturbation $\widehat{\varepsilon_{i,j}}(r,\phi)$ is defined as Rjeb et al. (2020b):

$$\widehat{\varepsilon_{i,j}}(r,\phi) = \varepsilon_{co} \Delta_n \begin{pmatrix} 0 & 1 & 0 \\ 1 & 0 & 0 \\ 0 & 0 & 0 \end{pmatrix} \tag{4.11}$$

Here, ε_{co} represents the dielectric permittivity in the core, and Δ_n refers to the birefringence. As shown in expression (4.11), the perturbation $\widehat{\varepsilon_{i,j}}(r,\phi))$ is independent of the azimuthal angle, meaning the azimuthal order is $k = 0$, and coupling occurs only between components of the same azimuthal order (n_μ and n_ϑ). Additionally, coupling is limited to the circular cross-polarized components ($\widehat{R} \leftrightarrow \widehat{L}$). For instance, $OAM_{l,m}^L$ and $OAM_{l,m}^R$ can couple through their dominant components, while no coupling takes place between $OAM_{-l,m}^L$ and $OAM_{l,m}^R$ modes. The IRC-FMF profile is integrated into the coupling coefficient calculations for stress birefringence and core ellipticity scenarios. The coupling coefficients for OAM modes are numerically computed and displayed in Fig. 4.14, with the results normalized to the maximum

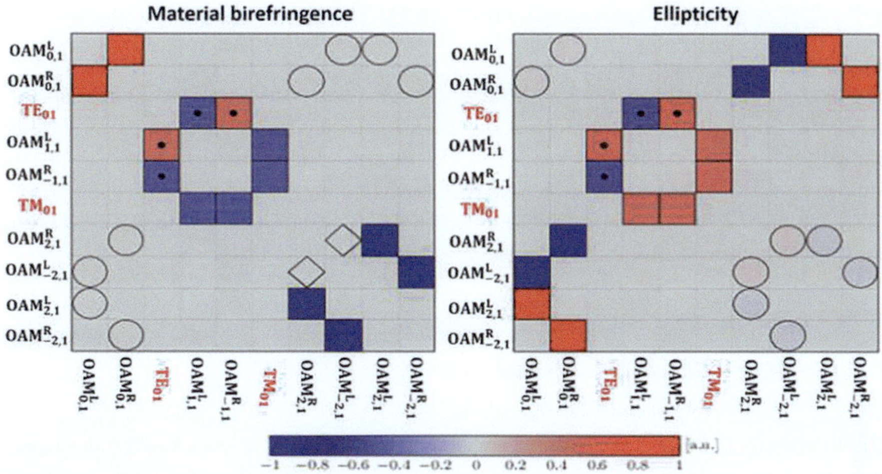

Fig. 4.14 Normalized magnitude of coupling coefficients evaluated for stress birefringence and core ellipticity: squares, circles and diamonds represent-coupling feasibility between dominant-dominant components, between secondary-secondary components, and between secondary-dominant (respectively dominant-secondary), respectively

value for each type of perturbation. Different markers are used to represent mode coupling: squares indicate dominant-dominant component coupling, diamonds denote dominant-secondary coupling, and circles represent secondary-secondary coupling. Dots are used to indicate the imaginary coupling coefficient. From Fig. 4.14, it is evident that stress birefringence and core ellipticity yield similar coupling results in OAM modes, albeit with differing strengths. This is due to both perturbations causing coupling in hybrid modes ($HE_{li,mi}$ and/or $EH_{li,mi}$), provided their orders satisfy the condition $\left|l_i - l_{j\neq i}\right| = 2$ (Rjeb et al. 2020b). Regarding magnitudes, for stress birefringence, the peak coupling occurs between the $OAM_{01}^{R/L}$ modes, reaching approximately $(1.9 \times 10^6) \times \Delta_n$ rad/m. In the case of core ellipticity, the maximum coupling is $(1. \times 10^4) \times \gamma$ rad/m, occurring between the $OAM_{01}^{R/L}$ and $OAM_{\pm21}^{R/L}$ modes. To assess these findings, the IRC-FMF results are compared with those from [Guerra et al., 2019]. The mode coupling observed in IRC-FMF is lower than in hollow-core fibers (e.g., ACF), due to the larger intermodal separation in IRC-FMF (minimum $\Delta n_{eff} = 1.27 \times 10^{-4}$) compared to hollow-core fibers, which have a value around 1×10^{-4}. This suggests that the IRC-FMF design demonstrates greater resilience to certain fiber perturbations. Furthermore, Fig. 4.14 illustrates that the coupling coefficients for mixed-strength components (secondary-dominant) and secondary-secondary components are negligible (uncolored circles and diamonds), while those for dominant-dominant components are substantial. This can be attributed to the high purity of the OAM modes in IRC-FMF, where secondary components play a minimal role.

4.7 Effects of Bending on IRC-FMF Performances

In practical applications, optical fibers are often subject to bending, which can significantly impact transmission quality. Bending can alter various factors such as intermodal separation, the purity of OAM modes, and result in power loss. This section examines the effects of bending on OAM channels in IRC-FMF, focusing on the deformation of OAM modes, inter-channel crosstalk (XT), mode walk-offs, and bending loss that occurs during propagation through IRC-FMF.

4.7.1 Bent IRC-FMF Model

The bending is caused by movement over a short distance due to localized stresses or lateral forces along the length. It causes a change of the refractive index profile of the fiber. Based on the standard conformal mapping approach, the refractive index of a bent fiber, noted as $n_{bent}(x, y)$, is expressed with respect to the refractive index of straight fiber, noted as $n_{straigth}(x, y)$, as follows (Rjeb et al. 2021c):

$$n_{bent}(x, y) = n_{straigth}(x, y)\left(1 + \frac{x}{1.4R_b}\right) \tag{4.12}$$

where the parameter R_b indicates the bend radius applied along the transverse x-axis direction. The bending degree increases as the bend radius R_b decreases. The term 1.40 presents the photoelastic effects on the refractive index profile induced by the local strain in bent fused silica fibers (Schulze et al. 2013). Figure 4.15 highlights a bent fiber model, where an optical fiber is bent around a bend post of radius Rb. The impact of bending on IRC-FMF supported modes are highlighted in Fig. 4.16a, b. Figure 4.16a depicts the bending effects on OAM modes phases ($l \neq 0$), while Fig. 4.16b depicts the normalized intensity of the transverse electric mode (TE_{01}).

Fig. 4.15 A bent fiber model

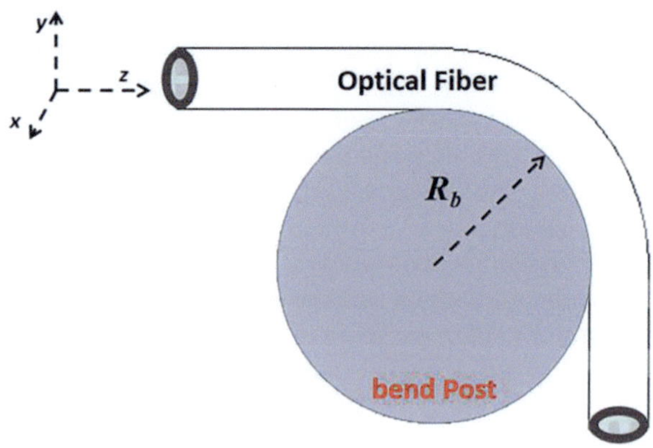

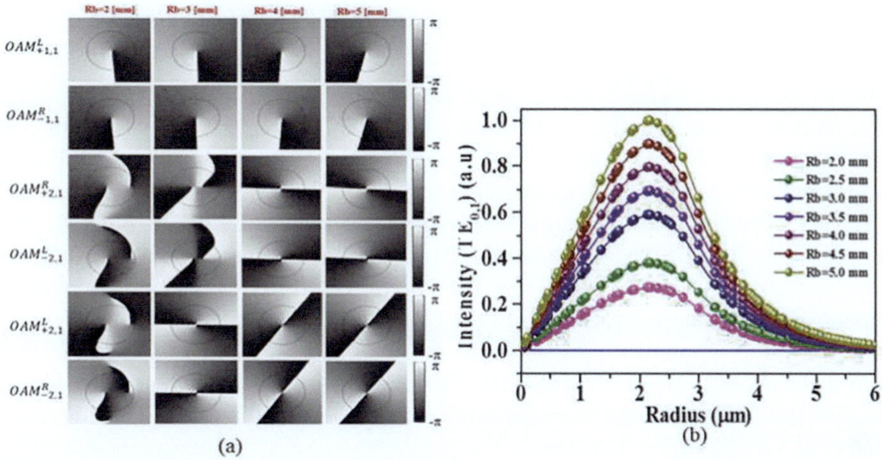

Fig. 4.16 **a** Impacts of bending on OAM modes phases ($l \neq 0$) supported in IRC-FMF. **b** effects of bending on the normalized intensity of TE_{01} mode

The selected bending radius R_b is spanning between 2 to 5 mm, for both Figures. At very small bending rang ($R_b < 2$ mm), only the first fundamental mode (i.e. OAM_{01}) is supported by the designed IRC-FMF. Since the bending affects the refractive index profile of the fiber, it tends to influence the supported fiber eigenmodes and as a results the OAM modes, since the later are constructed from the former modes. In details, when the fiber is bent, the power leaks from mode to another mode or from modes to the cladding. This is reflected by the decreasing of intensity of modes (shown in Fig. 4.16b) which will affects the effective area (A_{eff}) of modes hence increases the effects of nonlinearity along IRC-FMF transmission. Furthermore, the bending influences the circular symmetric patterns of true modes (eigenmodes), thereby, it directly influence the phase distributions of resulting OAM modes. These changes are reflected as distortions (i.e. deformities) in the OAM phase distribution (shown in Fig. 4.16a). Moreover, an exchange of phase is occurred between $OAM_{+l,m}$ and $OAM_{-l,m}$. This is clearly highlighted for example at $R_b = 4$ mm for modes $OAM_{+1,1}$/ $OAM_{-1,1}$. These effects have results on the data transmission operation over IRC-FMF. In the following, we examine the propagation performances metrics of OAM channels over a bent IRC-FMF. The investigated parameters are the number of OAM channels, the inter-channels separations, the mode walk-offs (L_{10ps}, $L_{2\pi}$) and the loss due to bending. Our numerical simulations have been carried out over a bending radius (R_b) ranging from 1 to 22 mm.

4.7.2 OAM Channel Number and Inter-Channels Crosstalk

The first parameter that could be affected by the bending is the number of OAM modes (i.e. channels), where under tight bending (small R_b); some modes are unlaunched due to the perturbation in the refractive index profile, which tend to leak power from fiber core to the cladding region (i.e. unguided). In Fig. 4.17, we depicts the effective indexes separation (Δn_{eff}) between successive modes versus R_b at $R_b = 4$ mm, the bent IRC-FMF supports all the 10 eigenmodes modes presented in its ideal case (i.e. straight case). Considering the Δn_{eff}, IRC-FMF keeps the modes well separated. This clearly reflected in Fig. 4.17 and the zoomed part therein, where all $\Delta n_{eff} > 10^{-4}$. Thus, IRC-FMF is resilient (i.e. tolerant) to bending (from $R_b = 4$ mm and beyond), hence, enables low channels crosstalk. The threshold value of $R_b = 4$ mm, is better than those reported in Rjeb et al. (2020c) ($R_b = 8$ mm) and Jiao et al. (2019) ($R_b = 5$ mm).

4.7.3 Modes Walk Offs

Other parameter that could be affected by the bending is the separation between odd and even eigenmodes that form the same an OAM mode. This change will give a mode walk-off during propagation over the IRC-FMF. Two types of mode walk-offs are occurred during propagation which are characterized by $L_{2\pi}$ and L_{10ps} where L is denoted for the length of walk-off. $L_{2\pi}$ is known as the distance when the two eigenmodes walked off to each other with a 2π optical cycle (i.e. a relative phase shift of 2π). L_{10ps} is defined as the distance after the two eigenmodes has a 10 *ps* temporal walk off. Considering 10-Gbaud/s signals, 10-*ps* temporal walk-off is equivalent

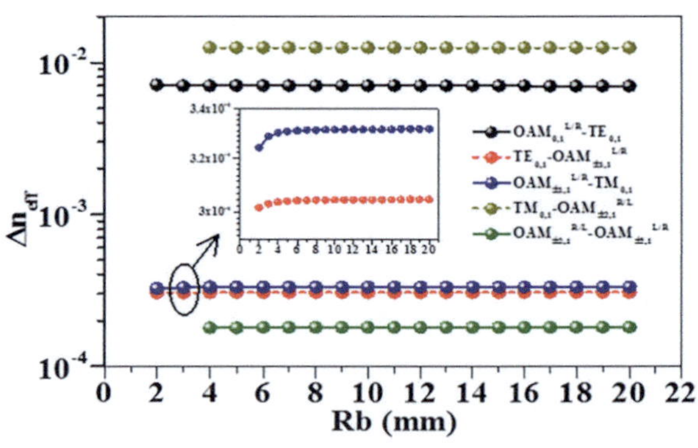

Fig. 4.17 Intermodal separation (Δn_{eff}) among successive modes supported by IRC-FMF versus bending radius R_b

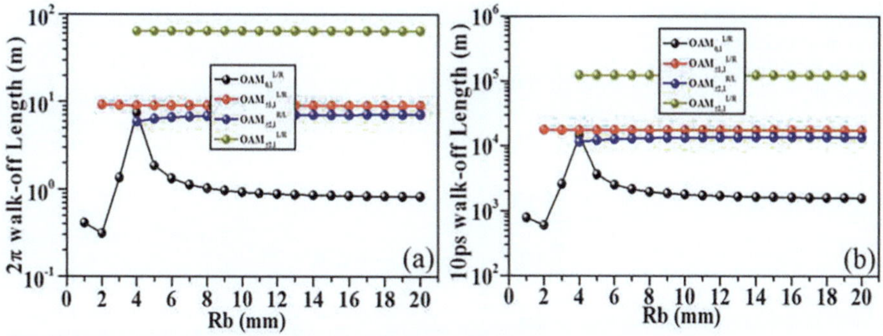

Fig. 4.18 **a** $L_{2\pi}$ of OAM modes supported in IRC-FMF versus R_b. **b** L_{10ps} of OAM modes supported in IRC-FMF versus R_b

to 10% of the symbol period (Yue et al. 2012). $L_{2\pi}$ and L_{10ps} are expressed as follows (Yue et al. 2012):

$$\begin{cases} L_{2\pi}(m) = \dfrac{\lambda}{HE_{l,m}^{even} HE_{l,m}^{odd}} = \dfrac{1.55 \times 10^{-6}}{HE_{l,m}^{even} HE_{l,m}^{odd}} \\ L_{10ps}(m) = \dfrac{c \times \Delta t}{HE_{l,m}^{even} HE_{l,m}^{odd}} = \dfrac{3 \times 10^{-3}}{HE_{l,m}^{even} HE_{l,m}^{odd}} \end{cases} \tag{4.13}$$

where the parameters λ, c, and Δt are used to denote the wavelength, the speed of light in vacuum, and the walk-off time ($\Delta t = 10ps$), respectively. $L_{2\pi}$ and $L_{10ps}$of our designed IRC-FMF are depicted in Fig. 4.18a, b, respectively. Higher modes have the greater $L_{2\pi}$ and L_{10ps} compared to other modes (i.e. lower modes). This is related to the small effective index separation occurred between even and odd eigenmodes of $HE_{3,1}$. This involves that $HE_{3,1}$ has more azimuthal periods in its transverse field distribution than lower modes. Considering the effects of bending radius R_b on the walk-offs distances, $L_{2\pi}$ and L_{10ps} are rather stable except those for fundamental mode. The minimum $L_{2\pi}$ and L_{10ps} are 0.3 and 600 m at $R_b = 2$ cm respectively. These values are consistent with those in recently reported OAM-fibers (Rjeb et al. 2020c; Wang et al. 2020).

4.7.4 Bending Inducing Confinement Loss

This sub-section describes the effects of bending on the occurred power loss (expressed in dB). In fact, due to the bending, the effective index of modes comprise an imaginary part that is related to the loss of power during propagation. This imaginary part (i.e. Im (n_{eff})) increases as the bending radius Rb increases. The

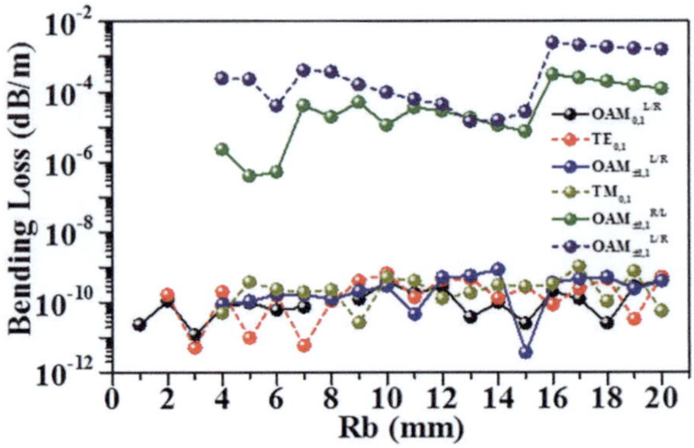

Fig. 4.19 Bending loss of supported modes in IRC-FMF versus bending radius R_b

$Im(n_{eff})$ is incorporated in the calculation of confinement loss (due to bending), as Rjeb et al. (2021c):

$$Loss[\text{dB/m}] = -20\log_{10}\exp\left(-\frac{2\pi Im(n_{eff})}{\lambda}\right) \tag{4.14}$$

where λ is the used wavelength. The bending loss of all modes versus R_b are shown in Fig. 4.19. Similarly as the walk-offs distances, higher modes observe the highest confinement loss (around 0.0001 dB/m) compared to lowest modes (around 10^{-10} dB/m). This is explained by the close proximity of higher modes to the cut-off (i.e. $n_{cladding}$) which tends to increase the susceptibility of the leakage of these modes into the cladding. Other cause behind that is related to the high index contrast (i.e. $\Delta n_{\max}$) that our designed IRC-FMF exhibits at the vicinity between the core and the cladding layers, which tend to raise the loss during propagation over the fiber. On the other side, lower modes exhibit lower confinement loss under tight bending radius range (e.g. R_b below 5 mm). These results are consistent with those reported in IPGI (Ung et al. 2014) and HTAN fiber in Rjeb et al. (2020c). In order to further decrease the bending loss, one solution is based on incorporating additional structures such as holes or trenches around the core. This tend to enhance the confinement of modes within the core fiber. Even though, the designed IRC-FMF seems resilient to bending effects with no need to add more complex structures to his refractive index profile.

4.8 Conclusion

In this chapter, we discussed one of the last new optical fiber that has been proposed and designed for OAM channels transmission. Various aspects that touch the designed IRC-FMF have been addressed. We started from the design and optimization process performed on the fiber and extract key parameters responsible for performances metrics enhancement. Several metrics are assessed including the number of available

OAM data channels, the separation among them, the differential delays, the chromatic dispersion, the purity, the intrinsic crosstalk the effective area and the nonlinearity. Other parameters are evaluated including the reliability of IRC-FMF to several perturbations such us ellipticity and birefringence. The effects of these imperfections are assessed in terms of OAM mode coupling. At last, the resilience of the designed IRC-FMF to a realistic environment including the bending is investigated and analyzed. At each stage of our studies, we evaluate and compare the obtained results to recently reported OAM-fibers used as benchmarks. These comparisons allowed as to confirm the outperforms of our designed fiber and its viability to became potential candidate for next generation optical communication systems. As a future work, the incorporation of IRC-FMF into quantum communication (especially in high dimensional quantum key distribution HD QKD) is among the perspectives of our analysis/ and exploration of new applications that could benefits from our designed fiber.

References

Bae, S., Jung, Y., Kim, B.G., Chung, Y.C.: Compensation of mode crosstalk in MDM system using digital optical phase conjugation. IEEE Photon. Technol. Lett. **31**(10):739–742 (2019)

Berdague, S., Facq, P.: Mode division multiplexing in optical fibers. Appl. Opt. **24**(11), 1950–1955 (1982)

Bozinovic, N., Yue, Y., Ren, Y., Tur, M., Kristensen, P., Huang, H., Ramachandran, S.: Terabit-scale orbital angular momentum mode division multiplexing in fibers. Science **340**(6140), 1545–1548 (2013)

Brunet, C., Vaity, P., Messaddeq, Y., LaRochelle, S., Rusch, L.A.: Design, fabrication and validation of an OAM fiber supporting 36 states. Opt. Express **22**(21), 26117–26127 (2014)

Brunet, C., Ung, B., Wang, L., Messaddeq, Y., LaRochelle, S., Rusch, L.A.: Design of a family of ring-core fibres for OAM transmission studies. Opt. Express **23**(8), 10553–10563 (2015)

Chebaane, S., Fathallah, H., Seleem, H., Machhout, M.: Proposed raised cosine FMF for dispersion management in next-generation optical networks. IEEE Photon. J. **8**(1), 1–6 (2016)

Chebaane, S., Fathallah, H., Seleem, H., Machhout, M.: Trenched raised cosine FMF for differential mode delay management in next generation optical networks. Opt. Commun. **408**, 15–20 (2018)

Chen, S., Wang, J.: Theoretical analyses on orbital angular momentum modes in conventional graded-index multimode fibre. Sci. Rep. **7**(1), 3990 (2017)

Chen, S., Wang, J.: Characterization of red/green/blue orbital angular momentum modes in conventional G.652 fiber. IEEE J. Quant. Electron. **53**(4), 1–14, Art no. 6800414 (2017)

Chralyvy, A.: The coming capacity crunch. In: 35th European Conference on Optical Communication, pp. 1–1 (2009)

Cisco VNI Global IP traffic forecast, San Diego, 2018–2023 (2018)

Essiambre, R., Tkach, R.: Capacity trends and limits of optical communication networks. In: Proceedings of the IEEE, vol. 100, no. 5, pp. 1035–1055 (2012)

Guerra, G., Lonardi, M., Galtarossa, A., Rusch, L.A., Bononi, A., Palmieri, L.: Analysis of modal coupling due to birefringence and ellipticity in strongly guiding ring-core OAM fibers. Opt. Express **27**(6), 8308–8326 (2019)

Jiao, X., Zhang, H., Zhang, X., Li, H., Wei, J., Wang, Z., Xi, L., Zhang, W., Tang, X.: Performance of circular photonic crystal fiber transmitting orbital angular momentum modes under macrobending. J. Opt. **21**(6), 065703 (2019)

Lataoui, J., Rjeb, A., Jaba, N.H., Fathallah, H., Machhout, M.: Multicore Raised Cosine Fibers for Next Generation Short & Medium Haul Interconnect (2021). Accepted for publication in Optical Fiber Technology (in press)

Lonardi, M., Guerra, G., Marcon, L., Nejad, R.M., Santagiustina, M., Galtarossa, A., Rusch, L.A., Bononi, A., Palmieri, L.: Mode coupling analysis of hollow ring-core fibers for OAM transmission. In: 2017 European Conference on Optical Communication (ECOC), pp. 1–3 (2017)

Mercy Kingsta, R., Shantha Selvakumari, R.: A review on coupled and uncoupled multicore fibers for future ultra-high capacity optical communication. Optik **199**, 163341 (2019)

Palmieri, L.: Coupling mechanism in multimode fibers. In: Next-Generation Optical Communication: Components, Sub-Systems, and Systems III, vol. 9009, p. 90090G. International Society for Optics and Photonics (2014)

Proakis, J.G., Salehi, M.: Digital Communications, 6th edn. McGraw-Hill, APA, Boston (2008)

Ramachandran, S., Kristensen, P., Yan, M.F.: Generation and propagation of radially polarized beams in optical fiber. Opt. Lett. **34**, 2525 (2009)

Ramachandran, S., Kristensen, P.: Optical vortices in fiber. Nanophotonics **2**(5–6), 455–474 (2013)

Richardson, D.J., Fini, J.M., Nelson, L.E.: Space-division multiplexing in optical fibres. Nat. Photon. **7**(5), 354–362 (2013)

Rjeb, A., Guerra, G., Issa, K., Fathallah, H., Chebaane, S., Machhout, M., Palmieri L., Galtarossa, A.: Inverse-raised-cosine fibers for next-generation orbital angular momentum systems. Opt. Commun. 124736 (2019)

Rjeb, A., Seleem, H., Fathallah, H., Machhout, M.: Design of 12 OAM-Graded index few mode fibers for next generation short haul interconnect transmission. Opt. Fiber Technol. **55**, 102148 (2020)

Rjeb, A., Fathallah, H., Machhout, M.: Orbital angular momentum mode coupling analysis due to ellipticity and birefringence in inverse-raised cosine fiber. In: 2020 17th International Multi-Conference on Systems, Signals & Devices (SSD), pp. 929–932 (2020b)

Rjeb, A., Fathallah, H., Khaled, I., Machhout, M., Alshebeili, S.A.: A novel hyperbolic tangent profile for optical fiber for next generation OAM-MDM systems. IEEE Access **8**, 226737–22675 (2020)

Rjeb, A., Fathallah, H., Issa, K., Machhout, M., Alshebeili, S.A.: Inverse hyperbolic tangent fiber for OAM mode/mode-group division multiplexing optical networks. IEEE J. Quant. Electron. (2021a)

Rjeb, A., Seleem, H., Fathallah, H., and Machhout, M.: A novel inverse gaussian profile for orbital angular momentum mode division multiplexing optical networks. In: 2021 18th International Multi-Conference on Systems, Signals & Devices (SSD), pp. 480–487. IEEE (2021b)

Rjeb, A., Fathallah, H., Machhout, M.: Analysis of orbital angular momentum channels in bent inverse raised cosine fiber. In: 2021 IEEE International Conference on Design & Test of Integrated Micro & Nano-Systems (DTS), pp. 1–4. IEEE (2021c)

Rusch, L.A., Rad, M., Allahverdyan, K., Fazal, I., Bernier, E.: Carrying data on the orbital angular momentum of light. IEEE Commun. Mag. **56**(2), 219–224 (2018)

Ryf, R., Fontaine, N.K., Mestre, M.A., Randel, S., Palou, X., Bolle, C., Gnauck, A.H., Chandrasekhar, S., Liu, X., Guan, B., Essiambre, R.-J., Winzer, P.J., Leon-Saval, S., Bland-Hawthorn, J., Delbue, R., Pupalaikis, P., Sureka, A., Sun, Y., Gruner-Nielsen, L., Jensen, R.V., Lingle, R.: 12 $\times$12 mimo transmission over 130−km few−mode fiber. Front. Opt. Laser Sci. **XXVIII**, FW6C.4 (2012a)

Ryf, R., Randel, S., Gnauck, A.H., Bolle, C., Sierra, A., Mumtaz, S., Esmaeelpour, M., Burrows, E.C., Essiambre, R.-J., Winzer, P.J., Peckham, D.W., McCurdy, A.H., Lingle, R.: Mode-division multiplexing over 96 km of few-mode fiber using coherent 6 $\times$ 6 MIMO processing. J. Lightwave Technol. **30**(4), 521–531 (2012b)

Saridis, G.M., Alexandropoulos, D., Zervas, G., Simeonidou, D.: Survey and evaluation of space division multiplexing: from technologies to optical networks. IEEE Commun. Surv. Tutor. **17**(4), 2136–2156 (2015)

Schulze, C., Lorenz, A., Flamm, D., Hartung, A., Schröter, S., Bartelt, H., Duparré, M.: Mode resolved bend loss in few-mode optical fibers. Opt. Exp. **21**(3), 3170–3181 (2013)

Sillard, P., Bigot-Astruc, M., Boivin, D., Maerten H., Provost, L.: Few-mode fiber for uncoupled mode-division multiplexing transmissions. In: 37th European Conference and Exhibition on Optical Communication, pp. 1–3 (2011)

Trichili, A., Park, K.H., Zghal, M., Ooi, B.S., Alouini, M.S.: Communicating using spatial mode multiplexing: Potentials, challenges, and perspectives. IEEE Commun. Surv. Tutor. **21**(4), 3175–3203 (2019)

Ung, B., Vaity, P., Wang, L., Messaddeq, Y., Rusch, L.A., LaRochelle, S.: Few-mode fiber with inverse-parabolic graded-index profile for transmission of OAM-carrying modes. Opt. Express **22**(15), 18044–18055 (2014)

Wang, J., Yang, J.Y., Fazal, I.M., Ahmed, N., Yan, Y., Huang, H., Willner, A.E.: Terabit free-space data transmission employing orbital angular momentum multiplexing. Nat. Photon. **6**(7), 488–496 (2012)

Wang, Y., Bao, C., Geng, W., Lu, Y., Fang, Y., Mao, B., Liu, Y.-G., Liu, B., Huang, H., Ren, Y., Pan, Z., Yue, Y.: Air-core ring fiber with >1000 radially fundamental OAM modes across O E S C and L bands. IEEE Access **8**, 68280–68287 (2020)

Yaman, F., Bai, N., Zhu, B., Wang, T., Li, G.: Long distance transmission in few-mode fibers. Opt. Express **18**(12), 13250–13257 (2010)

Yue, Y., Yan, Y., Ahmed, N., Yang, J.-Y., Zhang, L., Ren, Y., Huang, H., Birnbaum, K.M., Erkmen, B.I., Dolinar, S., Tur, M., Willner, A.E.: Mode properties and propagation effects of optical orbital angular momentum (OAM) modes in a ring fiber. IEEE Photon. J. **4**(2), 535–543 (2012)

Zhang, R., Tan, H., Zhang, J., Shen, L., Liu, J., Liu, Y., and Yu, S.: A novel ring-core fiber supporting MIMO-free 50km transmission over high-order OAM modes. In: 2019 Optical Fiber Communications Conference and Exhibition (OFC), pp. 1–3. IEEE (2019)

Zhu, G., Chen, Y., Du, C., Zhang, Y., Liu, J., Yu, S.: A graded index ring-core fiber supporting 22 OAM states. In: Opto-Electronics and Communications Conference (OECC) and Photonics Global Conference (PGC), pp. 1–3. IEEE (2017)

Chapter 5
Inverse Gaussian Fibers for Future OAM-MDM-WDM Optical Networking Systems

Alaaeddine Rjeb, Hussein Seleem, Habib Fathallah, and Mohsen Machhout

Abstract To meet the growing demand for transmission capacity and spectral efficiency in optical communication systems, significant attention has shifted toward utilizing the orbital angular momentum (OAM) of light for information encoding and transmission through optical fibers. This trend has spurred the development of advanced specialty fibers tailored to support OAM channels with improved performance characteristics. In this chapter, we focus on a class of optical fibers specifically engineered for OAM mode propagation. We examine various design approaches and review recent advancements in this area. Furthermore, we introduce our custom-designed OAM-supporting fibers, inverse Gaussian fibers (IGFs). A comprehensive numerical investigation is carried out to analyze the key structural and optical parameters of IGFs that enable stable OAM mode guidance. We propose and optimize four IGF configurations that demonstrate enhanced intermodal separation, achieving a minimum effective refractive index difference of $\Delta n_{eff} = 2.74 \times 10^{-4}$. This large separation helps suppress unwanted mode coupling and significantly reduces inter-channel crosstalk. Additionally, we evaluate the transmission characteristics of the designed IG-FMFs in terms of differential group delay and chromatic dispersion across the ITU-T C+L wavelength bands. Comparative results reveal that our IGF designs offer superior performance relative to state-of-the-art specialty fibers, positioning them as strong candidates for future high-capacity optical communica-

A. Rjeb (✉) · M. Machhout
Laboratory of Electronics and Microelectronics, Physics Department,
Faculty of Sciences of Monastir, Monastir, Tunisia
e-mail: alaaeddine.rjeb@gmail.com

H. Seleem
Department of Electronics and Electrical Communications, Faculty of Engineering,
Tanta University, Tanta, Egypt
e-mail: eltaibee@f-eng.tanta.edu.eg; hussein.eltaibee@pua.edu.eg

Electical Engineering Department, Faculty of Engineering, Pharos University in Alexandria,
Alexandria, Egypt

H. Fathallah
Laboratory of Artificial Intelligence and Data Engineering Applications,
Computer Department, Faculty of Sciences of Bizerte, Bizerte, Tunisia

N. Derbel and Q. Zhu (eds.), *Advances in Communication, Signal and Image Processing*,
https://doi.org/10.1007/978-981-95-0401-5_5

tion networks incorporating wavelength-division multiplexing (WDM). This chapter aims to provide valuable insights and serve as a stimulus for further research and development within the optical fiber design community.

Keywords Mode division multiplexing · Orbital angular momentum · Specialty few mode fibers design · Wavelength division multiplexing · Inverse gaussian fibers · Intermodal separation · Crosstalk reduction · Chromatic dispersion

5.1 Introduction

Fueled by the increase of number of connected users and devices (from internet of things (IOT) to internet of everything (IOET)), and the exponential demand for more data online services (from 5G to 6G), the global connectivity infrastructure needs improvements in terms of novel strategies & schemes and new communication technologies (Cisco VNI Global 2018). These solutions are considered as required targets that optical communication society (scientists and engineers) seek to explore and make them in a continuous race to cope with the upcoming capacity crunch (Chralyvy 2009).

An interesting direction of research for the optical fiber community is the harnessing of optical signal dimensions. These dimensions are used for enhancing the transmission capacity and enlarging the spectral efficiency (Essiambre and Tkach 2012). These dimensions are the time, the frequency (i.e. wavelength), the polarization, the phase and recently the space (Richardson et al. 2013). Recently, investing in the space dimension, known as space division multiplexing (SDM), has seen tremendous interest translated by the development and exploration of novel fibers and devices. SDM has been considered as an efficient long-term solution (the space is the only unexhausted dimension) to deal with the upcoming capacity crunch and to overcome the saturation of existing systems (Chralyvy 2009; Essiambre and Tkach 2012; Richardson et al. 2013). Interestingly, one of the promising scheme of SDM is consist in carrying data on optical fiber modes. These modes are required to be independents (i.e. orthogonal). Known as mode division multiplexing (MDM) (Ryf et al. 2012a), is an attractive approach for enhancing the transmission capacity. The number of available independent modes that co-propagate within the same fiber multiplies the overall capacity of the fiber transmission link. Hence as improving the number of modes (i.e. data channels), as the capacity increase (Saridis et al. 2015). Indeed, when the number of modes increases (multimode fibers), the interactions among them will increases leading to mode coupling that enhances the induced crosstalk (XT). The crosstalk deteriorates the quality of the transmission process (Bae et al. 2019). Therefore, to deal with crosstalk, a heavy and complex multiple input multiple output digital signal processing (MIMO-DSP) is required at the receiving stage (Ryf et al. 2012b). Moreover, due to the diversity of modes, this may leads to a differential mode delays (DMD). Few mode fibers (FMFs) are considered as appropriate solution to overcome the cited issues in MDM systems (Yaman et al. 2010). As a last trend in

MDM using FMFs, harnessing orbital angular momentum (OAM) of light to encode information instead of linear polarized modes (LP) became an attractive technique that has showed a great capability in improving both the overall optical channel capacity (beyond terabit to petabit) and the spectral efficiency (up to hundreds (b /s/Hz)) (Yao and Padgett 2011; Wang et al. 2014; Trichili et al. 2019; Wang et al. 2012). OAM modes are characterized by their inherent spatial orthogonality and their unbounded states (Rusch et al. 2018). Generally, two strategies supervising the transmission of OAM channels through an optical fiber are performed. The first is known as OAM-MDM (Rjeb et al. 2020a). This is based on carrying data on each independent OAM mode and transmit them within FMFs. In that regards, each OAM mode is considered as one data channel. Hence, the separation between modes is considered as the important parameter that characterize the number of available data channels. Intermodal separation of 10^{-4} guarantees the minimum interaction between modes and keep them well independent hence limits the reduced XT between data channels. In order to obtain that threshold level, maintaining the fiber design by adjusting fiber key parameters including the refractive index profile parameters is an essential option. On the other side, fiber imperfections and impediments could disturb the ideality of the fiber hence threat the independency between channels. Hence, the MIMO DSP will be required at the receiver side to decipher channels. Aiming to prevent the incorporation of heavy and complex MIMO DSP, the second strategy is consist in carrying data on OAM mode groups (MG) co-propagating over MMFs, known as OAM-mode group division multiplexing (OAM-MGDM) (Rjeb et al. 2021a; Chen and Wang 2017b). OAM modes within each MG are quit close (superimposed). Each OAM MG is considered as one data channels with very high bandwidth. Compared to OAM-MDM, the number of available data channels, employing OAM-MGDM technique, is limited to the number of MGs hence degrading the scalability of the MGDM system.

In General, in order to support one from both strategies, with the goal to handle robust OAM channels, specialty optical fibers have been proposed, designed, and even prototyped (Rjeb et al. 2019a, 2021b; Ramachandran et al. 2009; Gregg et al. 2013; Wang et al. 2020; Zhang et al. 2019b). These fibers are known as OAM-fibers. On the other side, harnessing OAM modes in classical fibers (i.e. conventional fibers) has shown high potential (Chen and Wang 2017a, b). Robust OAM channels include the all the target performances metrics that should fulfilled in order to establish a high potential transmission operation. These performances metrics include the high number of supported OAM modes within the same fiber, the high separation among them that tend to reduce/eliminate channels XT, the low and the flat chromatic dispersion (CD) of channels, the low differential delays between modes known as differential group delay (DGD) and the high OAM mode purity that limits the intrinsic XT within the same OAM mode. Moreover, The excited OAM modes should be of the first order. Hence, it is preferable to avoid exciting the higher radial orders modes because it causes trouble in multiplexing and demultiplexing operations due to the intensity and phase variety distribution. These metrics are considered as ID card of an OAM-fiber that should be fulfilled along a wide range of wavelengths covering at least the C and L bands.

This chapter explores optical fibers specifically engineered to support orbital angular momentum (OAM) modes, emphasizing innovative designs that have shown promise in boosting transmission capacity and spectral efficiency in optical communication systems. We first review the core structural aspects and refractive index profile characteristics of these advanced fibers, highlighting significant outcomes reported in recent literature. Motivated by prior work and theoretical insights, we introduce a new class of custom-designed OAMsupporting few-mode fibers (FMFs). These fibers are based on a refractive index distribution inspired by the Gaussian function and are termed inverse Gaussian few-mode fibers (IG-FMFs) (Rjeb et al. 2021c). The Gaussian function, widely known for its applications in various scientific fields–including statistics (as the probability density function of the normal distribution (Arora 2002)), signal processing (as the discrete Gaussian kernel (Kondor and Lafferty 2002)), image processing (e.g., for image blurring and smoothing (Haddad and Akansu 1991)), and mathematical physics (e.g., in solving the heat equation (Chirikjian 2009)), is adapted here as a foundation for designing novel refractive index profiles in optical fibers. These IG-FMFs are rigorously analyzed to determine their ability to support stable and isolated OAM modes.

The chapter is structured as follows. Section 5.2 provides a general overview of OAM-guiding fibers, summarizing typical design principles fiber engineers adopt to effectively support OAM propagation. Recent fiber structures and notable achievements in this field are reviewed. In Sect. 5.3, we introduce our proposed IG refractive index profiles and examine the influence of their defining parameters on mode confinement and effective index separation. This numerical analysis is conducted using the Finite Element Method (FEM) within the COMSOL Multiphysics environment at the standard telecom wavelength $\lambda = 1.55$ μm. This investigation allows us to identify optimized profile configurations that favor low-order mode support and large intermodal spacing, facilitating the selection of promising IG-FMF designs. Section 5.4 focuses on the performance evaluation of the selected IG-FMFs across the full C+L bands of the ITU-T grid (1530–1625 nm). We analyze key transmission characteristics such as intermodal effective index separation, chromatic dispersion, and differential group delay. The results are compared with those of recently developed OAM fibers, revealing the advantages of our IG-FMF designs for next-generation optical communication systems.

5.2 OAM Specialty Fibers: Designs and Recently Results

In this section, we highlights the design guidelines of OAM-fibers. Then, we present such designs and important achievements of them.

5.2.1 OAM Fibers Design Guidelines

An examination of recent advances in OAM-fiber technologies reveals that most reported designs exhibit shared structural and conceptual features, which are typically distilled into a set of design principles and practical guidelines (Rjeb et al. 2019a, 2021b; Ramachandran et al. 2009; Gregg et al. 2013; Wang et al. 2020; Zhang et al. 2019b). These principles are illustrated in Fig. 5.1. A fundamental requirement is that the fiber's refractive index distribution conforms to the donut-shaped intensity profile characteristic of OAM modes, commonly achieved through ring-shaped structures. For instance, the top panel of Fig. 5.1 displays the refractive index profile of a step-index ring-core fiber (RCF), defined by an inner radius a_1 and an outer radius a_2, with a high-index ring region spanning this range. This configuration aligns with the nature of OAM modes, which exhibit a central phase singularity and thus lack intensity at their cores. Since the OAM field distribution follows the refractive index profile, employing a central high-index core (as in traditional step- or gradedindex fibers) becomes incompatible with supporting stable OAM propagation. Another critical design criterion is ensuring a sufficiently large refractive index contrast between the core and the cladding layers (Zhang et al. 2015; Rjeb et al. 2019b). This enhances mode confinement and promotes the generation of true OAM states over conventional LP modes. Moreover, to preserve OAM purity, the refractive index profile should transition smoothly at the core-cladding interface. Abrupt changes can trigger spin-orbit coupling, leading to mode splitting: each OAM state degenerates into two components with opposite circular polarizations and differing topological charges. This results in unequal power distributions and intrinsic crosstalk that undermines signal integrity. Such issues can be mitigated by softening the gradient

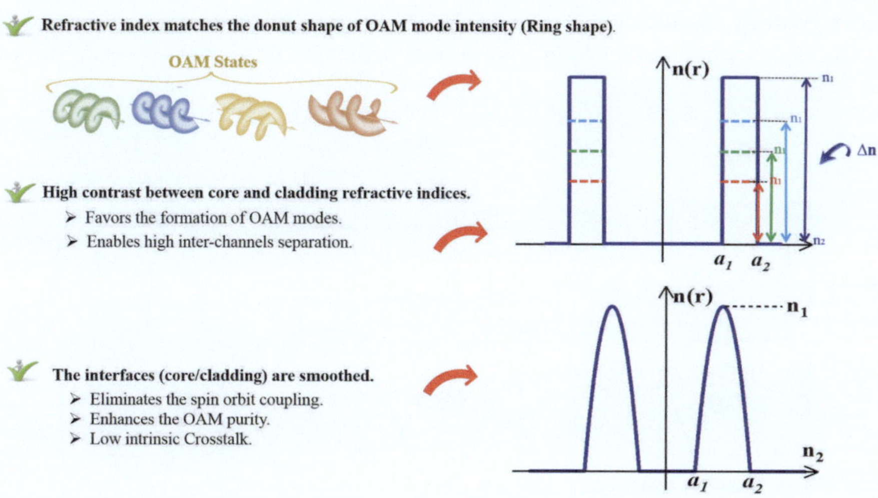

Fig. 5.1 OAM fibers design guidelines

at either the inner or outer boundary of the core region (Rjeb et al. 2020b). The lower panel of Fig. 5.1 shows the profile of a graded-index ring-core fiber (GIRCF), where both boundaries feature gradual transitions designed to enhance single-mode OAM propagation.

5.2.2 OAM Specialty Fibers Designs

Following the previously outlined design principles, many specialty optical fibers have been proposed and fabricated to support orbital angular momentum (OAM) transmission. These developments have been further facilitated by advancements in optical fiber fabrication techniques, including material engineering and novel processing methods, making the realization of complex refractive index profiles with high contrast significantly more feasible. In the following, we present an overview of the most prominent families of specialty fibers tailored for OAM applications.

One of the earliest fibers specifically designed for OAM transmission is the vortex fiber, introduced by Ramachandran and colleagues in (2009). This structure features a central core encircled by a lower-index trench and an outer high-index ring (vortex). It successfully guided several fundamental and cylindrical vector modes, including $LP_{0,1}$, $TE_{0,1}$, $TM_{0,1}$, as well as OAM mode groups over distances exceeding 20 m. Subsequent experiments extended this transmission distance to 1 km (Golowich et al. 2012; Bozinovic et al. 2011a, b), and a comprehensive summary of these results was compiled in Bozinovic et al. (2013). More recently, vortex fibers have been integrated into quantum communication systems, especially for quantum key distribution (QKD) (Sit et al. 2018). Another key category is the air-core fiber (ACF), also known as the hollow-core fiber, which benefits from its naturally high refractive index contrast. As illustrated in Fig. 5.2a, the ACF supports low-loss guidance of OAM modes within an air-filled core. Initial demonstrations showed stable propa-

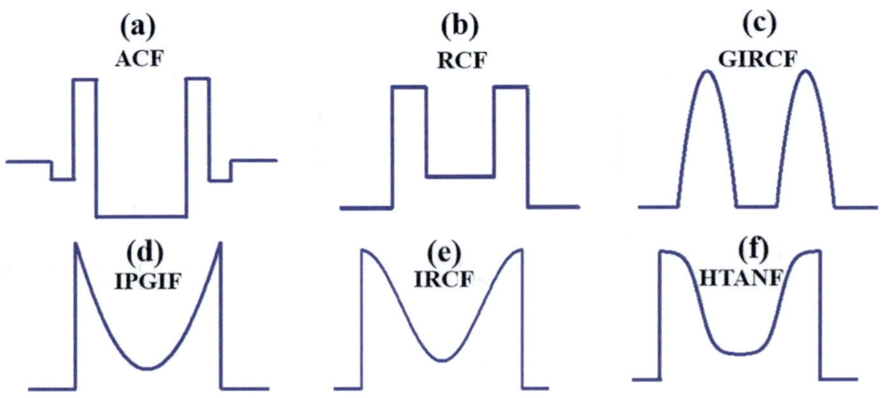

Fig. 5.2 Recently proposed and demonstrated OAM specialty fibers

gation of 12 distinct OAM modes over 2 m of ACF. In a subsequent achievement, dual-OAM mode transmission was achieved over 1 km (Gregg et al. 2013). A record-setting demonstration transmitted 36 distinct OAM modes (from $l = 0$ to $l = \pm 9$) through an ACF (Brunet et al. 2014a). ACFs have also been leveraged for secure quantum key distribution in recent studies (Cozzolino et al. 2018).

Some OAM-supporting designs adapt conventional fibers, with the ring-core fiber (RCF) being a notable example. The RCF's refractive index profile closely resembles the doughnut-shaped intensity pattern of an OAM beam, as shown in Fig. 5.2b, making it inherently suited for such modes. Analytical methods for solving the full-vector Maxwell equations in RCFs were developed by Brunet et al. (Zhang et al. 2019a), laying the groundwork for several experimental implementations (Zhang et al. 2019b). Since then, various studies have proposed novel RCF structures and reported OAM transmission experiments through them (Brunet et al. 2014b, 2015; Ramachandran et al. 2015; Chen et al. 2019). In a more refined design approach, graded-index ring-core fibers (GIRCFs) have been proposed. These fibers combine ring-like geometry with smooth refractive index transitions across the inner and outer boundaries of the ring, enhancing modal confinement and reducing crosstalk. Figure 5.2c illustrates a typical GIRCF structure. A notable implementation supported 22 OAM modes over 10 km, with insertion loss below 1 dB/km and crosstalk suppression better than 14 dB (Zhu et al. 2017). In another demonstration, a 32 Gbaud QPSK signal was transmitted over 80 data channels, achieving a capacity of 5.12 Tbit/s and a spectral efficiency of 9 bit/s/Hz across 10 and 18 km GIRCF spans (Zhu et al. 2018a). More recently, transmission of 12 Gbaud 8-QAM signals over 224 channels–using two OAM modes and 112 WDM channels– yielded an impressive 8.4 Tbit/s throughput without relying on MIMO DSP (Zhu et al. 2018b).

A further innovation came from the work of Ung and co-workers at Laval University's COPL, who introduced the inverse parabolic graded-index few-mode fiber (IPGIF) (Ung et al. 2014). This design features a hybrid refractive index profile: a smooth parabolic variation in the inner core, coupled with a sharp transition at the cladding boundary, as depicted in Fig. 5.2d. Such a structure enhances the refractive index gradient, increases transverse field confinement, and boosts mode discrimination. These characteristics enable effective separation of OAM modes, which has been confirmed experimentally through short-range (1 m) and long-range (1 km) mode-division multiplexing using 8 and 2 OAM states, respectively. Recently, a 10 m IPGIF link achieved 3.36 Tbit/s throughput using 4 OAM modes and 15 wavelength channels (Wang et al. 2017).

Recently, we numerically demonstrated the appropriateness of IPGIF to support high pure OAM modes that enable low intrinsic crosstalk (Rjeb et al. 2020b). Based on the same approach of design (i.e. hybrid structure), we propose an OAM specialty fiber where their refractive index profile is inspired from the Inverse Raised Cosine function (IRC) (Rjeb et al. 2019a). The refractive index profile of IRC fiber is shown in Fig. 5.2e. The designed IRC-FMF outperforms the IPGIF in terms of intermodal separation enabling reduces channel crosstalk. We demonstrated the high purity of supported OAM modes (Rjeb et al. 2019b). Recently, the tolerance of IRC-FMF to bending effects have been numerically demonstrated (Rjeb et al. 2021d).

Recently, the hyperbolic tangent (HTAN) function has been employed as a refractive index profile in the design of few-mode fibers (FMFs) tailored for OAM transmission (Rjeb et al. 2020a). The profile of the HTAN-based FMF, depicted in Fig. 5.2e, exhibits a steep gradient concentrated around its inflection point. This design feature yields a substantial effective index separation (3×10^{-4}) among cylindrical vector modes such as $TE_{0,1}$, $HE_{2,1}$, and $TM_{0,1}$, effectively minimizing intermodal crosstalk. This separation surpasses that previously achieved in related fiber designs reported in the literature. Additionally, the generation and analysis of OAM modes within this fiber confirm a mode purity of at least 99.9% corresponding to intrinsic crosstalk levels below –30 dB during propagation. Such results point to a highly robust mode transmission environment. Key performance indicators–including chromatic dispersion and differential mode delay–were further assessed across the C-band, demonstrating favorable characteristics. Numerical simulations have also shown that the HTAN-FMF exhibits bending tolerance, making it suitable for long-haul and short-reach communication links. Building upon this concept, we have recently extended the use of the HTAN function to design a novel MMF referred to as inverse HTAN MMF (IHTAN-MMF) (Rjeb et al. 2021a). This newly developed structure supports 14 distinct mode groups (MGs), outperforming conventional graded-index MMFs and many state-of-the-art OAM-capable fibers. A system level demonstration involving the multiplexing, transmission, and demultiplexing of 50 OAM data channels has been successfully implemented using compact, energy-efficient multiple-input multiple-output (MIMO) digital signal processing (DSP) modules, showcasing the potential of IHTAN-MMF for high-capacity optical networks.

5.3 The Proposed Inverse Gaussian Profile

Based on the OAM specialty fibers design recommendations provided above, and on the extensively literature review of various OAM fibers designs and obtained results, in this section, we present our proposed OAM fiber where their refractive index profile is inspired from the well-known Gaussian function. To our knowledge, the inverse Gaussian function has not previously been employed as a refractive index profile for optical fibers. Unlike conventional refractive index distributions discussed earlier, typically governed by a single profile-shaping parameter (often the shape parameter or profile exponent), the proposed inverse Gaussian fiber (IGF) introduces a novel dual-parameter control mechanism. Specifically, the profile is shaped by two independent parameters, denoted as α and P. Including two tunable shape parameters offers enhanced design flexibility, enabling the development of fiber structures that can better accommodate and optimize mode propagation characteristics.

5.3.1 Refractive Index Description

The refractive index profile of the proposed IGF is expressed as follows (Rjeb et al. 2021c):

$$n(r) = \begin{cases} n_1 - \Delta n_1 e^{-\alpha(\frac{r}{a})^P} & if\ 0 \leqslant r \leqslant a \\ n_3 & if\ r > a \end{cases} \tag{5.1}$$

where the parameters r, a, n_1, n_2, n_3 represent the radial coordinate, the core radius, the refractive index at the core-cladding interface (i.e. $r = a$), the refractive index at the fiber center (i.e. $r = 0$), and the refractive index of the cladding, respectively. The relative index difference is denoted by $\Delta = \frac{n_1 - n_2}{n_1}/$. The IG-FMF structure is governed by two key shape controlling parameters: the profile exponent α and the curvature parameter P. These parameters collectively regulate the refractive index transition from the fiber center to the cladding, effectively determining the interior smoothness of the ring region spanning $r = 0$ to $r = a$.

The corresponding refractive index profiles of IG-FMF are illustrated in Fig. 5.3a, b. Figure 5.3a shows the impact of varying α values–[5, 10, 15, 20, 30, 60]–on the index distribution, while Fig. 5.3b presents the effect of adjusting the curvature parameter P–[2, 4, 6, 8, 10]. For comparative analysis, we include the refractive index profiles of the inverse raised cosine FMF (IRC-FMF, with $\alpha_{IRC} = 1$) [20] and the inverse parabolic graded index FMF (IPGI-FMF, with $N = 4$) (Ung et al. 2014) as reference benchmarks. The hybrid nature of the IG-FMF profile, characterized by a smooth Gaussian-like gradient in the inner core and a sharp transition near the cladding, is strategically chosen to maximize intermodal index separation while preserving high mode purity for OAM modes. This dual benefit contributes to minimizing both interand intra-channel crosstalk. As observed in Fig. 5.3a, b, each parameter distinctly influences the profile geometry: when P is fixed and α increases, the core ring expands, resulting in a broader guiding region compared to IRC and

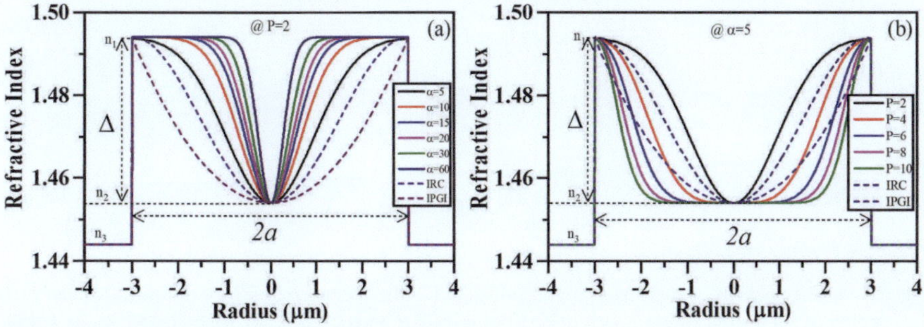

Fig. 5.3 Inverse Gaussian profiles: **a** for different values of α at fixed P, ($P = 2$), **b** for different values of P for fixed α, ($\alpha = 5$). The inverse raised cosine (IRC) profile and the inverse parabolic graded index (IPGI) profile (i.e. dashed lines) are added for aspect of comparison

IPGI profiles. Conversely, for a fixed α, increasing P compresses the ring, producing a narrower and more confined guiding layer.

5.3.2 Supported Modes (i.e. Hybrid Modes)

In order to conduct numerical investigation of each parameter impact on the supported eigenmodes in the proposed fiber, we incorporate the IG profile into Comsol Multiphysics software. The later uses Finite Element Method (FEM) to solve for modes. Preliminary simulations have been performed using fixed wavelength $\lambda = 1.55\ \mu$m and a uniform cladding index $n_3 = 1.444$ (i.e. pure Silica). Fiber eigenmodes are the exact solutions of the vectorial Maxwell equations and represent the modes that can propagate along the fiber without distortion. These fundamental modes are typically classified as $HE_{l,m}$, $EH_{l,m}$, $TE_{0,m}$ and $TM_{0,m}$, m where the indices l and m denote the azimuthal and radial mode numbers, respectively. The hybrid modes, $HE_{l,m}$ and $EH_{l,m}$, are inherently degenerate, each comprising both even and odd components.

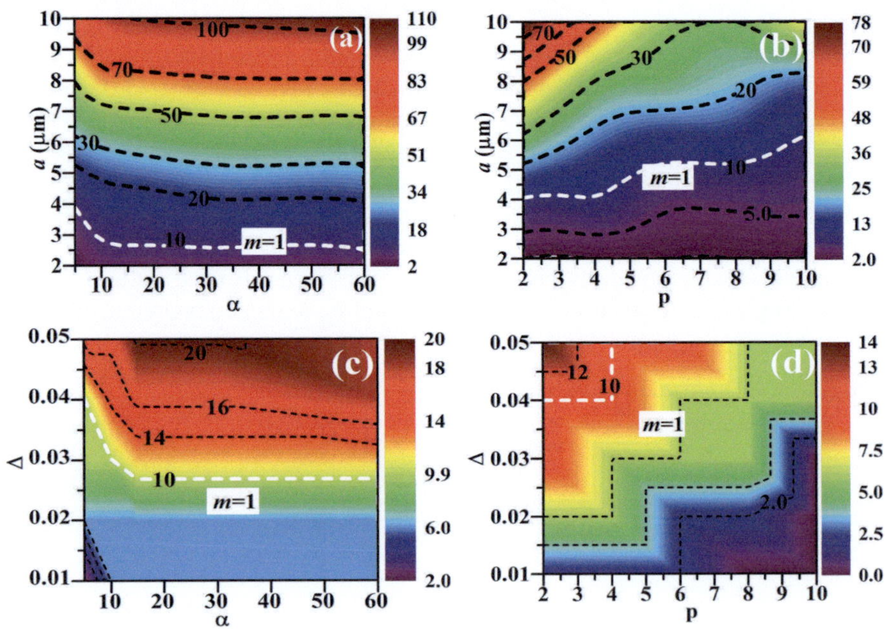

Fig. 5.4 **a** The number of supported Eigenmodes in IG fiber versus α and core radius (a) $(P = 2)$, **b** the number of supported Eigenmodes in IG fiber versus P and (a) $(\alpha = 2)$. $n_3 = 1.444$, $n_2 = 1.454$, $\Delta = 0.026$ $(n_1 = 1.494)$ and $\lambda = 1.55\ \mu$m. **c** the number of supported Eigenmodes in IG fiber versus α and Δ $(P = 2)$, **d** the number of supported Eigenmodes in IG fiber versus P and Δ, $(\alpha = 2)$ and for $n_3 = 1.444$, $n_2 = 1.454$, $a = 3\ \mu$m and $\lambda = 1.55\ \mu$m

A key objective of this study is to investigate how the shape parameters α and P influence the excitation and propagation of these eigenmodes. Figure 5.4a–d presents the outcomes of this analysis. Specifically, Fig. 5.4a, b show the number of supported modes as a function of the core radius a and the shape parameters α and P, respectively. Meanwhile, Fig. 5.4c, d illustrate the variation in the number of eigenmodes with respect to the refractive index contrast Δ and the same shape parameters α and P. From Fig. 5.4a, b, it can be observed that, for a fixed core radius (set to $a = 3\,\mu\text{m}$), increasing the profile exponent α results in a higher number of guided modes, whereas increasing the curvature parameter P leads to a reduction in the mode count. Similarly, Fig. 5.4c, d show that the number of supported modes rises with increasing Δ and α, but decreases as P increases. This trend aligns with physical intuition: increasing α broadens the effective index profile, thereby supporting more mode solutions, while higher values of P compress the guiding region, limiting modal support. It is also important to note that higher-order radial modes (i.e., those with $m > 1$) tend to exhibit intricate field distributions and contribute to generating higher radial-order OAM states, which can complicate multiplexing and demultiplexing procedures. For this reason, our analysis focuses only on the first ten eigenmodes. In Fig. 5.4a–d, white dashed lines indicate the boundary where all excited modes maintain a radial index of $m = 1$.

5.3.3 Intermodal Separation

This subsection is dedicated to investigate the intermodal separation responsible for the characterization of modes' interactions. Considering the first 10 low radial modes, the IG fiber supports $HE_{1,1}$, $TE_{0,1}$, $HE_{2,1}$, $TM_{0,1}$, $EH_{1,1}$, and $HE_{3,1}$. Each $HE_{l,m}$ and $EH_{l,m}$ is degenerate where each of which comprises odd and even mode. Thus, 10 eigenmodes are supported. Aiming to perform a straightforward comparison between our IG profiles and the benchmarks profiles (IPGI with shape parameter N = 4 (Ung et al. 2014) and IRC with shape parameter $\alpha_{IRC} = 1$ (Rjeb et al. 2019a)), we select the parameters as $a = 3\mu\text{m}$ and $n_1 = 1.494$, $n_3 = 1.444$. Figure 5.5 highlights the minimum intermodal separation (Min Δn_{eff}), calculated between two successive modes. The Min Δn_{eff} is evaluated versus the two shape parameters P and α. In Fig. 5.5, blue dashed line is corresponding to the Min Δn_{eff} demonstrated in IPGI-FMF (Ung et al. 2014). The green dashed line is corresponding to the Min Δn_{eff} reported in IRC-FMF (Rjeb et al. 2019a). Gold lines are corresponding to insufficient regions (i.e. parameters values) for supporting OAM modes where $\Delta n_{eff} < 10^{-4}$). In Fig. 5.5, the identified shaded region is corresponding to IG-FMFs that outperform both the IPGI-FMF and IRC-FMF in terms of Δn_{eff}, which limits the interaction between modes (i.e. reducing the channel crosstalk).

5.3.4 Designed IG-FMF and Supported Modes

Any point selected within the shaded region of Fig. 5.5a corresponds to an IG-FMF exhibiting enhanced intermodal separation. A magnified view of this region is presented in Fig. 5.5b, where four representative (α, P) pairs are selected to examine the influence of the shape parameters: $Fiber$#1 (10, 2), $Fiber$#2 (20, 2), $Fiber$#3 (40, 4), and $Fiber$#4 (60, 4). These configurations are chosen to systematically assess the roles of both α and P, and to gain deeper insights into how inner profile smoothness impacts various performance metrics associated with OAM modes. The selected parameter combinations span a meaningful range of cases, such as high α with low P, low α with high P, and intermediate configurations, thus enabling a comprehensive performance evaluation. The IG-FMFs described above support the following OAM and non-OAM eigenmodes: $OAM_{0,1}^{L/R}$) , $OAM_{1,1}^{L/R}$,$OAM_{\mp2,1}^{L/R}$ and $OAM_{\pm2,1}^{L/R}$, in addition to the $TE_{0,1}$ and $TM_{0,1}$ modes. These modes are categorized into three groups: G_1 includes the fundamental OAM mode $OAM_{0,1}^{L/R}$ 0,1 ; G_2 comprises $TE_{0,1}$, $OAM_{1,1}^{L/R}$, $TM_{0,1}$; and G_3 includes the higher-order OAM modes $OAM_{\mp2,1}^{L/R}$ and $OAM_{\pm2,1}^{L/R}$. Here, the superscripts L/R represent the spin angular momentum, left and right circular polarizations, while the subscripts l and m denote the azimuthal and radial indices, respectively. It is worth noting that the $TE_{0,1}$ and $TM_{0,1}$ modes do not contribute to the OAM mode set due to their significant modal separation in terms of effective refractive index, which precludes them from forming vectorial superpositions that carry orbital angular momentum.

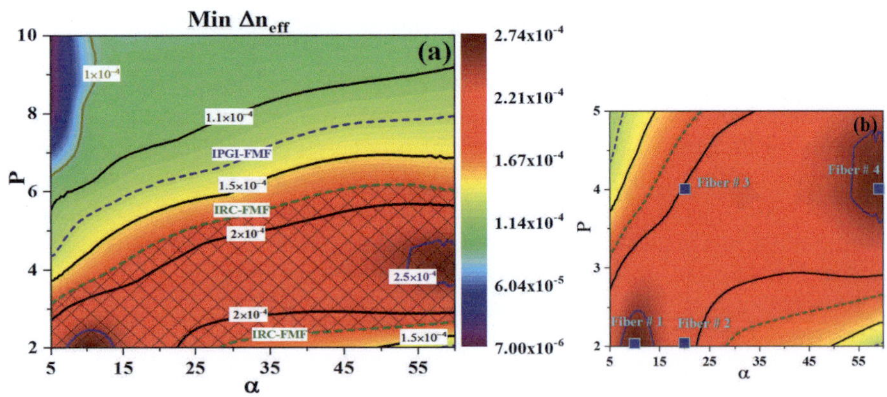

Fig. 5.5 **a** Minimum effective index separation between modes as a function of both shape parameters α and P. **b** zoomed part: the selected fibers. Other fiber parameters are kept as: $a = 3$ µm and $n_1 = 1.494$, $n_1 = 1.444$ and $\lambda = 1.55$µm

5.4 Performance Transmission Metrics of the IG-FMF over C+L Bands

In this section, we evaluate the performance of the four selected IG-FMF designs based on key transmission metrics, namely the intermodal effective index separation (Δn_{eff}), chromatic dispersion (CD), and differential group delay (DGD). The analysis spans the full C+L wavelength bands as specified by ITU-T, covering the spectral range from 1.530 to 1.625 μm. This spectral domain is particularly relevant for wavelength-division multiplexing (WDM) systems, where the fiber's capacity is determined by the product of the number of supported spatial modes (M) and the number of wavelength channels (D). In such a system, each spatial mode is an independent carrier for D WDM channels, which can be demultiplexed at the receiver side without mutual interference.

5.4.1 Intermodal Separation Across the C+L Bands

The effective refractive index differences (Δn_{eff}) between each pair of adjacent modes were computed for all four selected fibers and plotted over the entire C+L

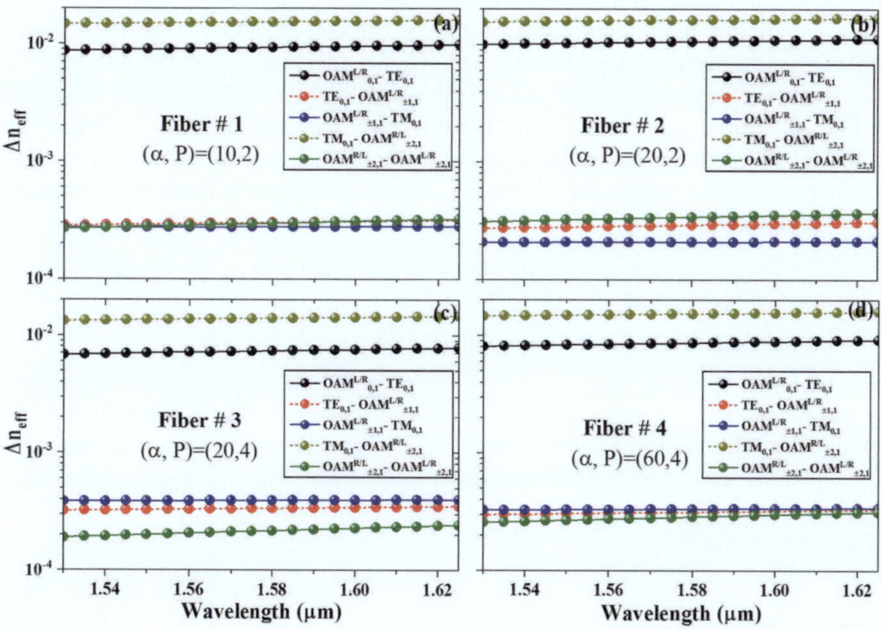

Fig. 5.6 Δn_{eff} in the designed IG-FMFs across C+L bands. **a** Fiber 1, **b** Fiber 2, **c** Fiber 3, and **d** Fiber 4

Table 5.1 Δn_{eff} in IG-FMFs compared to those in IRC-FMF (Rjeb et al. 2019a) and IPGI-FMF (Ung et al. 2014)

Δn_{eff}	#1	#2	#3	#4	IRC	IPGI
$G_1 - G_2$ ($\times 10^{-3}$)	8.93	10.24	7.02	8.26	6.96	6.61
$TE_{0,1} - OAM^{L/R}_{\pm 1,1}$ ($\times 10^{-4}$)	2.90	2.80	3.27	3.03	3.04	2.43
$OAM^{L/R}_{\pm 1,1} - TM_{0,1}$ ($\times 10^{-4}$)	2.74	2.09	3.88	3.29	3.31	2.44
$G_2 - G_3$ ($\times 10^{-3}$)	1.49	1.57	1.34	1.47	1.25	1.07
$OAM^{R/L}_{\pm 2,1} - OAM^{L/R}_{\pm 2,1}$ ($\times 10^{-4}$)	2.79	3.23	2.01	2.74	1.78	1.27

wavelength range. The resulting trends are illustrated in Fig. 5.6a–d. Across the spectrum, all modal pairs exhibit sufficient separation ($\Delta n_{eff} \geq 1 \times 10^{-4}$), indicating minimal intermodal crosstalk and ensuring reliable multiplexing performance. A comparative analysis between Fiber #1 and Fiber #2, having the same curvature parameter P but different profile constants α, reveals that increasing α generally leads to enhanced mode separation. An exception is observed for the Δn_{eff} values between the non-OAM modes ($TE_{0,1}$, $TM_{0,1}$) and $OAM^{L/R}_{1,1}$, where the increase is not consistent. Conversely, when comparing Fiber #2 and Fiber #3–which share the same α but differ in P–an increase in curvature parameter P typically reduces Δn_{eff}, again except the same mode pair involving non-OAM and $OAM^{L/R}_{1,1}$ modes. To benchmark performance, all obtained Δn_{eff} values were compared to those of the reference fibers, namely IRC-FMF (Rjeb et al. 2019a) and IPGI-FMF (Ung et al. 2014). As summarized in Table 5.1 for the central wavelength $\lambda = 1.55$ μm, the proposed IG-FMF designs exhibit superior intermodal separation, except in cases involving parasitic modes and $OAM^{L/R}_{1,1}$, where performance remains comparable to the benchmarks.

5.4.2 Chromatic Dispersion Across the C+L Bands

In this subsection, we evaluate the chromatic dispersion (CD) of supported modes within each selected IG-FMF. By definition, the chromatic dispersion is the spreading of a light pulse as it propagates through an optical fiber. This will tend to degrade the transmission quality over any fiber. The equation that describes the CD is as follows (Chen and Wang 2017b):

$$CD = \frac{-\lambda}{c} \frac{\partial^2 n_{eff}}{\partial \lambda^2} \tag{5.2}$$

Table 5.2 Comparison of chromatic dispersion (CD) of supported modes in the designed IG-FMFs at λ = 1.55 μm compared to those in IRC-FMF (Rjeb et al. 2019a) and IPGI-FMF (Ung et al. 2014)

Supported modes	#1	#2	#3	#4	IRC	IPGI
$OAM_{0,1}^{L/R}$	−1.92	4.72	−17.71	−7.697	−14.86	4.00
$TE_{0,1}$	14.70	17.37	5.62	13.39	0.25	12.00
$OAM_{\pm1,1}^{L/R}$	14.99	17.82	5.47	13.62	−0.08	11.83
$TM_{0,1}$	11.93	14.90	2.10	10.38	−3.17	8.86
$OAM_{\pm2,1}^{R/L}$	−35.56	−32.28	−44.44	−34.83	−74.09	−136.16
$OAM_{\pm2,1}^{L/R}$	−34.43	−31.53	−42.64	−33.31	−74.68	−138.20

where λ is the wavelength in vacuum, c is the light velocity in vacuum and n_{eff} is the effective index of supported mode. The CDs curves are plotted in Fig. 5.7a–d. Considering all the modes CD curves within each IG-FMF, the CD is flat and slowly decreases with wavelength except for those of higher modes: $OAM_{\pm2,1}^{R/L}$ and $OAM_{\pm2,1}^{L/R}$. The latter observe CDs that significantly increase with wavelength. This due to the closeness of considering modes to the cut off. Considering all the selected IG-FMF, The worst CD in each fiber: fiber #1: −67.41 [ps/(km nm)], fiber#2: −71.73 [ps/(km nm)], fiber #3: −82.63 [ps/(km nm)], fiber #4: −69.87 [ps/(km nm)]. Considering the impact of profile constant α on CDs, the CD increases with α except for and (fiber #1 vs fiber #2), this is explained by the direct effect of α on the profile index. When α increases, it tends to the profile to broad hence the higher modes index far away from the cutoff. This tends to decrease their CD values. Contrarily, the curvature P tends to increase the CD values of higher modes (fiber #2 vs. fiber #3). In order to compare the designed IG-FMFs with benchmarks fibers, Table 5.1 outlines the CDs of IG-FMFs and those of IPGI-FMF and IRC-FMF at $\lambda = 1.55\mu$m. The selected IG fibers outperforms in term of CDs of higher modes (i.e. G_3). On the other side, the CD of lower modes are reasonable since they are all lower than ITU-T Corning SMF-28 ® (i.e. 18 [ps/km nm]). This confirms the capability of IG-FMF to be potential candidate for next generation SDM-WDM communication systems using OAM modes bearing information (Table 5.2).

5.4.3 *Differential Group Delay Across the C+L Bands*

In this sub-section, we assessed the supported modes in terms of differential group delays (DGD) versus the C and L bands. The DGD increases the load of digital signal

Table 5.3 DGDs in IG-FMF at λ = 1.55 μm compared to those in IRC-FMF (Rjeb et al. 2019a) and IPGI-FMF (Ung et al. 2014)

Modes	Fiber #1	Fiber #2	Fiber #3	Fiber #4	IRC	IPGI
$OAM_{0,1}^{L/R}$	–	–	–	–	–	–
$TE_{0,1}$	31.94	31.11	30.63	31.75	29.33	25.3
$OAM_{\pm1,1}^{L/R}$	32.79	32.12	31.12	32.55	29.54	26.75
$TM_{0,1}$	32.41	31.68	30.73	32.25	28.82	26.4
$OAM_{\pm2,1}^{R/L}$	53.87	51.20	54.13	54.50	46.46	27.32
$OAM_{\pm2,1}^{L/R}$	56.06	53.38	56.28	56.76	48.46	28.10

processing (DSP) and power consumption. This DGD tardy the de-multiplexing process at the receiver side. The group delay (τ_g) is given by the following expression (Rjeb et al. 2020c):

$$\tau_g = \frac{1}{c}(n_{eff} - \frac{\partial n_{eff}}{\partial \lambda} \tag{5.3}$$

where λ is the wavelength in vacuum, c is the light velocity in vacuum, and n_{eff} is the mode effective index. All the DGDs are calculated considering the delay of fundamental mode (Rjeb et al. 2020c). All the DGDs of designed IG-FMF versus the C+L bands are shown in Fig. 5.7a–d. Considering one designed IG-FMF, the DGDs of lower modes belonging to n_{eff} and G_2 slowly increase with wavelength while the DGDs of higher modes belonging to G_3 significantly decrease with wavelength. Considering all the designed IGFMFs, the worst DGDs are fiber #1: 51.96 [ps/m], fiber #2: 49.06 [ps/m], fiber #3: 52.65 [ps/m], and fiber #4: 51.96 [ps/m]. Considering the impact of α on DGDs values, through a comparison between fiber #1 and fiber #2, the DGD decreases with the exponent α. Contrarily, considering fiber #2 and fiber #3, the DGD increases with the curvature P. This is true for all supported modes except for higher modes where their DGDs deceases with α and increases with P. Thanks to the direct dependency between the intermodal separation and the DGD range, the fourth fiber observes the highest DGDs values since it possess the highest separation among modes. The obtained DGDs for our designed IG-FMF are compared with those reported in IRC-FMF and those of IPGI-FMF, where the results at $\lambda = 1.55$ μm are shown in Table 5.3. IG-FMFs performs comparably DGD with those in IRC-FMF and IPGI-FMF except for DGDs values of higher modes.

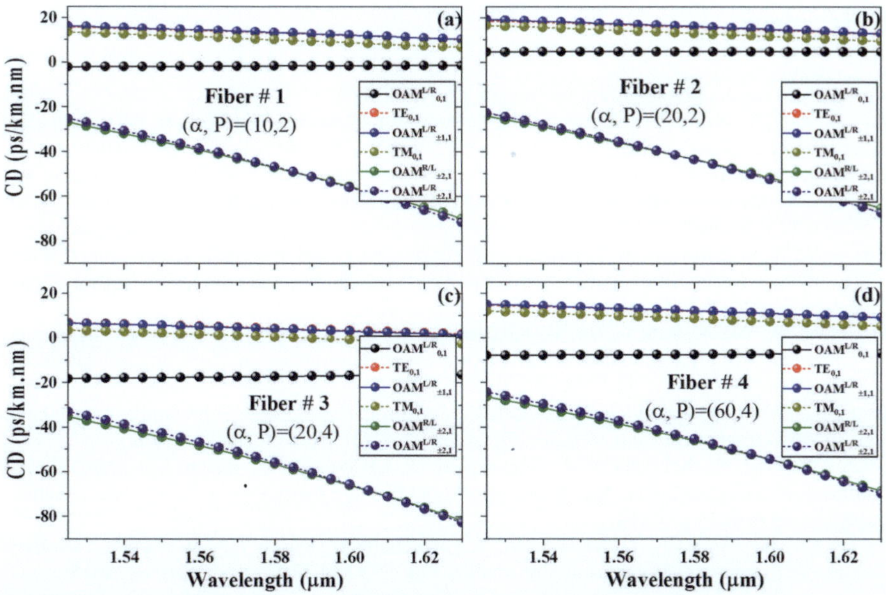

Fig. 5.7 DGD in IG-FMFs versus C+L bands. **a** Fiber 1, **b** Fiber 2, **c** Fiber 3, and **d** Fiber 4

5.5 Conclusion

This chapter focuses on developing specialty optical fibers specifically tailored for the efficient transmission of orbital angular momentum (OAM) channels. We began by discussing the key design principles of OAM-supporting fibers, referencing notable experimental demonstrations and major advancements that have emerged in recent years. Despite the progress, the field remains highly active, with considerable scope for further innovation. We adopted these design principles and introduced a novel refractive index profile, inverse Gaussian (IG) fibers, engineered to support stable OAM mode propagation. Through comprehensive parametric sweeps, we delineated a performance-optimized region where IG fibers achieve improved intermodal separation, with a minimum Δn_{eff} of 2.74×10^{-4}—a notable enhancement over existing designs such as IRC and IPGI fibers. From this optimal region, four distinct IG-FMF structures were selected and thoroughly analyzed across the C+L telecommunications bands. Key performance indicators–including modal separation, chromatic dispersion (CD), and differential group delay (DGD)–were evaluated. Comparative analysis with benchmark designs reported in the literature confirms that the proposed IG-FMFs hold strong potential for future integration in advanced optical networks utilizing OAM multiplexing in conjunction with wavelength-division multiplexing (WDM).

References

Arora, J.S.: Introduction to Optimum Design. Elsevier (2004)

Bae, S., Jung, Y., Kim B.G., Chung, Y.C.: Compensation of mode crosstalk in MDM system using digital optical phase conjugation. IEEE Photon. Technol. Lett. **31**(10), 739–742 (2019)

Bozinovic, N., Kristensen, P., Ramachandran, S.: Long-range fiber transmission of photons with orbital angular momentum. In: CLEO: 2011 Laser Applications to Photonic Applications. Optical Society of America (2011a)

Bozinovic, N., Ramachandran, S., Brodsky, M., Kristensen, P.: Record length transmission of entangled photons with orbital angular momentum (vortices). In: Frontiers in Optics. Optical Society of America (2011b)

Bozinovic, N., Yue, Y., Ren, Y., Tur, M., Kristensen, P., Huang, H., Willner, A.E., Ramachandran, S.: Terabit scale orbital angular momentum mode division multiplexing in fibers. Science **340**(6140), 1545–1548 (2013)

Brunet, C., Vaity, P., Messaddeq, Y., LaRochelle, S., Rusch, L.A.: Design, fabrication and validation of an OAM fiber supporting 36 states. Opt. Express **22**(21), 26117–26127 (2014a)

Brunet, C., Ung, B., Bélanger, P.-A., Messaddeq, Y., LaRochelle, S., Rusch, L.A.: Vector mode analysis of ring-core fibers: design tools for spatial division multiplexing. J. Lightwave Technol. **32**(23), 4046–4057 (2014b)

Brunet, C., Ung, B., Wang, L., Messaddeq, Y., LaRochelle, S., Rusch, L.A.: Design of a family of ring-core fibres for OAM transmission studies. Opt. Express **23**(8), 10553–10563 (2015)

Chirikjian, G.S.: Gaussian distributions and the heat equation. In: Stochastic Models, Information Theory, and Lie Groups, vol. 1, pp. 31–61. Birkhäuser Boston (2009)

Chralyvy, A.: The coming capacity crunch. In: 35th European Conference on Optical Communication, pp. 1–1 (2009)

Chen, S., Wang, J.: Characterization of red/green/blue orbital angular momentum modes in conventional G. 652 fiber. IEEE J. Quant. Electron. **53**(4), 1–14 (2017a)

Chen, S., Wang, J.: Theoretical analyses on orbital angular momentum modes in conventional graded-index multimode fibre. Sci. Rep. **7**(1), 3990 (2017b)

Chen, S., Li, S., Fang, L., Wang, A., Wang, J.: OAM mode multiplexing in weakly guiding ring-core fiber with simplified MIMO-DSP. Opt. Express **27**(26), 38049–38060 (2019)

Cisco VNI Global IP traffic forecast, San Diego, 2018–2023 (2018)

Cozzolino, D., Bacco, D., Da Lio, B., Inaerslev, K., Ding, Y., Dalgaard, K., Kristensen, P., Galili, M., Rottwitt, K., Ramachandran, S., Oxenløwe, L.K.: Fiber-based high-dimensional quantum key distribution with twisted photons. In: 2018 Conference on Lasers and Electro-Optics Pacific Rim (CLEO-PR), pp. 1–2. IEEE (2018)

Essiambre, R., Tkach, R.: Capacity trends and limits of optical communication networks. In: Proceedings of the IEEE, vol. 100, no. 5, pp. 1035–1055 (2012)

Golowich, S., Kristensen, P., Bozinovic, N., Gregg, P., Ramachandran, S.: Fibers supporting orbital angular momentum states for information capacity scaling. In: Frontiers in Optics 2012/Laser Science XXVIII, OSA Technical Digest, Optica Publishing Group, Paper FW2D.2 (2012)

Gregg, P., Kristensen, P., Golowich, S.E., Olsen, J.Ø., Steinvurzel, P., Ramachandran S.: Stable transmission of 12 OAM states in an air-core fiber. In: Conference on Lasers and Electro-Optics (CLEO), Paper CTu2K2 (2013)

Haddad, R.A., Akansu, A.N.: A class of fast Gaussian binomial filters for speech and image processing. IEEE Trans. Signal Process. **39**(3), 723–727 (1991)

Kondor, R.I., Lafferty, J.: Diffusion kernels on graphs and other discrete structures. In: 19th International Conference on Machine Learning, pp. 315–22 (2002)

Ramachandran, S., Kristensen, P., Yan, M.F.: Generation and propagation of radially polarized beams in optical fiber. Opt. Lett. **34**, 2525 (2009)

Ramachandran, S., Gregg, P., Kristensen, P., Golowich, S.E.: On the scalability of ring fiber designs for OAM multiplexing. Opt. Express **23**(3), 3721–3730 (2015)

Richardson, D.J., Fini, J.M., Nelson, L.E.: Space-division multiplexing in optical fibres. Nat. Photon. **7**(5), 354–362 (2013)

Rjeb, A., Guerra, G., Issa, K., Fathallah, H., Chebaane, S., Machhout, M., Palmieri L., Galtarossa, A.: Inverse-raised-cosine fibers for next-generation orbital angular momentum systems. Opt. Commun. 124736 (2019a)

Rjeb, A., Fathallah, H., Machhout, M.: Investigation of orbital angular momentum mode purity in Inverse Raised Cosine fiber. In: ICO & IUPAP-C17 Topical Meeting on OPTIcs and Applications to Sustainable Development (OPTISUD), Carthage, Tunis, Tunisia, SPIE, 4–7 September (2019b)

Rjeb, A., Fathallah, H., Khaled, I., Machhout, M., Alshebeili, S.A.: A novel hyperbolic tangent profile for optical fiber for next generation OAM-MDM systems. IEEE Access **8**, 226737–22675 (2020a)

Rjeb, A., Fathallah, H., Machhout, M.: Numerical investigation of orbital angular momentum mode purity in inverse parabolic graded index fiber. In: 17th International Multi-Conference on Systems, Signals & Devices (SSD'20), pp. 925–928. IEEE (2020b)

Rjeb, A., Seleem, H., Fathallah, H., Machhout, M.: Design of 12 OAM-graded index few mode fibers for next generation short haul interconnect transmission. Opt. Fiber Technol. **55**, 102148 (2020c)

Rjeb, A., Fathallah, H., Issa, K., Machhout, M., Alshebeili, S.A.: Inverse hyperbolic tangent fiber for OAM mode/mode-group division multiplexing optical networks. IEEE J. Quant. Electron. (2021a)

Rjeb, A., Fathallah, H., Machhout, M.: OAM modes in optical fibers for next generation space division multiplexing (SDM) systems. In: Fiber Optics—Technology and Applications, Guillermo Huerta-Cuellar, IntechOpen, November (2021b)

Rjeb, A., Seleem, H., Fathallah, H., Machhout, M.: A novel inverse gaussian profile for orbital angular momentum mode division multiplexing optical networks. In: 2021 18th International Multi-Conference on Systems, Signals & Devices (SSD), pp. 480–487. IEEE (2021c)

Rjeb, A., Fathallah, H., Machhout, M.: Analysis of orbital angular momentum channels in bent inverse raised cosine fiber. In: 2021 IEEE International Conference on Design & Test of Integrated Micro & Nano-Systems (DTS), pp. 1–4. IEEE (2021d)

Rusch, L.A., Rad, M., Allahverdyan, K., Fazal, I., Bernier, E.: Carrying data on the orbital angular momentum of light. IEEE Commun. Mag. **56**(2), 219–224 (2018)

Ryf, R., Randel, S., Gnauck, A.H., Bolle, C., Sierra, A., Mumtaz, S., Peckham, D.W.: Mode−division multiplexing over 96 km of few−mode fiber using coherent 6 × 6 MIMO processing. J. Lightwave Technol. **30**(4), 521–531 (2012a)

Ryf, R., Fontaine, N.K., Mestre, M.A., Randel, S., Palou, X., Bolle, C., Gnauck, A.H., Chandrasekhar, S., Liu, X., Guan, B., Essiambre, R.-J., Winzer, P.J., Leon-Saval, S., Bland-Hawthorn, J., Delbue, R., Pupalaikis, P., Sureka, A., Sun, Y., Gruner-Nielsen, L., Jensen, R.V., Lingle, R.: 12 ×12 MIMO transmission over 130−km few−mode fiber. In: Proceedings of FIO: Laser Science XXVIII, p. FW6C.4 (2012b)

Saridis, G.M., Alexandropoulos, D., Zervas, G., Simeonidou, D.: Survey and evaluation of space division multiplexing: from technologies to optical networks. IEEE Commun. Surv. Tutor. **17**(4), 2136–2156 (2015)

Sit, A., Fickler, R., Alsaiari, F., Bouchard, F., Larocque, H., Gregg, P., Yan, L., Boyd, R.W., Ramachandran, S., Karimi, E.: Quantum cryptography with structured photons through a vortex fiber. Opt. Lett. **43**(17), 4108–4111 (2018)

Trichili, A., Park, K.H., Zghal, M., Ooi, B.S., Alouini, M.S.: Communicating using spatial mode multiplexing: potentials, challenges, and perspectives. IEEE Commun. Surv. Tutor. **21**(4), 3175–3203 (2019)

Ung, B., Vaity, P., Wang, L., Messaddeq, Y., Rusch, L.A., LaRochelle, S.: Few-mode fiber with inverse-parabolic graded-index profile for transmission of OAM-carrying modes. Opt. Express **22**(15), 18044–18055 (2014)

Wang, J., Yang, J.Y., Fazal, I.M., Ahmed, N., Yan, Y., Huang, H., Willner, A.E.: Terabit free-space data transmission employing orbital angular momentum multiplexing. Nat. Photon. **6**(7), 488–496 (2012)

Wang, J., Li, S., Luo, M., Liu, J., Zhu, L., Li, C., Xie, D., Yang, Q., Yu, S., Sun, J., Zhang, X., Shieh, W., Willner A.E.: N-dimentional multiplexing link with 1.036-Pbit/s transmission capacity and 112.6-bit/s/Hz spectral efficiency using OFDM-8 QAM signals over 368 WDM pol muxed 26 OAM modes. In: 40th European Conference on Optical Communication, September, Paper No. 4.5.1. IEEE (2014)

Wang, X., Yan, S., Zhu, J., Ou, Y., Hu, Z., Messaddeq, Y., LaRochelle, S., Rusch, L.A., Simeonidou, D., Yu, S.: 3.36-Tbit/s OAM and wavelength multiplexed transmission over an inverse-parabolic graded index fiber. In: Conference on Lasers and Electro-Optics CLEO 2017, San Jose, p. SW4I.3 (2017)

Wang, Y., Bao, C., Geng, W., Lu, Y., Fang, Y., Mao, B., Liu, Y.-G., Liu, B., Huang, H., Ren, Y., Pan, Z., Yue, Y.: Air-core ring fiber with > 1000 radially fundamental OAM modes across O, E, S, C, and L bands. IEEE Access **8**, 68280–68287 (2020)

Yao, A.M., Padgett, M.J.: Orbital angular momentum: origins, behavior and applications. Adv. Opt. Photon. **3**, 161–204 (2011)

Yaman, F., Bai, N., Zhu, B., Wang, T., Li, G.: Long distance transmission in few-mode fibers. Opt. Express **18**(12), 13250–13257 (2010)

Zhang, Z., Gan, J., Heng, X., Wu, Y., Li, Q., Qian, Q., Yang, Z.: Optical fiber design with orbital angular momentum light purity higher than 99.9. Opt. Express **23**(23), 29331–29341 (2015)

Zhang, R., Tan, H., Zhang, J., Shen, L., Liu, J., Liu, Y., Yu, S.: A novel ring-core fiber supporting MIMO-free 50 km transmission over high-order OAM modes. In: 2019 Optical Fiber Communications Conference and Exhibition (OFC), pp. 1–3. IEEE (2019a)

Zhang, R., Tan, H., Zhang, J., Shen, L., Liu, J., Liu, Y., Zhang, L., Yu, S.: A novel ring-core fiber supporting MIMO-free 50km transmission over high-order OAM modes. In: Optical Fiber Communications Conference and Exhibition (OFC) pp. 1–3. IEEE (2019b)

Zhu, G., Chen, Y., Du, C., Zhang, Y., Liu, J., Yu, S.: A graded index ring-core fiber supporting 22 OAM states. In: Opto-Electronics and Communications Conference (OECC) and Photonics Global Conference (PGC), pp. 1–3. IEEE (2017)

Zhu, G., Hu, Z., Wu, X., Du, C., Luo, W., Chen, Y., Cai, X., Liu, J., Zhu, J., Yu, S.: Scalable mode division multiplexed transmission over a 10-km ring-core fiber using high-order orbital angular momentum modes. Opt. Express **26**, 594–604 (2018)

Zhu, L., Zhu, G., Wang, A., Wang, L., Ai, J., Chen, S., Du, C., Liu, J., Yu, S., Wang, J.: 18 km low-crosstalk OAM+WDM transmission with 224 individual channels enabled by a ring-core fiber with large high-order mode group separation. Opt. Lett. **43**(8), 1890–1893 (2018)

Chapter 6
Performance of GLRT-LTD CFAR Processor in Correlated Pareto Clutter

Taha Hocine Kerbaa, Amar Mezache, and Houcine Oudira

Abstract The Pareto Type II distribution has proven to be a robust statistical model for capturing the characteristics of high-resolution sea clutter, particularly under low grazing angle conditions. The application of the Generalized Likelihood Ratio Test Linear Threshold Detector (GLRT-LTD) within such clutter environments necessitates precise estimation of clutter parameters to maintain optimal detection performance. In this study, we conduct a comprehensive evaluation of several parameter estimation techniques-namely Maximum Likelihood Estimation (MLE), integer-order moments (HOME), fractional-order moments (FOME), and the $z \log z$ estimator-under conditions of correlated Pareto-distributed clutter. Through extensive Monte Carlo simulations, we demonstrate that while MLE and $z \log z$ estimators yield comparably accurate results in general, the $z \log z$ method provides superior performance in scenarios involving strong correlation among clutter samples. This highlights the estimator's robustness and computational efficiency, particularly in challenging detection environments where correlation effects are non-negligible.

Keywords Pareto type II clutter · Parameter estimation · MLE · $z \log(z)$ · Moments · GLRT-LTD · CFAR detection

T. H. Kerbaa (✉)
Département d'Electronique and Laboratoire d'Analyse des Signaux et Systèmes, Université Mohamed-Boudiaf, M'sila, Algeria
e-mail: tahahocine.kerbaa@univ-msila.dz

A. Mezache
Département d'Electronique, Université Mohamed-Boudiaf M'sila, Laboratoire SISCOM, Université de Constantine, Constantine, Algeria
e-mail: amar.mezache@univ-msila.dz

H. Oudira
Département d'Electronique, Université Mohamed- Boudiaf Laboratoire de Génie Electrique (LGE), M'sila, Algeria
e-mail: houcine.oudira@univ-msila.dz

N. Derbel and Q. Zhu (eds.), *Advances in Communication, Signal and Image Processing*,
https://doi.org/10.1007/978-981-95-0401-5_6

6.1 Introduction

The continuous advancement of digital signal processing systems has significantly transformed the landscape of modern radar technology. These innovations have enabled radar systems to achieve remarkable improvements in accuracy, robustness, and adaptability, thus facilitating reliable operation in increasingly intricate electromagnetic environments (Gini 2021). As radars become more intelligent and versatile, capable of addressing a wide array of surveillance and tracking tasks, they simultaneously encounter growing complexity in the surrounding environment. One of the principal challenges faced in such settings is the reliable estimation of interference characteristics-particularly in non-homogeneous and non-stationary scenarios-where conventional assumptions may no longer hold.

Among the central objectives of radar detection theory is the discrimination of fluctuating target signals against a background of clutter and noise. This task must be accomplished while ensuring a high probability of detection (P_D) and maintaining a suitably low probability of false alarm (P_{FA}). However, in practical environments such as maritime surveillance or low-grazing-angle operations, the received echoes are often submerged in noise and diffuse clutter, thereby complicating detection performance. The integration of an accurate and well-adapted clutter model, derived from prior knowledge or on-site estimation, can significantly enhance detection outcomes under these conditions (Rosenberg et al. 2019).

Extensive empirical studies have revealed that high-resolution radar data frequently exhibit impulsive or spiky behavior, especially when observing sea surfaces. This phenomenon results in amplitude distributions with heavy tails, diverging notably from the Gaussian assumption typically employed in early radar detection models. Consequently, alternative statistical models-such as the log-normal, Weibull, Pareto, and K-distributions-have been widely proposed to describe radar clutter under high-resolution or low-grazing-angle regimes (Farina et al. 1997; Lampropoulos et al. 1999; Sangston et al. 1999). These distributions better capture the variability and statistical behavior of sea clutter and are essential in formulating realistic detection strategies.

Furthermore, clutter is rarely uncorrelated in practical radar returns. Spatial and temporal correlation can arise due to physical characteristics of the propagation medium, motion-induced coherence, or even sensor array geometry. This introduces additional challenges in both modeling and detection. The prevalent assumption of independent and identically distributed (i.i.d.) clutter samples-while analytically convenient-has been shown to produce suboptimal results when applied to correlated environments. Foundational works such as Cantrell (1986), Farina et al. (1987) theory addressed these issues by proposing models that account for non-Gaussian correlation structures. These approaches build upon the multivariate statistical framework introduced by Martinez et al. (1984), which was initially developed for modeling general multivariate non-Gaussian vectors and later extended to handle complex-valued radar measurements, such as those encountered in log-normal and Weibull clutter.

n adaptive radar systems, particularly those employing Constant False Alarm Rate (CFAR) techniques-either coherent or non-coherent-the task of estimating clutter parameters becomes paramount. The Pareto Type II distribution, due to its heavy-tailed nature and analytical tractability, is an attractive model for sea clutter. However, successful deployment of CFAR algorithms like the Generalized Likelihood Ratio Test with Linear Threshold Detector (GLRT-LTD) critically depends on accurate estimation of the distribution's shape and scale parameters. Various estimation techniques have been proposed to this end, particularly for the uncorrelated case (Balleri et al. 2007; Bocquet 2015a, b).

In Balleri et al. (2007), a Maximum Likelihood Estimation (MLE) procedure was introduced, offering an effective estimation of the scale parameter, with the shape parameter computed in a forward manner. Despite its theoretical advantages, MLE often entails a high computational cost, particularly when applied to large datasets or online processing. In a complementary line of work, Bocquet (2015b) evaluated multiple approaches: the Higher Order Moments Estimator (HOME), which is computationally efficient but sensitive to outliers in spiky clutter environments; the Constrained Maximum Likelihood Estimator (CMLE), which improves estimation precision but at the expense of increased numerical complexity; and the $z\log(z)$ estimator, which leverages the generalized exponential integral function to provide a balanced solution. The latter exhibits near-MLE performance while offering reduced computational requirements. A closed-form expression for this estimator in multi-look configurations was later provided in Bocquet (2015a), making it particularly suitable for real-time applications where both accuracy and efficiency are required. However, all these methods may exhibit statistical variability in small-sample regimes due to the inherent uncertainty of moment-based statistics.

Beyond parameter estimation, several GLRT-based CFAR detectors have been proposed in the literature to address detection in complex, non-Gaussian clutter environments (Gini and Greco 2002; Sangston et al. 2012; Weinberg et al. 2019). These detectors typically require estimation of the clutter covariance matrix, which can be approached through iterative or non-iterative techniques. For example, Gini and Greco (2002) considered adaptive detection in correlated K-distributed clutter and proposed an Approximate Maximum Likelihood (AML) solution for estimating the covariance matrix under unknown statistics. Meanwhile, Sangston et al. (2012) introduced a coherent detection framework for compound-Gaussian clutter with inverse gamma texture, presenting an optimal matched filter detector whose threshold adapts to the data via a quadratic statistic-coined the GLRT-LTD. More recently, a Bayesian detection scheme tailored for uncorrelated Pareto Type II clutter was proposed in Weinberg et al. (2019), although it relies heavily on numerical integration, which can be computationally prohibitive in real-time applications.

While the GLRT-LTD has been analyzed under the assumption of known clutter parameters (Sangston et al. 2012), practical applications necessitate the estimation of these parameters from observed data. In the same study, the authors pro-posed an Adaptive GLRT (AGLRT) where parameter estimation was achieved via the method of moments. However, the sensitivity of this estimator to the order of moments-particularly in the presence of impulsive clutter-renders it less suitable for real-time or embedded systems. Moreover, these estimation methods are often tailored to specific clutter scenarios, limiting their generalizability.

Despite the growing body of literature, a comprehensive comparison of the impact of various parameter estimators-specifically integer-order moments (HOME), fractional-order moments (FOME), $z \log(z)$, and MLE-on the performance of the GLRT-LTD detector under correlated Pareto clutter has not yet been undertaken. This chapter addresses this gap by conducting a detailed investigation into how these estimation methods influence detection probability across a range of signal-to-clutter ratios (SCRs) and correlation levels, using synthetically generated clutter data.

Through extensive simulation, we show that while MLE and $z \log(z)$ estimators tend to offer similar detection performance under general conditions, the $z \log(z)$ method consistently demonstrates superior resilience and accuracy when clutter is highly correlated. This makes it an attractive candidate for practical deployment in advanced radar systems operating under non-ideal environmental conditions.

The structure of the remainder of this chapter is as follows. Section 6.2 details the methodology for generating correlated Pareto Type II clutter samples. Section 6.3 presents the mathematical formulations used for estimating the distribution parameters, including MLE, HOME, FOME, and the $z \log(z)$ method. Section 6.4 provides an overview of the GLRT-LTD detection algorithm. In Sect. 6.5, we present numerical results comparing estimation accuracy and detection performance across different estimation strategies. Finally, Sect. 6.6 concludes the chapter and dis-cusses potential directions for future research.

6.2 Pareto Distribution and Related Correlated Clutter Generator

This section provides an overview of the Pareto Type II distribution, which is widely employed for modeling high-resolution sea clutter, particularly in radar signal processing. Furthermore, we discuss established methodologies for generating correlated clutter samples based on this statistical model. These methods serve as critical tools for developing and evaluating advanced detection algorithms under realistic operational conditions.

6.2.1 Pareto Distribution

In statistical radar modeling, sea clutter is often represented as a compound process, where the received signal amplitude is modeled as a combination of two components: a fast-varying speckle term and a slowly varying texture component. Mathematically, the probability density function (PDF) of the compound process can be obtained by integrating the conditional PDF of the speckle, denoted by $p(x|\tau)$, over the PDF of the texture, $p(\tau)$, where $\tau > 0$ represents the local power or texture variable:

$$p(x) = \int_0^\infty p(x|\tau)\, p(\tau)\, d\tau \tag{6.1}$$

For the specific case of single-look, independent Pareto Type II clutter, the conditional speckle amplitude x follows a Rayleigh distribution, while the texture τ is modeled by an inverse gamma distribution (Gini and Greco 2002; Weinberg et al. 2019):

$$\begin{cases} p(x|\tau) = \dfrac{2x}{\tau} \exp\left(-\dfrac{x^2}{\tau}\right) \\ p(\tau) = \dfrac{1}{\Gamma(\lambda)} \left(\dfrac{\lambda}{\eta}\right)^{\lambda} \dfrac{1}{\tau^{\lambda+1}} \exp\left(-\dfrac{\lambda}{\eta\tau}\right) \end{cases} \tag{6.2}$$

Here, λ and η are the shape and scale parameters, respectively, and $\Gamma(.)$ denotes the gamma function. Substituting 6.2 into 6.1 leads to the closed-form expression for the Pareto Type II distribution:

$$p(x) = \frac{2x\eta}{\left(\frac{\eta}{\lambda}\lambda + 1\right)} \tag{6.3}$$

Assuming $\lambda > \dfrac{k}{2}$, the kth moment of the amplitude x is given by:

$$\mu_k = \left(\frac{\lambda}{\eta}\right)^{\frac{k}{2}} \frac{\Gamma\left(1 + \frac{k}{2}\right) \Gamma\left(\lambda - \frac{k}{2}\right)}{\Gamma(\lambda)} \tag{6.4}$$

When a square-law detector is applied i.e., the detected power is $Z = X^2$-the resulting PDF and k^{th} moment become:

$$p(z) = \frac{\lambda b^\lambda}{(z + b)^{\lambda+1}} \tag{6.5}$$

and

$$\mu_k = \left(\frac{\lambda}{\eta}\right)^k \frac{\Gamma(1 + k)\, \Gamma(\lambda - k)}{\Gamma(\lambda)} \tag{6.6}$$

where $b = \dfrac{\eta}{\lambda}$ represents the scale parameter in the transformed domain. In this analysis, the system noise contribution is assumed to be negligible, focusing exclusively on the speckle and texture components.

6.2.2 *Generation Approaches for Correlated Multivariate Model*

Given the scarcity and controlled access to real radar sea clutter measurements, especially under various operational environments, simulated data has become a key enabler in radar algorithm development. It allows for reproducibility and precise manipulation of clutter characteristics, particularly the correlation properties that significantly impact detection performance.

In compound clutter models, the speckle component typically exhibits rapid fluctuations and is well-approximated by a Gaussian process, while the texture, representing the local power, varies more slowly. Hence, to replicate the realistic behavior of sea clutter, the data generation process must incorporate temporal and/or spatial correlation in both components.

6.2.2.1 Memoryless Nonlinear Transformation (MNLT) Approach

The Memoryless Nonlinear Transformation (MNLT) technique constructs a correlated non-Gaussian process from a correlated Gaussian process. A zero-mean, unit-variance Gaussian sequence is first generated (commonly via Fourier synthesis), then transformed using a nonlinear mapping that reshapes its marginal distribution to match a target PDF (Bocquet 2013).

In this method, the error function is employed to transform the Gaussian process into a uniform distribution, which is then mapped to the desired distribution using the inverse cumulative distribution function (CDF). When the target distribution is inverse gamma, the resulting transformed process follows a Pareto Type II model when combined with a Rayleigh speckle component.

The general transformation for a sequence Φ with inverse gamma CDF F_d is given by:

$$\Phi(x) = F_d^{-1}\left[\frac{1}{2}\operatorname{erfc}\left(\frac{x}{\sqrt{2}}\right)\right] \tag{6.7}$$

Here, $erfc(.)$ denotes the complementary error function. Efficient numerical algorithms for computing F_d^{-1} are detailed in Bocquet (2013). The correlation properties of the transformed sequence can be understood through the Hermite polynomial expansion, which relates the output autocorrelation to the input Gaussian process autocorrelation $R_G(t)$:

$$\langle \Phi(0)\Phi(t)\rangle = \frac{1}{\pi}\sum_{n=0}^{\infty}\frac{R_G(t)^n}{2^n n!}I_n^2 \tag{6.8}$$

where:

$$I_n = \int_{-\infty}^{+\infty} \exp(-x^2)H_n(x)F_d^{-1}\left[\frac{1}{2}\operatorname{erfc}\left(\frac{x}{\sqrt{2}}\right)\right]dx \tag{6.9}$$

The series converges rapidly for larger values of λ, but becomes increasingly slow for smaller λ due to the divergence of the inverse gamma variance as $\lambda \leq 2$. The normalized correlation function for the inverse gamma process is expressed as:

$$R_{\gamma}(t) = \frac{\langle \Phi(0)\Phi(t) \rangle - \langle \Phi \rangle^2}{\langle \Phi^2 \rangle - \langle \Phi \rangle^2} \tag{6.10}$$

It is important to note that the MNLT method does not simultaneously preserve both the marginal PDF and the power spectral density. Additionally, the nonlinear mapping may violate the Hermitian symmetry required in the covariance matrix, resulting in imperfect emulation of correlation structure in multivariate data.

6.2.2.2 Spherically Invariant Random Vector (SIRV) Approach

A widely adopted and analytically tractable method for modeling non-Gaussian clutter in radar systems is the use of Spherically Invariant Random Processes (SIRPs), which generate Spherically Invariant Random Vectors (SIRVs) upon realization. This modeling framework is particularly effective for simulating compound-Gaussian clutter with heavy tails and is especially suitable for high-resolution sea clutter environments (Sangston et al. 2012).

In the SIRV formulation, the radar return vector c is modeled as the product of two independent random components:

$$\boldsymbol{c} = \sqrt{\tau}\,\boldsymbol{x} \tag{6.11}$$

where $\boldsymbol{x} \sim \mathcal{CN}(0, \mathbf{M})$ is a zero-mean complex Gaussian speckle vector with a positive definite Hermitian covariance matrix $\mathbf{M}$, and τ is a non-negative random scalar known as the texture, which controls the local power of the clutter. The trace of $\mathbf{M}$ is normalized such that $Tr(\mathbf{M}) = m$, where m denotes the dimensionality of the clutter vector. The speckle and texture components are assumed statistically in-dependent.

This model enables the generation of correlated Pareto Type II clutter through the steps summarized in Algorithm 1, which outlines the process of synthesizing SIRV-based clutter vectors with prescribed correlation and marginal statistical properties.

In this procedure, the speckle covariance matrix $\mathbf{M}_x$ is constructed as a Toeplitz matrix to enforce correlation with exponential decay governed by ρ. The Cholesky factorization enables generation of correlated Gaussian vectors. The texture variable τ is drawn from an inverse-gamma distribution, ensuring that the resulting compound-Gaussian process yields a Pareto Type II marginal distribution.

This approach offers several advantages over transformation-based methods, such as the Memoryless Nonlinear Transformation (MNLT), including exact control over both the second-order statistics (through $\mathbf{M}$) and the marginal distribution (through τ). Moreover, the SIRV model naturally preserves the Hermitian property of the

Algorithm 1: Synthesis of Correlated Pareto-Distributed Clutter using the SIRV Model

Require: Number of pulses N, observation window size m, number of Monte Carlo trials n, number of clutter vectors nb

1: Generate random correlation coefficient $\rho \in [0, 1]$
2: Generate random values for shape parameter λ and scale parameter η
3: **for** $i = 1$ to nb **do**
4: Construct speckle covariance matrix: $\mathbf{M}_x \leftarrow \text{toeplitz}(\rho^m)$
5: Compute Cholesky factor: $\mathbf{L} \leftarrow \text{chol}(\mathbf{M}_x)$
6: Generate complex Gaussian speckle:

$$\mathbf{x} \leftarrow \mathbf{L} \cdot \frac{1}{\sqrt{2}} (\text{randn}(m, N) + j \cdot \text{randn}(m, N))$$

7: Generate inverse-gamma texture variable:

$$\tau \leftarrow \frac{1}{\text{gamrnd}(\lambda, \eta, 1, N)}$$

8: Compute clutter vector:

$$\mathbf{c} \leftarrow \sqrt{\tau} \cdot \mathbf{x}$$

9: **end for**

Ensure: Correlated Pareto-distributed clutter matrix $\mathbf{c}$ of size $N \times m$

covariance matrix and supports multidimensional simulation, making it ideal for testing CFAR detectors and adaptive signal processing algorithms under realistic clutter conditions.

6.3 Estimators of Pareto Type II Clutter

Accurate estimation of distribution parameters is a fundamental prerequisite for achieving robust detection performance in radar systems employing adaptive CFAR (Constant False Alarm Rate) algorithms. Since these detectors operate under uncertain and dynamic conditions, they rely heavily on precise evaluation of clutter model parameters-particularly the shape (λ) and scale (η) parameters-from local data samples within a reference window (Cantrell 1986).

In the context of Pareto Type II distributed clutter, a range of estimation strategies have been proposed in the literature. This section presents and analyzes four prominent methods for estimating the clutter parameters: the Higher Order Moments Estimator (HOME), the Fractional Order Moments Estimator (FOME), the $z \log z$ estimator, and the classical Maximum Likelihood Estimator (MLE). While the MLE is derived directly from the theoretical distribution (Eq. 6.3), the other estimators are constructed by leveraging statistical moment relationships as expressed in Eq. (6.4).

6.3.1 Higher Order Moments Estimator (HOME)

The HOME approach is based on matching theoretical and empirical moments of the observed data. Let $\{x_i\}_{i=1}^N$ denote a set of independent samples assumed to follow a Pareto Type II distribution. The shape and scale parameters of the texture component can be estimated by equating the theoretical and empirical second- and fourth-order moments:

$$\begin{cases} r_\mu = \dfrac{\widehat{\mu}_4}{\widehat{\mu}_2^2} \\ \widehat{\lambda} = \dfrac{2r_\mu - 1}{r_\mu - 1} \\ \widehat{\eta} = \dfrac{\widehat{\mu}_2}{\widehat{\mu}_4}\left(2r_\mu - 1\right) \end{cases} \tag{6.12}$$

Here:

$$\widehat{\mu}_k = \frac{1}{N}\sum_{i=1}^{N} x_i^k$$

denotes the kth sample moment. This method is particularly appealing due to its computational simplicity. However, its applicability is theoretically restricted to cases where $\lambda > 2$. To extend its usability to heavier-tailed clutter (i.e., lower λ), the second-order moment can be replaced with a lower-order moment, such as $\widehat{\mu}_{1/2}$, as proposed in Balleri et al. (2007).

6.3.2 Fractional Order Moments Estimator (FOME)

The FOME method enhances the flexibility of moment-based estimation by combining fractional-order statistical moments. In particular, moments of order 1 and 1/2 are used to derive a set of nonlinear equations that relate the sample moments to the model parameters. This technique is valid for shape parameters $\lambda > 1$, and typically requires numerical methods to solve for $\widehat{\lambda}$:

$$\begin{cases} \dfrac{\Gamma\left(\dfrac{3}{2}\right)^2 \widehat{\mu}_{\frac{1}{2}}}{\Gamma\left(\dfrac{5}{4}\right)\widehat{\mu}_1^2} = \dfrac{\Gamma\left(\widehat{\lambda} - \frac{1}{4}\right)}{\Gamma\left(\widehat{\lambda}\right)\Gamma\left(\widehat{\lambda} - \frac{1}{2}\right)^2} \\ \widehat{\eta} = \widehat{\lambda}\left(\dfrac{\Gamma\left(\frac{5}{4}\right)\Gamma\left(\widehat{\lambda} - \frac{1}{4}\right)}{\widehat{\mu}_1 \Gamma\left(\widehat{\lambda}\right)}\right)^4 \end{cases} \tag{6.13}$$

The strength of this estimator lies in its applicability to heavy-tailed data and its improved numerical stability in comparison to higher-order moment matching.

Nonetheless, solving the implicit nonlinear equation for $\widehat{\lambda}$ typically necessitates iterative root-finding algorithms.

6.3.3 $z\log(z)$ Estimator

The $z\log z$ estimator, introduced by Bocquet (2013), provides an efficient closed-form estimation strategy particularly suitable for data processed using envelope detectors. This method involves computing the expectations $\langle \log z^2 \rangle$ and $\langle z^2 \log z^2 \rangle$ to derive a direct relationship for estimating the shape and scale parameters:

$$\begin{cases} \widehat{\lambda} = 1 + \left(\dfrac{\langle z^2 \log(z^2) \rangle}{\langle z^2 \rangle} - \langle \log(z^2) \rangle - 1 \right)^{-1} \\ \widehat{\eta} = \widehat{\lambda} \left(\dfrac{\widehat{\mu}_1 \Gamma\left(\widehat{\lambda}\right)}{\Gamma\left(\frac{5}{4}\right) \Gamma\left(\widehat{\lambda} - \frac{1}{4}\right)} \right)^{-4} \end{cases} \tag{6.14}$$

This estimator effectively bridges the trade-off between computational complexity and accuracy. It has been shown to yield near-optimal results compared to MLE under correlated and spiky clutter conditions, with a significantly lower computational cost. However, the accuracy of this method can degrade in low-SNR scenarios or when the data set is very limited.

6.3.4 Maximum Likelihood Estimator (MLE)

The MLE is often considered the gold standard for parameter estimation due to its asymptotic efficiency. For the Pareto Type II model, the log-likelihood function is derived from the PDF in Eq. (6.3). The estimation is typically performed by maximizing the likelihood with respect to the scale-to-shape ratio $\widehat{b} = \dfrac{\widehat{\eta}}{\widehat{\lambda}}$. The optimal $\widehat{b}$ is found by solving the following nonlinear equation (Balleri et al. 2007):

$$\frac{N \sum_{i=1}^{N} \dfrac{z_i^2}{\widehat{b} z_i^2 + 1}}{N - \widehat{b} \sum_{i=1}^{N} \dfrac{z_i^2}{\widehat{b} z_i^2 + 1}} - \sum_{i=1}^{N} \log\left(\widehat{b} z_i^2 + 1\right) = 0 \tag{6.15}$$

Once $\widehat{b}$ is determined numerically, the shape parameter $\widehat{\lambda}$ can be calculated explicitly as:

$$\widehat{\lambda} = \frac{N}{\widehat{b} \sum_{i=1}^{N} \frac{z_i^2}{\widehat{b} z_i^2 + 1}} - 1 \tag{6.16}$$

The MLE approach typically achieves the lowest estimation error, particularly as the number of samples increases. However, it is computationally more intensive than moment-based or closed-form alternatives and may be impractical for real-time radar systems with strict latency constraints.

In summary, each of the presented estimators offers a unique balance between complexity, accuracy, and applicability to different radar scenarios. While the MLE is theoretically optimal, it may be unsuitable for real-time applications. In contrast, the HOME and FOME estimators provide simplicity and flexibility, albeit at the cost of accuracy under certain conditions. The $z \log z$ method offers a compelling compromise and is particularly effective in correlated and heavy-tailed clutter environments-making it a favorable candidate for CFAR-based detection systems in modern maritime and surveillance radars.

6.4 GLRT-LTD Detector

In radar signal processing, the problem of target detection is typically formulated as a binary hypothesis testing scenario. The objective is to determine whether a target is present in a given observation window based on the received signal. Let the received complex-valued observation vector be defined as $z = z_I + jz_Q = [z_1, z_2, \ldots, z_m]^T$, where z_I et z_Q represent the in-phase and quadrature components, respectively, of the complex radar return. The dimension m corresponds to the number of coherent pulses (or snapshots) collected during the Coherent Processing Interval (CPI).

The binary hypothesis testing problem can then be expressed as:

$$\begin{cases} H_0 : z = c & \text{(clutter} - \text{only)} \\ H_1 : z = s + c & \text{(target} + \text{clutter)} \end{cases} \tag{6.17}$$

Here, c denotes the clutter component, and s is the known target signal. Under this model, the goal is to determine whether the observed vector z consists purely of clutter or includes a target contribution.

According to the Neyman-Pearson criterion, the optimal test statistic is based on the likelihood ratio:

$$\Lambda(z) = \frac{p_{z|H_1}(z|H_1)}{p_{z|H_0}(z|H_0)} \gtrless_{H_0}^{H_1} e^T \tag{6.18}$$

In Eq. (6.18), $p_{z|H_i}$ denotes the probability density function (PDF) of the observed vector under hypothesis H_i, and T is a threshold chosen to achieve a desired probability of false alarm PFA. To derive the likelihood expressions, we assume that the clutter follows a compound-Gaussian model with a known form of texture distribution. Under this assumption, and assuming the target signature s is known, the conditional PDF under each hypothesis is expressed as Sangston et al. (2012):

$$p_{Z|H_i}(Z|H_i) = \frac{1}{\pi^m |M|} h_m(q_i(z)); i = 1, 2 \tag{6.19}$$

where $\mathbf{M}$ is the clutter covariance matrix, and the function $h_m(.)$ is given by:

$$h_m(q_i(z)) = \int_0^{\infty} \alpha^m \, exp(-\alpha q) \, p_\alpha(\alpha) \, d\alpha \tag{6.20}$$

Here, $p_\alpha(\alpha)$ denotes the distribution of the texture variable α. The quadratic forms q_0 and q_1 represent energy terms under H_0 and H_1, respectively:

$$\begin{cases} q_0(z) = Z^H M^{-1} Z \\ q_1(z) = q_0(z - s) \end{cases} \tag{6.21}$$

The optimal decision rule is obtained by comparing the difference $q_0 - q_1$ with a threshold function $f_{opt}(q_0, T)$: The density $p_\alpha(\alpha)$ is given in Sangston et al. (2012) [and therein]. It has been shown that the optimal detector can be written as:

$$q_0 - q_1 \gtrless_{H_0}^{H_1} f_{opt}(q_0, T) \tag{6.22}$$

he threshold function f_{opt} depends on the clutter model and is given in terms of the inverse of the function $h_m(.)$:

$$f_{opt}(q_0, T) = q_0 - h_m^{-1}(e^T h_m \left(e^T h_m(q_0)\right) \tag{6.23}$$

Further simplification of this test under the compound-Gaussian clutter model leads to the Generalized Likelihood Ratio Test with a Linear Threshold Detector (GLRT-LTD), whose decision rule is expressed as:

$$q_0 - q_1 = \frac{\left|\mathbf{p}^H \mathbf{M}^{-1} \mathbf{z}\right|^2}{\mathbf{p}^H \mathbf{M}^{-1} \mathbf{p}} \gtrless_{\mathbf{H}_0}^{\mathbf{H}_1} \left(1 - e^{-T/m+\lambda}\right) \left(\frac{\lambda}{\eta} + q_0\right) \tag{6.24}$$

In Eq. (6.24), p is the known steering vector aligned with the Doppler frequency of the target, $\mathbf{M}$ is the estimated covariance matrix of the clutter, and $q_0 = z_H \mathbf{M}^{-1} z$ corresponds to the clutter-normalized energy of the test vector. The right-hand side incorporates the Pareto clutter parameters λ and η, and the detection threshold T. The false alarm probability associated with this detector, under the assumption of a large number of independent secondary data samples ($K \longrightarrow \infty$), is given by:

$$P_{FA} = exp\left(-T \, \frac{m + \lambda - 1}{m + \lambda}\right) \tag{6.25}$$

This expression highlights the dependence of P_{FA} on both the number of coherent pulses m and the shape parameter λ of the Pareto clutter model.

The covariance matrix **M**, which is critical to the detector's performance, is typically estimated from neighboring range cells assumed to be free of targets (i.e., secondary data). Common estimation methods include the Sample Covariance Matrix (SCM) and the Normalized Sample Covariance Matrix (NSCM), as defined in Gini and Greco (2002):

$$\widehat{M}_{SCM} = \frac{1}{K} z_k \, z^H \tag{6.26}$$

and:

$$\widehat{M}_{NSCM} = \frac{1}{K} \sum_{k=1}^{K} z_k \, z^H \tag{6.27}$$

These estimators provide a practical means to obtain **M** from real data, especially under homogeneous conditions. However, in heterogeneous environments, robust covariance estimation becomes essential to ensure CFAR performance.

To deploy the GLRT-LTD detector in operational radar systems, accurate values of the Pareto distribution parameters λ and η must be estimated using the procedures detailed in Sect. 6.3. The choice of estimation method directly influences the detector's sensitivity and reliability, particularly under conditions of high clutter correlation or limited sample availability. As such, the integration of efficient estimation techniques with the GLRT-LTD framework forms a critical component of adaptive detection strategies in non-Gaussian environments.

6.5 Numerical Illustration

To evaluate the practical performance and robustness of the GLRT-LTD detector in the presence of correlated Pareto Type II clutter, a two-stage comparative analysis is conducted. This analysis focuses on two key aspects: the accuracy of parameter estimation and the impact of the chosen estimators on the target detection capabilities of the GLRT-LTD framework.

In the first stage, we assess the estimation accuracy of the shape and scale parameters of the Pareto clutter distribution using the Mean Square Error (MSE) as the performance metric. The four estimators introduced in Sect. 6.3 namely, HOME, FOME, $z \log z$, and MLE-are applied to estimate the shape parameter λ under different correlation regimes. In the second stage, we evaluate the detection performance of the GLRT-LTD detector in terms of the probability of detection (P_D) and its CFAR properties, using synthetically generated clutter data with varying correlation levels.

All simulations are conducted using Monte Carlo methods with (= 1000 trials. The simulation parameters are set as follows: the number of reference cells is $N = 24$, the number of pulses within a coherent processing interval is $m = 16$, and the

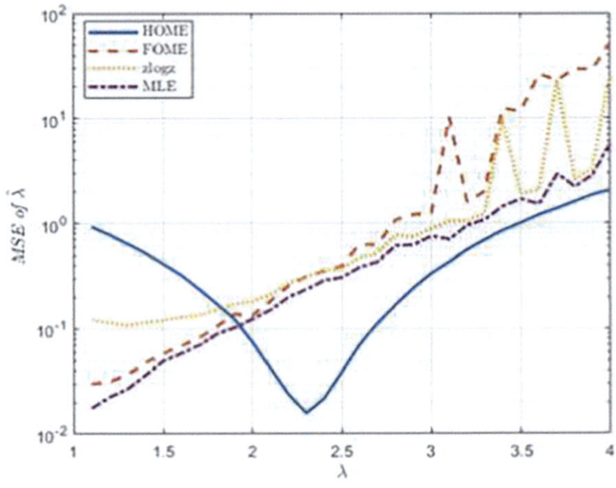

Fig. 6.1 Shape parameter estimation performance using the HOME, FOME, $z\log(z)$ and MLE methods for $m = 16$, $\rho = 0.01$ and $N = 24$

correlation coefficient ρ is varied in the interval $[0, \ 1]$. The shape parameter λ is examined over the range $1.1 < \lambda < 4$.

Figure 6.1 presents the estimation performance of the shape parameter for weakly correlated clutter data ($\rho = 0.01$), which represents a nearly uncorrelated scenario. The results reveal that for highly spiky clutter (i.e., $\lambda < 2$), the FOME and MLE estimators outperform their counterparts in terms of MSE. Conversely, at higher values of λ, the HOME method exhibits superior performance. However, it is important to emphasize that HOME is only valid when $\lambda > 2$, as it relies on higher-order moments that may not exist or be stable for heavy-tailed distributions. In practical applications, this constraint may limit its usefulness.

The estimation performance under a high-correlation scenario ($\rho = 0.9$) is illustrated in Fig. 6.2. While FOME and MLE maintain slight superiority in the low λ regime, the $z\log z$ and HOME methods provide more accurate estimates when $\lambda > 1.5$. The comparison between Figs. 6.1 and 6.2 underscores the influence of the correlation coefficient on estimator performance. These fluctuations are attributed to the fact that most moment-based estimators are derived under the assumption of uncorrelated clutter, leading to performance degradation when applied to strongly correlated data.

The second part of the evaluation focuses on the CFAR properties of the GLRT-LTD detector. Figures 6.3 and 6.4 display the probability of false alarm (PFA) as a function of the shape parameter λ and threshold factor T under uncorrelated and correlated scenarios, respectively. These figures include both the ideal detector-assumed to operate with perfectly known parameters and the adaptive detector, which estimates parameters using the $z\log z$ method.

As shown in Fig. 6.3 for $\rho = 0.01$, the adaptive GLRT-LTD detector pre-serves the CFAR texture property, with performance closely matching the theoretical model. A value of $T = 8.55$dB corresponds to a $P_{FA} = 10^{-3}$. Figure 6.4 confirms similar CFAR behavior in the highly correlated case ($\rho = 0.9$), demonstrating that the

Fig. 6.2 Shape parameter estimation performance using the HOME, FOME, $z\log(z)$ and MLE methods for $m = 16$, $\rho = 0.9$ and $N = 24$

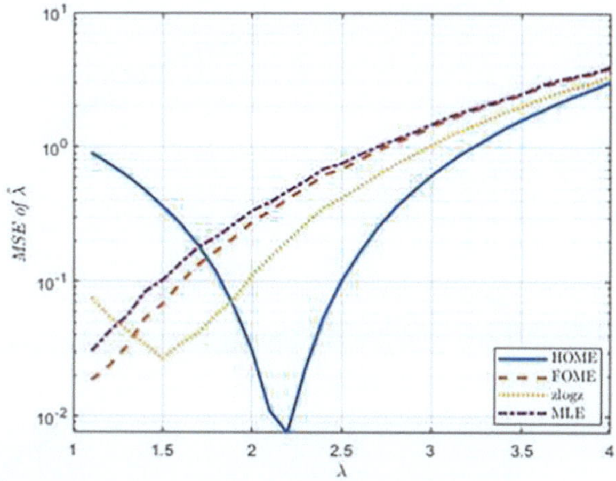

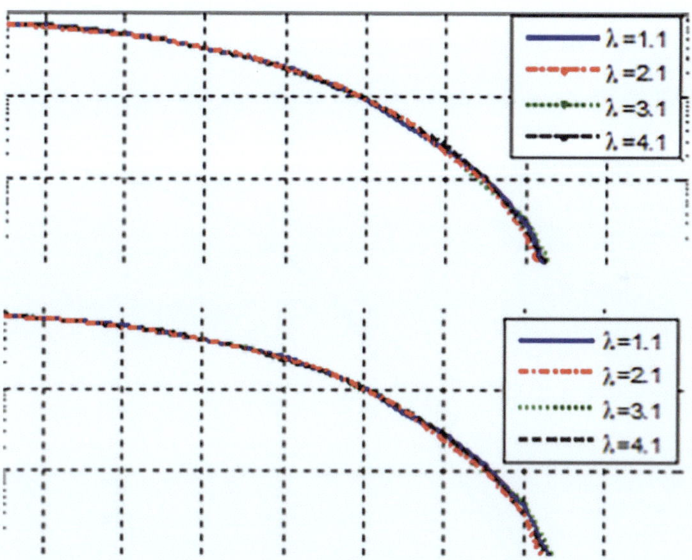

Fig. 6.3 CFAR texture property of the GLRT-LTD detector for $m = 16$, $\rho = 0.01$ and $N = 24$

GLRT-LTD maintains its false alarm regulation capability across varying degrees of correlation.

Figures 6.5 and 6.6 depict the detection probability (P_D) as a function of the signal-to-clutter ratio (SCR) for both uncorrelated and fully correlated clutter scenarios. For $\rho = 0.01$ (Fig. 6.5), the MLE and $z\log z$ estimators yield similar detection curves, validating the accuracy of the closed-form estimator. In the highly correlated case ($\rho = 0.9$), Fig. 6.6 reveals a significant performance advantage for the $z\log z$ method, confirming its robustness in environments with strong texture dependence.

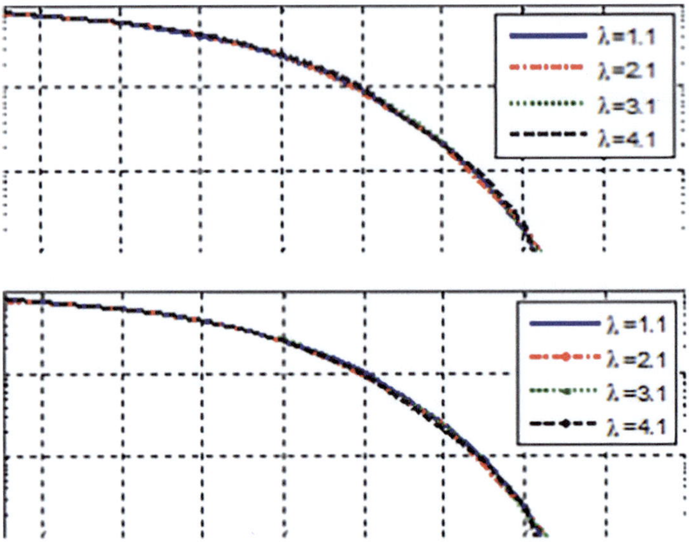

Fig. 6.4 CFAR texture property of the GLRT-LTD detector for $m = 16$, $\rho = 0.9$ and $N = 24$

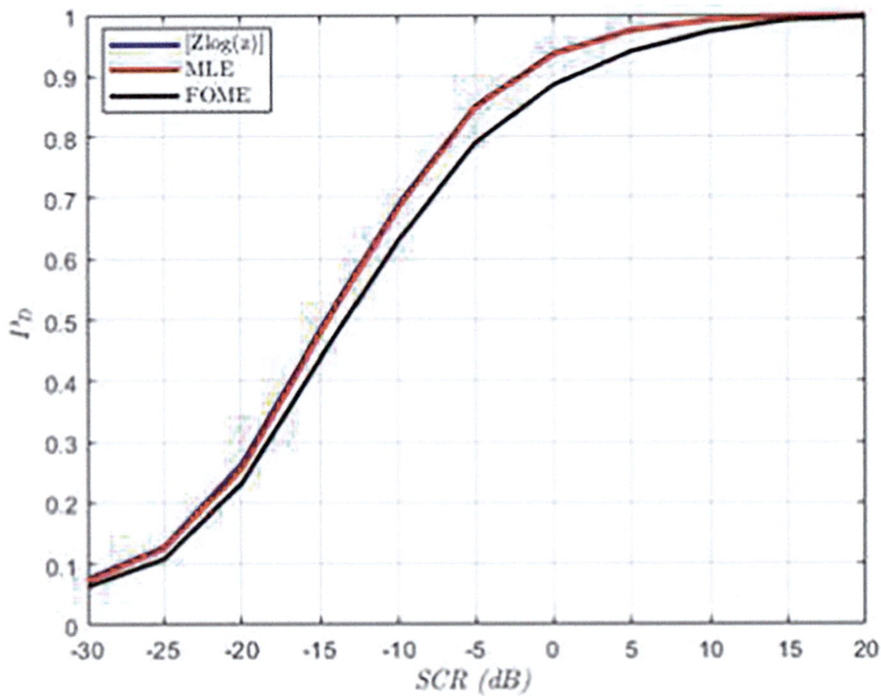

Fig. 6.5 Detection performance of the GLRT-LTD algorithm with $\lambda = 1.1$, $m = 16$, $\rho = 0.01$ and $N = 24$

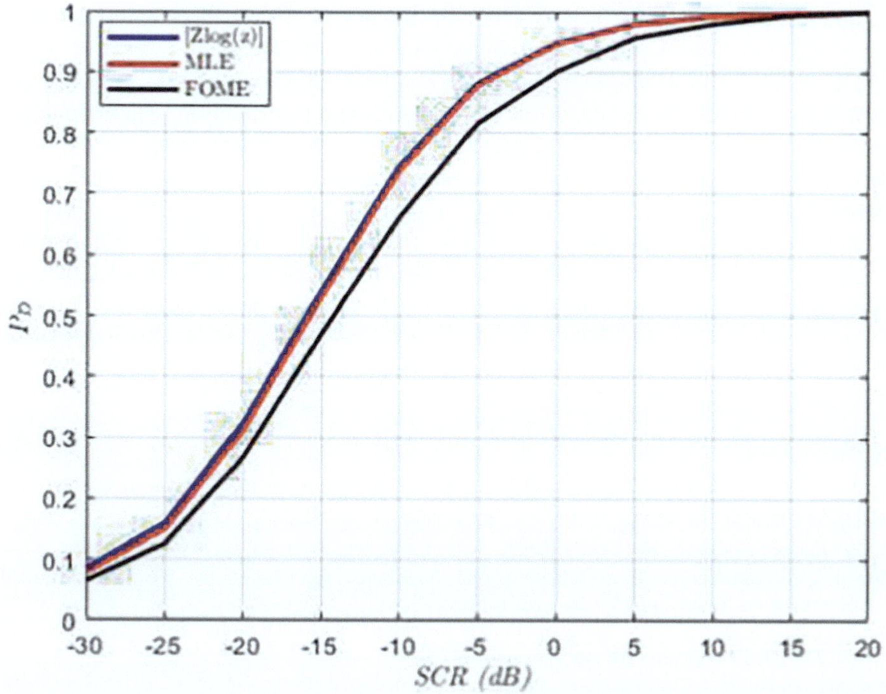

Fig. 6.6 Detection performance of the GLRT-LTD algorithm with $\lambda = 1.1$, $m = 16$, $\rho = 0.9$ and $N = 24$

6.6 Conclusions

This chapter presented an in-depth study of parameter estimation and adaptive detection in the context of radar signal processing with correlated Pareto Type II sea clutter. Emphasis was placed on evaluating the impact of various estimation strategies on the detection performance of the GLRT-LTD algorithm, particularly in heavy-tailed and correlated environments.

Several estimators were investigated-namely, the Higher Order Moments Estimator (HOME), Fractional Order Moments Estimator (FOME), the $z \log z$ method, and the Maximum Likelihood Estimator (MLE). Through extensive Monte Carlo simulations, it was demonstrated that the accuracy of parameter estimation is strongly influenced by the degree of correlation in the clutter. In low-correlation scenarios, FOME and MLE performed reliably across different shape parameter regimes. However, in high-correlation environments, the $z \log z$ estimator outper-formed all others, offering a compelling trade-off between computational efficiency and robustness.

Moreover, the GLRT-LTD detector was shown to preserve CFAR properties-both in texture and covariance structure-even when operating with adaptively estimated parameters. This is a crucial characteristic for practical radar applications, where environmental conditions are rarely homogeneous or stationary.

Detection performance analysis further confirmed the effectiveness of the $z \log z$ approach, particularly in scenarios with severe clutter correlation. These findings position the $z \log z$ method as a strong candidate for integration into real-time radar detection systems, offering a robust and efficient alternative to traditional MLE-based frameworks.

Future research could explore the integration of these estimators within more complex radar system architectures, including multichannel, polarimetric, or cognitive radar platforms, as well as their extension to other heavy-tailed models and Bayesian inference-based detectors.

References

Balleri, A., Nehorai, A., Wang, J.: Maximum likelihood estimation for compound-gaussian clutter with inverse gamma texture. IEEE Trans. Aerosp. Electron. Syst. **43**(2), 775–779 (2007)

Bocquet, S.: Simulation of correlated pareto distributed sea clutter. In: 2013 International Conference on Radar, pp. 258–261. IEEE (2013)

Bocquet, S.: Closed-form parameter estimators for pareto distributed clutter with noise. Electron. Lett. **51**(23), 1924–1926 (2015a)

Bocquet, S.: Parameter estimation for pareto and k distributed clutter with noise. IET Radar Sonar Navig. **9**(1), 104–113 (2015b)

Cantrell, B.: Radar detection in non-gaussian, correlated clutter. Naval Res. Lab. Rep. **9015**, 21 (1986)

Farina, A., Russo, A., Scannapieco, F., Barbarossa, S.: Theory of radar detection in coherent weibull clutter. In: IEE Proceedings F (Communications, Radar and Signal Processing), vol. 134, pp. 174–190. IET (1987)

Farina, A., Gini, F., Greco, M., Verrazzani, L.: High resolution sea clutter data: statistical analysis of recorded live data. IEE Proc.-Radar Sonar Navig. **144**(3), 121–130 (1997)

Gini, F., Greco, M.: Covariance matrix estimation for CFAR detection in correlated heavy tailed clutter. Signal Process. **82**(12), 1847–1859 (2002)

Gini, F.: Grand challenges in radar signal processing. Front. Signal Process. **1**, 664232 (2021)

Lampropoulos, G., Drosopoulos, A., Rey, N., et al.: High resolution radar clutter statistics. IEEE Trans. Aerosp. Electron. Syst. **35**(1), 43–60 (1999)

Martinez, A., Swaszek, P., Thomas, J.: Locally optimal detection in multivariate non-gaussian noise. IEEE Trans. Inf. Theory **30**(6), 815–822 (1984)

Rosenberg, L., Watts, S., Greco, M.S.: Modeling the statistics of microwave radar sea clutter. IEEE Aerosp. Electron. Syst. Mag. **34**(10), 44–75 (2019)

Sangston, K.J., Gini, F., Greco, M.V., Farina, A.: Structures for radar detection in compound gaussian clutter. IEEE Trans. Aerosp. Electron. Syst. **35**(2), 445–458 (1999)

Sangston, K.J., Gini, F., Greco, M.S.: Coherent radar target detection in heavy-tailed compound-gaussian clutter. IEEE Trans. Aerosp. Electron. Syst. **48**(1), 64–77 (2012)

Weinberg, G.V., Howard, S.D., Tran, C.: Bayesian framework for detector development in pareto distributed clutter. IET Radar Sonar Navig. **13**(9), 1548–1555 (2019)

Chapter 7
Design and Hardware Implementation of a Novel Protocol for Digital Video in GSM Interfaces

Dalenda Bouzidi, Youssef Oudhini, and Fahmi Ghozzi

Abstract Recently, In terms of broadcasting and transferring the procedures related to this area of contemporary technology, as well as in terms of its fundamental processing and coding, digital television has taken a while to gain traction. It is well recognized that the challenges associated with fundamental digital processing methods and digital image storage are the primary cause of this observed delay. It has taken a while for digital television to gain traction, both in terms of basic processing and coding as well as in terms of broadcasting and transferring the procedures related to this area of contemporary technology. It is well recognized that the primary cause of this observed delay is the challenges associated with digital image storage and fundamental digital processing methods. To enable digital picture reception on handheld devices, a few options are suggested. Digital Multimedia Broadcasting DMB (Asia), Media Flo (USA), and Digital Video Broadcasting-Handheld DVB-H (Europe) are the three primary standards that have been authorized. They are using the same notional TV or DAB platform transport stream. In this work, we provide a novel way to integrate digital video into GSM interfaces without altering any of the standard's specifications. The novel suggested technique therefore makes it possible to accomplish this goal by utilizing GSM channels directly. The primary goal is to use the H265 advanced standard to reduce the bit rate coming from the video encoder by inserting a novel design known as a "Brewer-buffer". To create the RF channels, the outputs of this newly suggested element will be routed to the GSM norm's time division multiple access TDMA frames. Our receiver has to have the

D. Bouzidi (✉)
SM@RTS, Digital Research Center of Sfax, University of Sfax, Sfax, Tunisia
e-mail: dalendabouzidi91@gmail.com

Y. Oudhini
National School of Electronics and Telecommunications of Sfax, Sakiet Ezzit, Tunisia

F. Ghozzi
Laboratory of Signals, Systems, Artificial Intelligence and Networks (SM@RTS), National School of Electronics and Telecommunications of Sfax, Sakiet Ezzit, Tunisia
e-mail: Fahmi.ghozzi@enetcom.usf.tn

N. Derbel and Q. Zhu (eds.), *Advances in Communication, Signal and Image Processing*,
https://doi.org/10.1007/978-981-95-0401-5_7

symmetric component "Invert brewerbuffer" in order to restore the digital video. In fact, we will also suggest a new receiver architecture for this handheld video reception application. We illustrate the basic architecture and define the uniqueness of the suggested approach.

Keywords DVB-H · Media Flo · DMB · GSM · Brewer-buffer · Invert brewer-buffer

7.1 Introduction

The first use of video signal digitization was in the production industry, where it allowed for more intricate processing to be done without sacrificing signal quality (slow motion, special effects, overlay, etc.). Subsequently, studios began exchanging television shows over so-called contribution links using digital technology. Then, in order to send digital video signals via telecommunications arteries that offer bit rates of 34–140 Mbit/s, it became necessary to compress them to a gross bit rate of 216 Mbit/s. The source encoder decoder and the program component multiplexer are the two types of equipment that have been constructed. Once these processing operations are finished, the quality requirements are lower, making it possible to push the compression to bit rates between 4 and 8 Mbit/s. However, at the production level, it may be challenging to compress at bit rates lower than 34 Mbit/s due to the quality requirements of the processing operations carried out. The MPEG standardization group established this goal, and the standard of the same name was born out of their findings. Programs could then be sent to users in digital format. Therefore, transmitting the digital shows on already-existing channels constituted the second phase. Each broadcast media has its own set of so-called digital modulation techniques that have been researched and developed. It is evident that multiple programs can be carried on a single channel thanks to various modulation techniques. It seems required to add a program multiplexer to the transport segment that groups this possibility in order to take full advantage of it.

The production industry was the first to digitize video signals because it made it possible to perform more intricate processing without sacrificing signal quality (slow motion, special effects, overlay, etc.). Subsequently, studios began exchanging television shows over so-called contribution links using digital technology. The need then arose to compress digital video signals to a raw bit rate of 216 Mbit/s in order to transmit them over telecommunications arteries offering bit rates of 34–140 Mbit/s. Two types of equipment have been built: the program component multiplexer and the source decoder encoder. The quality requirements for the processing operations that are carried out make it difficult to compress at bit rates lower than 34 Mbit/s at the production level. However, after these processing operations are finished, the quality requirements are lowered, making it feasible to push the compression at bit rates between 4 and 8 Mbit/s. This was the objective of the MPEG standards group, whose findings led to the creation of the eponymous standard. After that, consumers could receive the programs in digital format. Therefore, airing the digital programming on already-existing channels was the second stage. Each broadcasting medium has its own set of so-called digital modulation techniques that have been researched and developed. It is evident that these modulation techniques enable the transmission of several programs per channel. It seems required to add a program multiplexer to the transport segment that groups this possibilities together in order to take full advantage of it. The process of digitizing an image involves assigning a specific number of points, each represented by multiple bits, to the television stream. A television standard deals with the parameters of television digitization. This standard defines the (SDTV) as 625 L (576 active lines) with 720 points per line in the European system at 50 frames per second. Another higher resolution standard is high definition television (HDTV), which has a 16:9 format and 1920 dots per line and 1152 lines. This paper's remaining sections are arranged as follows: We provide a quick comparison of various digital handheld receivers in Sect. 7.2. The suggested solution and the fundamental architecture are then presented in Sect. 7.3. We show some implementation results of the primary circuits that are suggested in this work in Sect. 7.4. The Quartus FPGA implementation is shown in Sect. 7.5. Lastly, we would draw a conclusion and offer some viewpoints.

7.2 Emerging Standards for Mobile Picture Reception

The three major families of digital mobile television will be compared in the lines that follow. No one is flawless; (ETSI 2015) lists both benefits and drawbacks. The most crucial point we make at the outset of this study is that all three systems employ the same transmitter and the frame of a nominal digital TV.

7.2.1 DVB-H: European Standard

The new digital broadcasting standard for sending broadcast content to handheld terminal devices is called "digital video broadcasting-Handheld". It is adapted for portable receivers (mobile stations) and is based on the DVBT standard for digital terrestrial television. Therefore, DVB-H uses several technologies for the transmission system in order to save battery power in the receiver and provides multimedia services via digital terrestrial broadcasting networks to handheld mobile stations. To lower the average power consumption and facilitate a smooth and flawless hand-over, the transmission system is regularly shut off (ITU 2010; Lee and Kwak 2005). The fundamental idea behind the creation of the DVB-H stream, its incorporation into the DVBT nominal frame, and its transmission by the same issuer as the nominal television are all shown in Fig. 7.1.

The DVB-T system's drawback is that in order to access the TV channel, the entire data stream must be decrypted. The short battery life of mobile gadgets is their

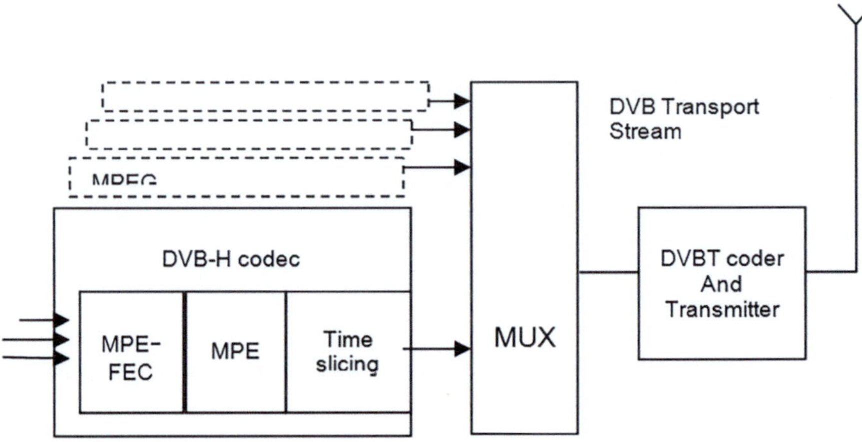

Fig. 7.1 Basic principle of the DVB-H frame

primary issue. Time slicing is used to lower the device's power usage. Additionally, it facilitates smooth service transition (ETSI 2016). Time division multiplexing (TDM) is the foundation of time slicing. The plan is to use a greater bit rate than standard streaming systems to transfer data in bursts. The receiver is informed of the following time (delta t) within the current burst. The receiver is idle and does not receive any data in between bursts. The memory of the device must be larger than the burst size. The burst bit-rate is the bit-rate used during transmission of a time sliced burst, and constant bit-rate is the average when time slicing is not used. This figure shows that as the burst bit-rate increases, the receiver's power savings also increase. The streaming of the digital flow of the DVB-H transmission placed into a nominal DVB-T frame is displayed in Fig. 7.2.

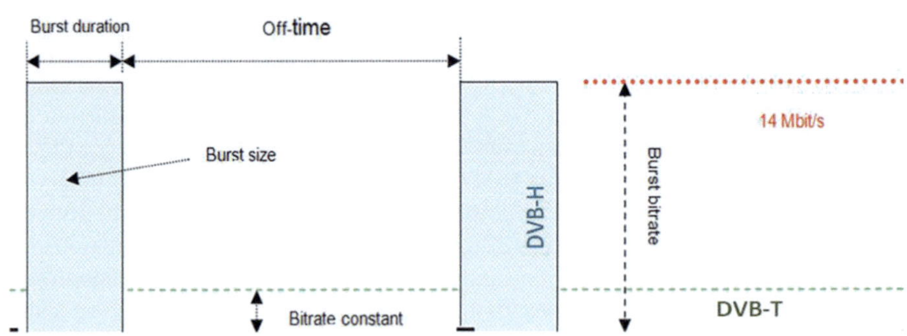

Fig. 7.2 Time slicing

7.2.2 Media Flo: American Standard

Qualcomm created the Media Flo technology, which allows data, video, and music to be transmitted to portable devices like cell phones and personal televisions (TVs).The FLO TV brand is the service that this technology offers in the United States (ETSI 2016). It is available in the site web (www.etri.re.kr). Table 7.1 shows Media Flo specifications.

The 3G telecommunications network is utilized to enable users' access to the network because Media Flo lacks uplinks (Up-Link). The following Fig. 7.3 illustrates the general architecture of the Media Flo system and the addition of 3G function in the nominal deployment.

Table 7.1 Specification of digital Flo

Modulation	OFDM-QPSK/16-QAM
Bandwidth	6 MHz
Channel coding	Turbo-Reed solomon
Maximum resolution	OVGA (240 * 320p)
Video frame rate	30 fps
Video coding	H264 (AVC/10)
Audio coding	HE-AAC
Channel switching time	1.5 s

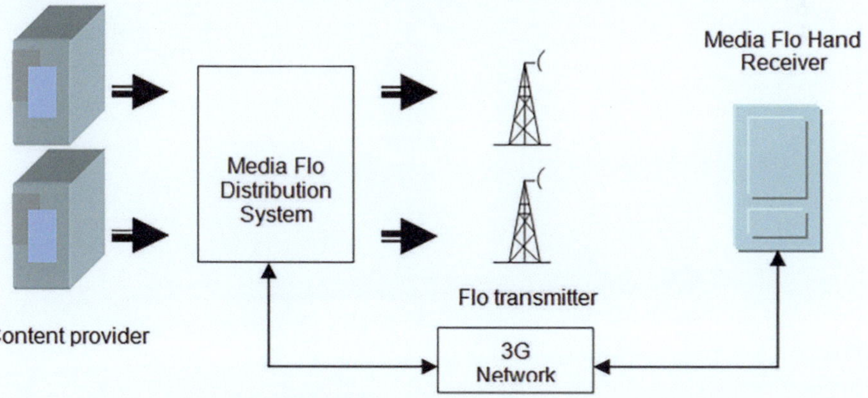

Fig. 7.3 Basic architecture of media Flo

7.2.3 T-DMB

Terrestrial Digital Multimedia Broadcasting is a digital technology for radio transmission that was created in South Korea as part of a national multimedia initiative. It allows for the transmission of sound and video as well as data to mobile devices like phones (Oudhini et al. 2014). This technology's primary benefit over DVBH is the quantity of free channels it offers. It was derived from DAB (Digital Audio Broadcasting). A DMB, often known as a pocket TV or TV-in-hand, is a mobile TV that offers data and audio services. The DMB service, as its name implies, is based on the DAB (Eureka-147), with the addition of video services components (AVC, BSAC, and MPEG4-BIFS) more recently (ITU 2010). We present the T-DMB system's design in a two-dimensional array (Fig. 7.4).

Following this concise overview of the three digital transmission techniques for the portable image effect, we compile all of the transmission parameters in each of the three scenarios into a comparison summary Table 7.2.

7.3 Proposed Solution

We will attempt to provide a novel way to send a video image over the GSM standard's radio interface in this section of the article. According to its definition, the GSM

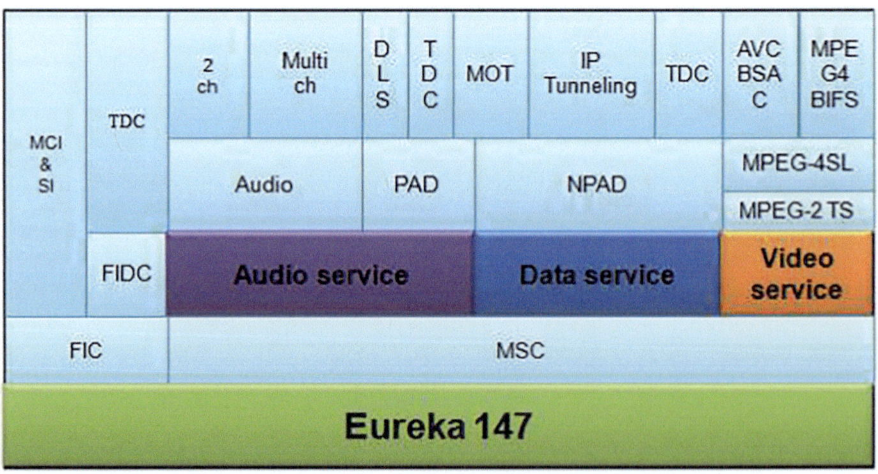

Fig. 7.4 Summary structure EUREKA 147 standard: frame DMB. *Legend* MCI: Multiplex configuration information; MOT: Multimedia object transfer; SI: Service Information; AVC: Advanced video coding; TDC: Transport Data Channel; BSAC: bit sliced arithmetic coding; FIDC: Fast information data channel; BIFS: Binary format for scene; FIC: Fast information channel; DLS: Dynamic label service; MSC: Main service channel NPAD: Non program associated data

Table 7.2 Comparative statement

	T-DMB	DVB-H	Media-FLO
Platform	Eureka147	DVB-T	N/A
Modulation	n/4DQPSK	QPSK	QPSK/16QAM
Transmission	OFDM	OFDM	OFDM(4K)
Error control	RS+ conv.code	RS+ conv.code	RS
Video compression	MPEG4AVC/H264	MPEG4AVC/H264	MPEG4AVC/H264
Audio compression	MPEG4-1	MPEG4AAC+	AAC+
Bandwidth	1.536 MHz	5-6-7-8 MHz	6 MHz
Frequency	174–216 MHz	470–838 MHz	

standard was initially utilized for telephone communication rather than video. In order to accomplish this, we must compress the video stream as much as possible using the standard MPEG, H264, and H265 formats and build the interfaces required to graft the video image into the physical channels and GSM logic. Therefore, it is important to present the specifications of this standard, which will host our suggested notion in some manner. This mobile communications standard will be covered in more details in the section that follows.

In this section of the paper, we will try to present an original solution to transmit the video image on the radio interface of the GSM standard. As defined, the GSM standard is used at the beginning to convey telephone and not video. To achieve this goal, we must make a maximum compression of the video stream using the standard MPEG (Hanhart et al. 2013), H264 (Gouyet and Mahieu 2009) and H265 (Mpeg 2012), and create the necessary interfaces to be able to graft the video image in the physical channels and GSM logic. So it appears necessary to introduce the specifications of this standard which will host in some ways our proposed idea. The following section will elaborate on this standard for mobile communications.

7.3.1 Brief Overview on the GSM Standard

A radio system needs a spectrum for function (Lagrange et al. 1997). The band dedicated to the GSM system and its extension DCS1800 is given Table 7.3.

The main constraint to be overcome in this architectural level, relation to cutting the allocated spectrum, i.e. time shot/frequency to result in channels that can properly support the transmission. Three techniques in this regard are retained: Sharing frequencies (FDMA: frequency division multiple access), Time sharing (TDMA: time division multiple Access), Sharing by codes (CDMA: Code Division Multiple Access). The method "multiple access" or more accurately by multiplexing involves assigning carrier frequencies and to divide the time axis into intervals (bursts). This distribution must provide the possibility for the radio channel (Abou Chakra 2004)

Table 7.3 Comparison chart between GSM and DCS

	GSM900	DSC1800
Band	890–915 (Up link)/	1710–1785 (Up link)/
MHz	935–960 (Down link)	1805–1880 (Down link)
2 × Simplex	2×25 MHz	2×75 MHz
Duplex gap	45 MHz	95 MHz

to support the transmission easily. The access method of the GSM standard is the TDMA: multiple access by time division which consists in subdividing the time slots regularly (slot). At the mobile terminal MS, the transmission and reception are shifted in time by a three slots, so the mobile receives and transmits at different times, to better manage this synchronization, the uplink TDMA frame is shifted also 3× TSLOTS. This feature makes it very simple duplex filtering in each mobile and its role will be reduced to reject the signal from another nearby station broadcasting for a mobile reception phase as illustrated in Fig. 7.5 (ITU 2010).

Figure 7.6 illustrates a general overview of our proposed solution which consists of the offering by an image coding block animated using standard codec H264-5.

This standard is an open standard known under license by the entire world and supports compression techniques most effective video currently exploited and reduces the size of bulky files without compromising image quality.The bit stream resulting from this compression will undergo a "fragmentation" which converts it into a sequence of 40 bits (vectors) may be grafted in the time intervals "slots" available. This TDMA frame device that performs fragmentation the linear flow of the digital image is called in our application: Tampon-brewer. Its modeling is performed by a VHDL description. To deploy our proposed solution, we need a set of six carrier frequencies in the downward direction BTS to MS including five reserved for 40 rounds of the digital image. The sixth frequency will be reserved for "his" accompanying the image and other subsequent applications. The following paragraph gives

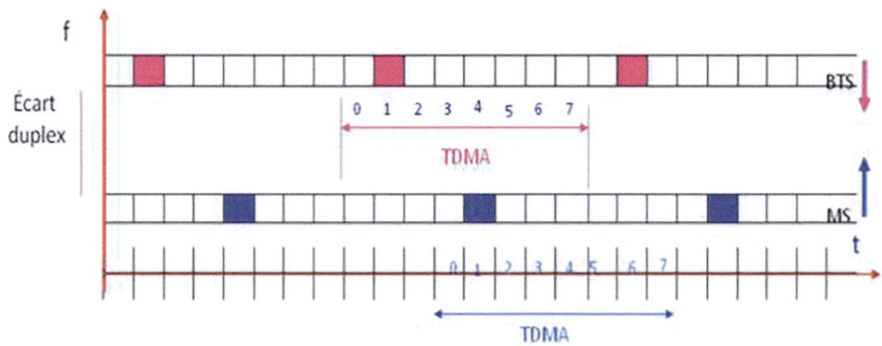

Fig. 7.5 Duplex physical channel

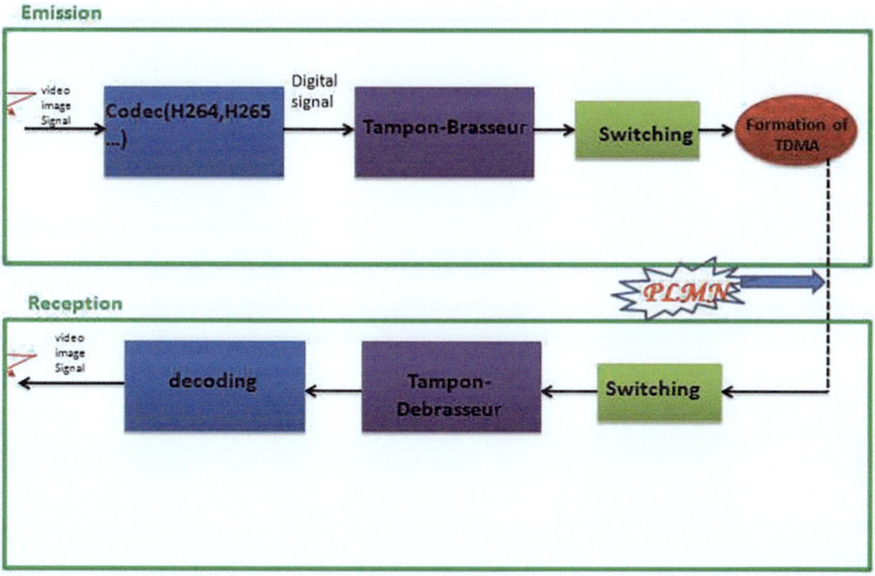

Fig. 7.6 General architecture of the proposed solution

basic explanation of this idea of fragmentation of the bit stream of the image and the formation of "bursts" to fill the TDMA frames in question.

7.3.2 Implementation

In order to comply with the GSM standard, the plan is to operate the "fragmentation" and interfere with the source stream H264: 13 kbit/s is equal to 260 bits in 20 milliseconds. A rate of 512 Kbit/s is the best option for transmitting a moving digital image in line with the compression ratio H26x: Examine the flow. 512 Kbit/s is as follows: 512 bits/ms = 512000 bits/s We need 20 ms in order to join the GSM standard:We get: It is evident that $512 \times 20 = 10240$ bits/20 ms exceeds the GSM 260 bit/20 ms limit (1). How should I proceed? We provide an alternative setup: After compression and the addition of service bits, our H264 stream of images will attack a specific circuit for this application (Brewer-Buffer) in Fig. 7.7 that fragments and distributes signal elements in 40 outings. This subdivision of the flow is organized in synchronism with the specific clock to insert each "element" in the TDMA slots.

We need 40 bursts (TDMA) to "establish" the 40 parts, so we need to find more spectrum to support the 40 relevant items. Going back to (1), $512 \times 20 = 10240$ pieces $10240/40 = 256$ bits/20 ms. The elementary flow is less than the value required by the GSM standard (260 bits/20 ms), so it is likely to be accepted by the GSM radio frequency modulation circuits. This new system should provide 40 input stream vectors

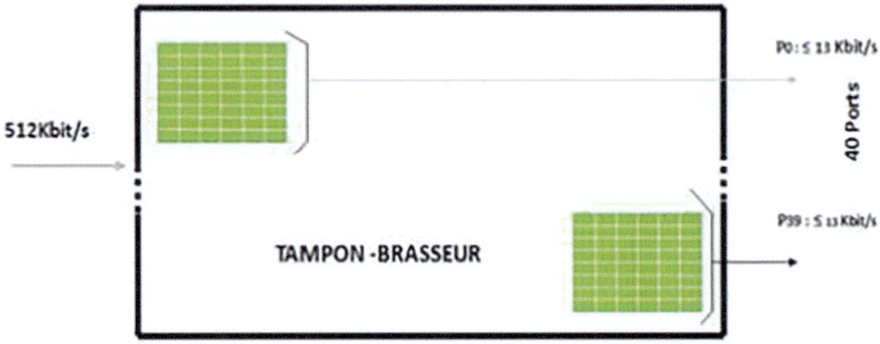

Fig. 7.7 Illustration of buffer brewer 1/40

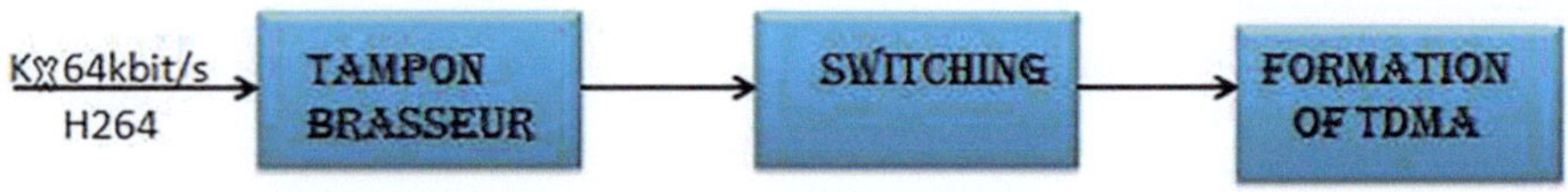

Fig. 7.8 Diagram of the proposed management

linearly H264 on its outputs. This specific distribution will cause the elementary flow rate to decrease and complies with the GSM standard, which permits 13 kHz in all frame time slots (TDMA). According to the concept shown in Fig. 7.8, the BTS (base station) is responsible for managing this function and allocating broadcast channels.

To insert these 40 "digital burst" through the GSM radio channel, we need in this case to five frequencies. But for more flexibility transmission take care six frequencies in the download direction; the 40 outputs of the video image will be transported by five frequencies, and the sixth frequency for transporting sound and signals. As shown in the sequential diagram (see below), we opted cyclically SFH (Slow Frequency Hopping) (Lagrange et al. 1997; Moreau 1995), this type of procedure does not need to run such an algorithm, but the process will be generated using a single synchronous clock ($N = 5$).

7.4 Hardware Description and Simulation

7.4.1 Brewer-Buffer

The way of the "Brewer-Buffer" block operates is shown in the accompanying chart (Fig. 7.9). The video image must be transmitted using five TDMA frames, with the sixth frame set aside for signaling and free slots. After examining this flowchart and verifying the previously suggested conditions (reset, clk, countslot, and counttrame) with:

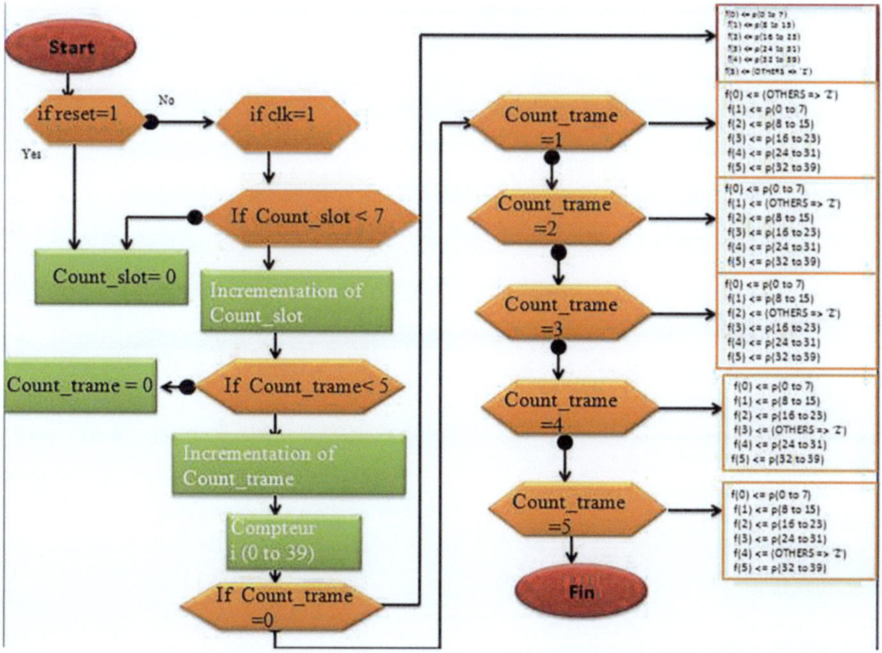

Fig. 7.9 "Brewer-buffer" chart

- count_slot is a slot counter which ranges from 0 to 7.
- count_trame is a trame counter 0–5. It happens after a counter i which varies from 0 to 39 to perform the test in each TDMA frame as six frequencies ($f(0)$, $f(1)$, $f(2)$, $f(3)$, $f(4)$, $f(5)$). Each TDMA frame, a single frequency we find that receives this character "others $\leq Z$" to say that the frame is reserved for signaling (Oudhini et al. 2015) and not to the transmission of a video image as illustrated by Fig. 7.10.

The simulation results of the "BrewerBuffer", which shows the state of several signals to each clock pulse (clk = 1), are displayed in Fig. 7.11.

We take into account the prerequisites that guarantee this interface's functionality, which we have previously mentioned inside the company: Six frequencies ($f_0 \ldots f_5$) are available; each frequency has a set of 8-bit slots or high impedance Z that are set aside for free slots and signaling. The 40-bit output component, which is linked to the six frequencies indicated, replicates the information injected into the input elements in time with the clock.

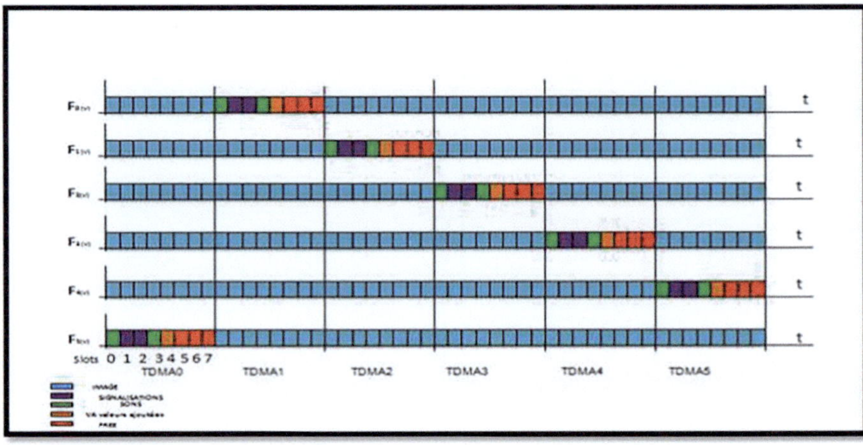

Fig. 7.10 Summary timing diagram

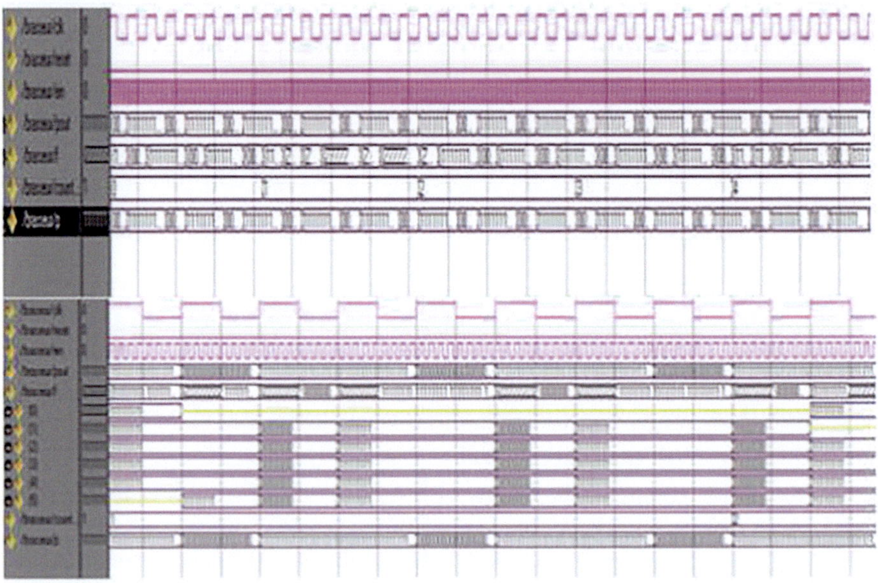

Fig. 7.11 Simulation of "Brewer-Buffer"

7.4.2 *Invert Brewer-Buffer*

Examining this "Invert brewer-buffer" chart, we will compare the same terms with the Invert Brewer buffer, but we add a counter i that meets the following requirements:

$$i <= count\_trame * 8 + count\_slot \;\; where\; i < 40 \tag{7.1}$$

Last but not least, this block's primary function is to "Invert BrewerBuffer" (pout less than or equal to ein) by increasing the frames and slots (Fig. 7.12). A simulation of the recapitulation "invert brewer-buffer" is shown in Fig. 7.13, explaining how the interface will function at each clock pulse (clk = 1), as we see, the entry ein (40-bit) was forced in order to have at the output of this debrasseur a serial bits during the course of 6 frames (0 … 5).

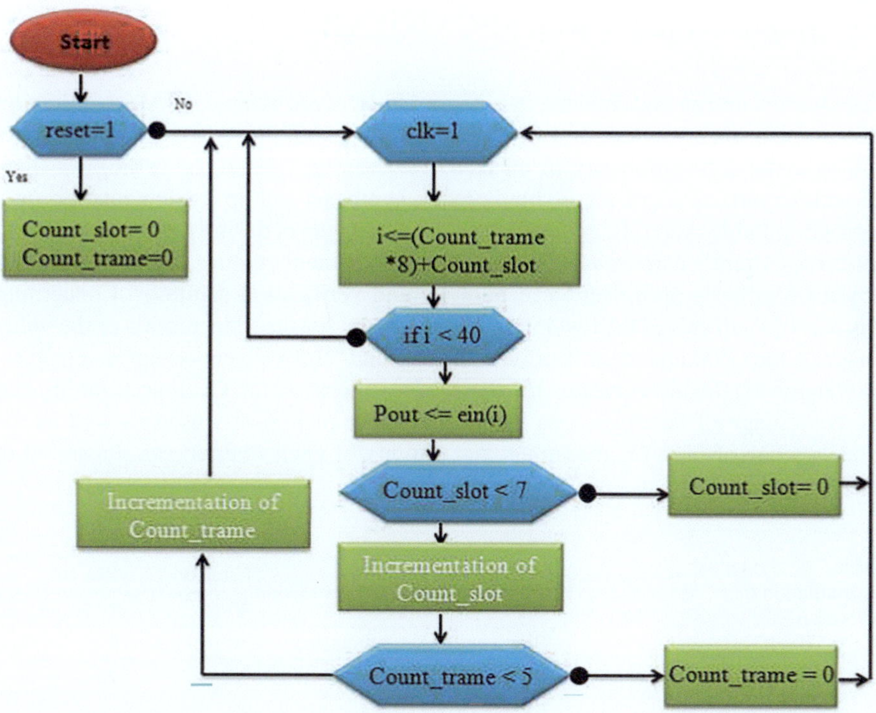

Fig. 7.12 "Invert Brewer-Buffer" chart

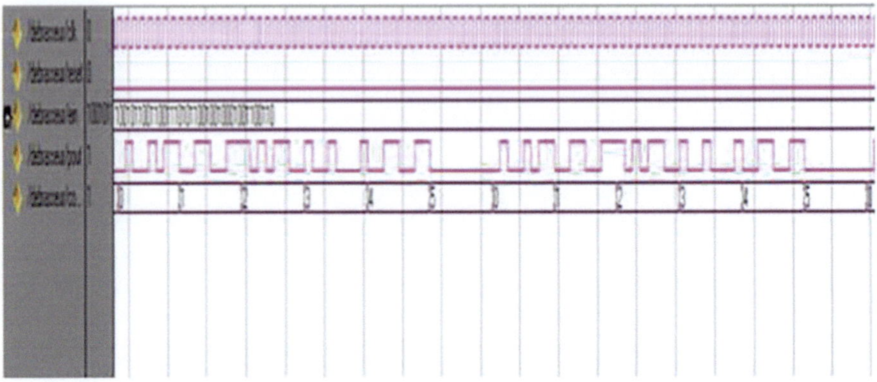

Fig. 7.13 Simulation results of "Invert brewer-buffer"

7.5 Implementation with FPGA: QUARTUS

The Quartus design and implementation tool is software offered by Altera, allowing the complete management of a CPLD or FPGA design flow. This software allows you to make a graphic entry or an HDL Description (VHDL or verilog) of digital architecture, to carry out a simulation, a synthesis and an implementation on a reprogrammable target. It includes a suite of system-level design features, providing access to Altera's extensive IP library and a placement-routing engine integrating Physics Synthesis optimization technology and verification solutions. Concerning this part of synthesis of the two interfaces necessary for the transmission of the video image on the GSM channels "Buffer-Brasseur" and "Buffer-Debrasseur" using Quartus version 10.0 which gives us the result of synthesis on the Cyclone II family and the percentage of occupation as the total number of registers used as well as the total number of pins for the inputs and outputs of each interface as illustrated in Tables 7.4 and 7.5.

Table 7.4 Synthetis recaputilation of "brewer-buffer"

	Number	Percent
Total logic elements	40	1
Total combinatory functions	0	0
Total logic registres	40	1
Total registres	40	
Virtual pins	0	
Total bit of memory	0	0
Multiplieurs	0	0

Table 7.5 Synthetis recaputilation of "brewer-buffer"

	Number	Percent
Total logic elements	37	1
Total combinatory functions	37	0
Total logic registres	7	1
Total registres	7	
Virtual pins	43	14
Total bit of memory	0	0
Multiplieurs	0	0

7.6 Conclusion

The three main mobile handheld systems are using the same transport stream of nominal TV platform. The complexity of modulation, the spectrum load and the limited power consumption in the receiver are real constrains. Using GSM logic and physical channels, we have suggested an alternative method for integrating digital motion pictures. Therefore, the current setup of the physical and logical channels of the mobile network, specifically the GSM standard, as well as the positive outcomes obtained in the field of image compression by injecting the two required blocks "BrewerBuffer" and "Invert Brewer-Buffer" for the transmission of the video image, will undoubtedly serve as inspiration for our suggested solution. We were interested in the study of the material design and the simulation of two proposed blocks based on the VHDL language as we synthesized these two interfaces using "Quartus version 10.0". Given the genuine opportunities presented by this basic concept, it would be worthwhile to consider a new arrangement of radio frequency channels specifically for GSM transmission. To enhance the number of channels and, thus, the number of subscribers per BTS, we can consider polarizing RF channels alternately between H and V. The nominal TV platform's transport stream is shared by the three main handheld mobile systems. The actual barriers to the implementation of the three standards are the receiver's energy consumption limit, spectrum size, and modulation complexity. In order to inject the digital image using the GSM's logical and physical channels, we have suggested an alternative method. By suggesting a novel RF channel allocation method, our solution offers an achievable perspective on the network saturation issue. After the switch-off of analog broadcasting on the UHF bands in the near future, this process considers the new scenario of releasing a somewhat easy frequency band. In this instance, we will be able to devote additional frequency space that is, radio frequency to the profile of our application, which is consuming physical material.

References

Abou Chakra, S.: The radio local loop and direct demodulation of broadband signals at 26 GHz, Doctoral Thesis (2004). pastel.archives-ouvertes.fr/pastel-00001988

ETSI.: Digital video broadcasting DVB, transmission system for handheld terminals DVB-H **15**(10), 300–744 (2015). http://www.etsi.org/standards-search

ETSI.: Radio-broadcasting systems-digital audio broadcasting DAB Portable and fixed receivers, **16**(10), 300–401 (2016)

Gouyet, J., Mahieu, F.: MPEG 4 advanced video coding H264: Institue National de l'Audiovisuel France (2009)

Hanhart, P., Rerabek, M., Korshunov, P., Ebrahimi, T.: AhG4: subjective evaluation of HEVC intra coding for still image compression. JCT-VC (2013)

ITU.R.: Radio data system for Automatic tuning and other applications in FM radio receivers for use with pilot-tone system, BS. 643–3 (2017)

ITU.R.: Methodology for the subjective assessment of the quality of television picture **06**(02), 11–500 (2010)

Lagrange, X., Godlewski, P., Tabbane, S.: Réseaux GSM-DCS. Edition Hermes, Paris (1997)

Lee, S., Kwak, D.: TV in your cell phone, the introduction of digital multimedia broadcasting in Korea. TPRC (2005)

Moreau, N.: Signal compression techniques. technique and scientific collection of telecommunications, CENT-ENST. Edition Masson, vol. xii, 274 p., Paris (1995)

Mpeg, L.: MPEG LA announces call for patents essential to high efficiency video coding (HEVC) (2012)

Oudhini, Y., Ghozzi, F., Sbita, L.: Dynamic telemetry network for remote photovoltaic stations, ICGE'2014, ENET'Com, Sfax, Tunisia. IEEE (2014). 978-1-4799-3602-1/14/.00

Oudhini, Y., Ghozzi, F., Sbita, L.: Collection and organization of telemetry data in an isolated photovoltaic field. Int. J. Sci. Res. Eng. Technol. (IJSET) **3**(2), 114–122 (2015). ISSN:2356-5608. Copyright IPCO'2015

Chapter 8
EEG Signals for Motor Imagery Left and Right Hand Movements Recognition

Aicha Reffad and Kamel Mebarkia

Abstract Since the past twenty years, a new type of interface has developed in a spectacular way: the Brain-Computer Interfaces (BCI). These interfaces enable users to control computers solely with their thoughts, generating various forms of brainwave activity. Indeed, BCIs are a new means of communication for people with total or partial paralysis. A key challenge in BCI system design involves automatically interpreting the user's brain activity patterns. Most BCIs are noninvasive systems that rely on the electroencephalogram (EEG) and use one of three mental tasks, among them motor imagery (MI). This chapter focuses on motor imagery-based BCI systems utilizing EEG signal processing. In conventional BCI systems, the detection and interpretation of brain signals typically depend on machine learning classifiers, which require carefully extracted EEG features for accurate operation. In this study, we specifically investigate motor imagery (MI) classification for left- and right-hand movements using EEG signals. Our novel approach introduces two key feature extraction methods. (1) Wavelet Transform (WT) for time-frequency analysis. (2) Mel-Frequency Cepstral Coefficients (MFCC) for spectral representation. These features serve as input to a Support Vector Machine (SVM) classifier, with a Genetic Algorithm (GA) optimizing feature selection. We rigorously evaluate the system's classification performance against state-of-the-art benchmarks, demonstrating significant improvements in MI recognition accuracy.

Keywords Electroencephalogram (EEG) · Brain computer interface (BCI) · Motor imagery (MI) · Features extraction · Classification algorithm · Genetic algorithm

---

A. Reffad
Electrotechnics Department, Faculty of Technology, Setif 1 University Setif, Setif, Algeria

K. Mebarkia (✉)
Electronics Department, Faculty of Technology, Setif 1 University Setif, Setif, Algeria
e-mail: kamel.mebarkia@rwth-aachen.de

N. Derbel and Q. Zhu (eds.), *Advances in Communication, Signal and Image Processing*,
https://doi.org/10.1007/978-981-95-0401-5_8

8.1 Introduction

Brain-Computer Interfaces (BCIs) are designed to create a direct communication pathway between neural activity and external devices, enhancing independence for individuals with disabilities (Yuan et al. 2018). These systems decode various brain signals-including cognitive processes, motor intentions, and emotional states-typically captured via electroencephalography (EEG). The translated signals then generate commands for assistive technologies, such as speech-generating software, motorized wheelchairs, or neuroprosthetic limbs (Lun et al. 2020). EEG records electrical potentials produced by the cerebral cortex. Its primary signals originate from the synchronized firing of large populations of pyramidal neurons in cortical layers.

EEG signals exhibit non-stationary characteristics with millisecond-scale temporal resolution (Khasnobish et al. 2010). These bio-signals typically demonstrate a voltage amplitude: from 0.5 to 100 μV and frequency spectrum from 0.5 100 Hz, categorized into five clinically significant bands (Teplan 2002). Depending on the brain states, several rhythms (waves) can be distinguished. The gamma rhythm (40 Hz) is associated with mental activity, perception, problem solving, and awareness. The daily activities and anxiety reflect beta waves (12–28 Hz). The alpha waves (from 8 13 Hz) are typical for state of relaxation, especially with closed eyes. Deep meditation, intense dreams, and intense emotions are associated with theta waves (4–7 Hz). The lowest frequency waves (from about 0.5 3 Hz) which represent the delta waves occur during a deep sleep. These waves are also common for young children and for certain types of brain damage (Teplan 2002). Furthermore, cortical activity exhibits distinct spatial organization, with specific brain functions localized to particular regions. This spatial distribution can be systematically mapped using standardized EEG electrode placement systems, most notably the international 10–20 system (Malmivuo and Plonsey 1995), which provides anatomical reference points for consistent electrode positioning and covers all major cerebral lobes.

BCI systems are categorized based on their signal acquisition approaches. Non-invasive BCIs acquisition measures electrical potentials with sensors placed on the scalp surface, called Electroencephalography (EEG) or detects magnetic fields, called Magnetoencephalography (MEG). In Semi-invasive BCIs acquisition, electrodes placed on the dura mater record cortical surface potentials with higher spatial resolution than EEG called Electrocorticography (ECoG). The invasive BCIs acquisition requires neurosurgery to implant micro-electrode arrays in cortical tissue. This approach provides single-neuron resolution with highest signal quality. Current human BCI research predominantly utilizes non-invasive approaches (Ofner et al. 2019), focusing on decoding user intent from EEG patterns while balancing safety and performance considerations. These approaches utilize three distinct neurophysiological phenomena: Steady-State Visual Evoked Potentials (SSVEP) based on frequency-locked cortical responses to visual stimuli (typically 5–30 Hz), P300 Event-Related Potential exploiting the endogenous component occurring 300 ms after the stimulus and Motor Imagery (MI) based on sensorimotor rhythm modulation during kinesthetic movement imagination (Wolpaw et al. 2002). MI is defined as:

a cognitive and dynamic capacity involving the cerebral representation of an action, without its actual motor execution. An imagined movement, in a healthy subject as well as a paralyzed subject, produces brain activity similar to the activity produced during a real movement. For the brain, mentally representing a movement is almost like realizing it. Decoding this brain activity presents significant technical challenges for EEG-driven BCI systems. The signals measuring the cerebral activity of patients during the imagination of a movement, or during a mental imagery task, can be used in the context of functional rehabilitation or for the control of external devices (prosthesis, wheelchair.) (Singh et al. 2021).

Recent advances in machine learning have demonstrated robust performance in MI task classification. Contemporary motor imagery classification systems employ a standardized processing framework consisting of two key stages: feature extraction to capture discriminative patterns in EEG signals and supervised classification of movement intent. Notable feature extraction methods include: time-frequency representations (Brodu et al. 2012), spatial filtering techniques (Lotte et al. 2007), nonlinear dynamics analysis (Zhong et al. 2008). These two-stages approach (feature and classifier) has become standard in MI-BCI systems since its validation by Zhang et al. (2013). Feature selection and extraction methodologies serve two critical functions in BCI systems: (1) eliminating irrelevant/redundant features, and (2) enhancing discriminative neurophysiological patterns. This preprocessing stage directly determines classification accuracy and consequently impacts overall system performance (Brodu et al. 2012).

The inherent non-linearity and non-stationarity of EEG signals continue to pose significant challenges for feature extraction (Chaudhari and Galiyawala 2017). Researchers have developed multiple approaches to address these difficulties: conventional feature extraction methods, spatial filtering approaches and wavelet-based techniques. Conventional feature extraction methods are based on spectral features like power spectral density (PSD), band power (BP) (Zhong et al. 2008; Brodu et al. 2012) or on parametric models like autoregressive (AR) coefficients (Pfurtscheller et al. 1998) and on time-frequency analysis like short-time Fourier transform (STFT) (Coyle et al. 2005). Spatial filtering approaches involve the common spatial pattern (CSP) algorithm which has emerged as particularly effective for MI-BCI systems (Blankertz et al. 2008). For binary classification tasks (e.g., left vs. right hand MI), CSP identifies optimal spatial filters that maximize variance for one class, minimize variance for the opposing class and operate in high-dimensional EEG space (Yu et al. 2019). Wavelet Transform (WT) has proven particularly suitable for EEG analysis due to its ability to handle non-stationary signal characteristics. Recent advances include :WT-LBP hybrid methods (Yuan et al. 2018), flexible analytic wavelet transform for limb MI discrimination (Chaudhary et al. 2020) and various applications across BCI research (Wang et al. 2013; Lotte et al. 2018; Al-Qerem et al. 2020). Chaudhary et al. (2020) introduced an innovative classification framework employing flexible analytic wavelet transform (FAWT) to discriminate between right-hand and right-foot motor imagery tasks in EEG signals. This approach specifically addressed the challenge of separating anatomically proximal limb representations in the sensor motor cortex.

The Mel frequency cepstral coefficients (MFCC) have been used for speech recognition applications and are known to provide a robust characterization of the temporal and spectral properties of the speech signal.

MFCCs have emerged as a powerful feature extraction technique for EEG classification, demonstrating particular efficacy in capturing non-linear signal characteristics (Golmohammadi et al. 2019). Building on this foundation, our study implements a approach combining two steps. Feature extraction step using WT and MFFC coefficients and classification step based on support vector machine (SVM) classifier (Lotte et al. 2007; Bashashati et al. 2015; Lotte et al. 2018) and genetic algorithm (GA) optimization (Brodu et al. 2011). SVM classifier is selected for its proven robustness in BCI applications and GA optimization is chosen for discriminative feature selection. Hence, this framework specifically targets the discrimination of left versus right hand MI tasks through EEG signal analysis.

The remaining of this chapter is structured as follows. In Sect. 8.2, we describe the BCI system and its elements. The SVM classifier is presented in Sect. 8.3. In Sect. 8.4, the proposed features that are derived from the WT and the MFCC are defined and described. Dataset and classification results are the subject of Sect. 8.5. Finally, a conclusion of this work is presented in Sect. 8.6.

8.2 BCI System

Modern brain-computer interface systems implement a sequential processing pipeline consisting of five critical stages: signal acquisition, signal preprocessing, feature extraction, classification and conversion of classified outputs to device commands (Golmohammadi et al. 2019; Lun et al. 2020).

The functional architecture of a typical BCI system is illustrated in Fig. 8.1, depicting the sequential signal processing stages from neural data acquisition to command execution.

Scalp-acquired EEG signals undergo specialized processing to decode the subject's intentional states through pattern recognition of neural activity markers. Thus, BCI constitutes an adaptive human-machine system that acquires neural activity via EEG to translate brain patterns into device commands and enables direct computer control for motor-impaired users (Lotte et al. 2007).

8.2.1 Signal Acquisition

Scalp-mounted electrodes detect electrical potentials (typically 5–100 μV) generated by cortical neural activity. These are the result of electrical exchanges between neurons. This method is inexpensive, non-invasive and provides good temporal resolution. Therefore, most research in BCI focuses on the EEG as a measure of brain activity. The measurement of cerebral activity by EEG leads to the acquisition of

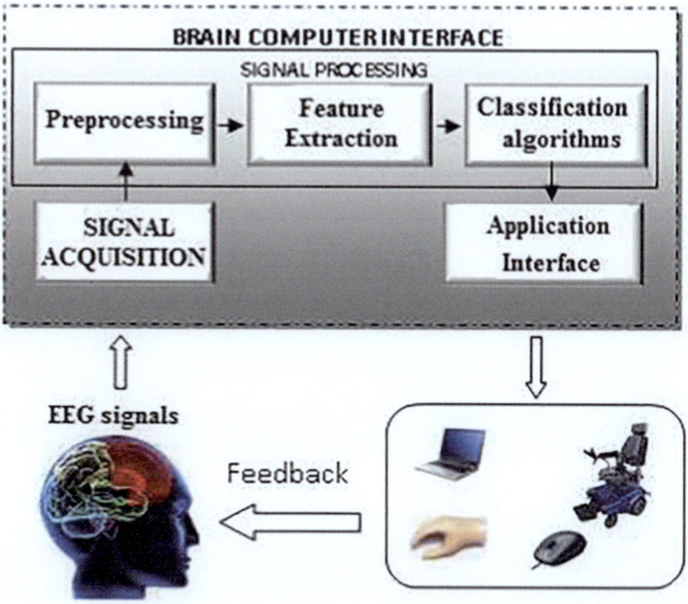

Fig. 8.1 Schematic representation of a brain-computer interface architecture

a considerable amount of data. Indeed, the signals are recorded with a number of electrodes ranging from 2 to 128 electrodes with a frequency, traditionally, varying between 100 and 1000 Hz (Lotte 2012). Electrode placement requires conductive gel to enhance signal acquisition by improving electrical contact between the scalp and sensors. Figure 8.1 illustrates the standardized 10–20 electrode placement system for EEG recordings, demonstrating optimal sensor positioning across cranial region (Fig. 8.2).

8.2.2 *Preprocessing*

The collected signals require preprocessing to eliminate noise and enhance the useful information they may contain. This is necessary because EEG signals are highly prone to interference, such as electrical noise or muscle activity from the eyes and face. To perform this preprocessing, different space-time filters are used as well as artifact filters. While spatio-temporal filters aim to reduce general noise, artifact filters focus on some specific type of noise, caused by well-identified sources. The mentioned filters make it possible to obtain a better signal to noise ratio (SNS). Consequently, with high signal quality , the BCI system will be able to identify the desired response more efficiently (Lotte 2012).

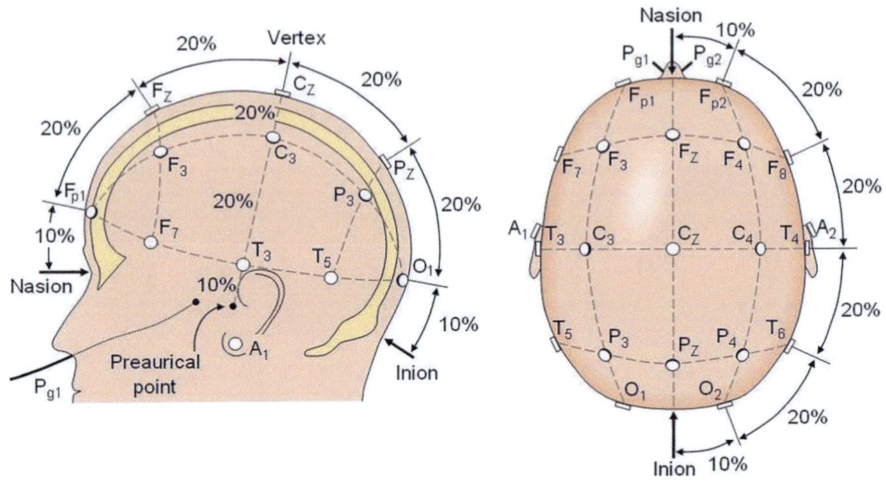

Fig. 8.2 The standardized 10–20 electrodes placement (Malmivuo and Plonsey 1995)

8.2.3 Feature Extraction

Due to the complexity of EEG signals, meaningful information cannot be extracted through visual inspection alone. Instead, signal processing techniques must be applied to identify and extract informative features. To optimize performance, these features should represent key signal properties while reducing the data dimensionality. They are aggregated in the form of a vector called the vector of features. There are many methods for features extraction, some of them are cited in the introduction. These feature extraction methods reduce the complexity of the signal by representing it by a set of features and thus facilitate the classification step. A high number of features can increase classifier latency, slowing down the classification process (Brodu et al. 2012). To address this issue, dimensionality reduction techniques such as principal component analysis (PCA) and independent component analysis (ICA) have been employed (Wang and Jung 2012). Additionally, feature selection through optimization provides an effective alternative to reduce feature dimensionality.

8.2.4 Classification

The goal of this step is to automatically classify the extracted feature vector (Duda et al. 2001), assigning it to a category that corresponds to the user's mental task in the brain-computer interface. Classification is performed using algorithms called classifiers. Classifiers are able to learn to identify the class of a feature vector using training sets. These sets are composed of features vectors labeled with their class membership. The training sessions make it possible to form a learning base for the classification of the different mental states. It is on this data that different data processing methods are tested and compared, as well as the different pairs of tasks to be separated. Classification helps to find out what type of mental task the subject is performing. Identifying good classifiers to associate the extracted features with the corresponding mental state is the objective of this phase. Thus, proper classifier selection is crucial for enhancing BCI performance. Various classifiers have been employed for mental task identification, including Linear Discriminant Analysis (LDA), Multilayer Perceptron (MLP), k-Nearest Neighbors (KNN), Hidden Markov Models (HMM), Support Vector Machines (SVM) with different kernel functions, and Fuzzy classifiers (Bashashati et al. 2015). Research has demonstrated that classifier ensembles typically outperform single-classifier approaches (Mebarkia and Reffad 2019).

8.2.5 *Translation Into an Order*

Once the mental state is identified, a corresponding command is assigned to enable user control of applications like text editors, robotic systems, or prostheses (Mattar et al. 2017). It is important to note here that in a classic BCI, the EEG signals are analyzed over time windows of lengths close to one second. A good quality BCI can allow an information rate (number of binary symbols sent per minute thanks to the BCI) of the order of 20 bits/min (this measurement takes into account the probability of recognition error of mental control) (Wolpaw et al. 2002).

8.3 SVM Classifier

The Support Vector Machine is a supervised learning method widely used for classification and regression tasks in data analysis and pattern recognition. Due to its robustness, it has become a prominent algorithm for EEG-based pattern recognition (Lotte et al. 2018). The core principle of SVM involves finding an optimal hyperplane that maximizes the margin between two classes (see Fig. 8.3). In a 2D feature space, this hyperplane is positioned to ensure the greatest possible separation between class-representing feature vectors. Maximizing this margin enhances the model's generalization capability and improves classification accuracy (Burges 1998). The support vectors, which are the closest data points to the hyperplane, ultimately define its position and orientation. These vectors are critical because they determine the decision boundary between classes.

When the data is nonlinear, then nonlinear SVM is used in which the original input space is mapped to higher-dimensional feature space by the use of kernel functions on training sets. Kernel functions transform the input feature space into a higher-dimensional space, enabling linear separation of non-linearly separable data (Mebarkia and Reffad 2019). The most commonly used kernel functions include radial basis function (RBF), polynomial and sigmoid.

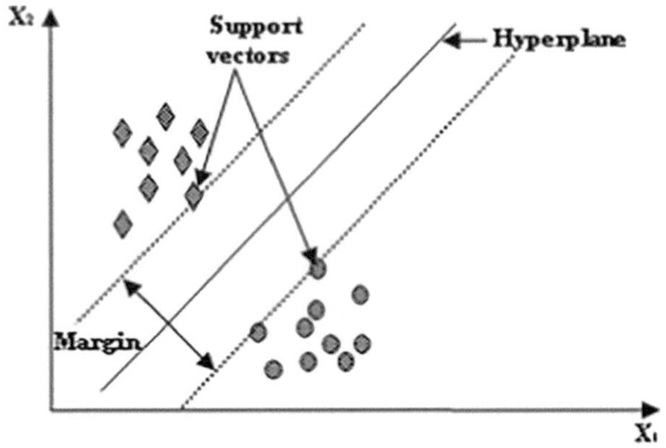

Fig. 8.3 In SVM classification, the algorithm constructs an optimal hyperplane that maximizes the separation margin between two classes (represented by "x" and "O" data points). The support vectors (critical training samples lying on the boundary hyperplanes of each class) define this margin and determine the decision boundary (Burges 1998)

8.4 Proposed Features

Figure 8.4 presents the block diagram for classifying left- versus right-hand MI tasks based on EEG signal features. The system processes signals SC3 and SC4 from channels C3 and C4, extracting relevant features that are then fed into a pre-trained SVM classifier (using labeled training data) to predict the intended hand movement. This classification output enables the BCI system to interface with electronic actuators and execute the desired actions.

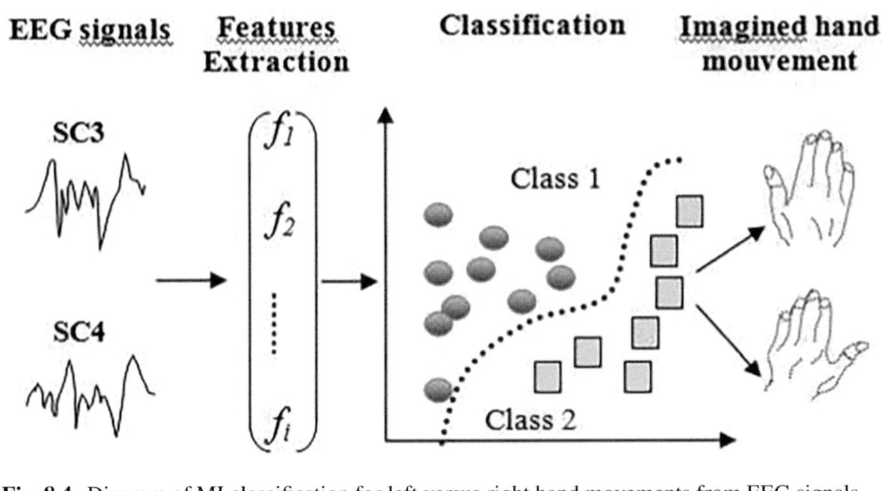

Fig. 8.4 Diagram of MI classification for left versus right hand movements from EEG signals

The feature extraction stage represents the most critical operation in any classification problem. Particular attention must be given to feature selection, as the chosen features must be highly discriminative between classes and must contain maximal relevant information. In pattern recognition systems, reducing the feature space dimensionality is essential to accelerate the classification process, to enhance generalization capability and to improve computational efficiency.The primary goal of feature selection is to identify an optimal subset from the original feature space that maximizes inter-class separability while minimizing redundant or irrelevant information. Effective features should provide clear discrimination between different mental states.

8.4.1 Features Based on Wavelet Transform

Due to the non-stationary nature of EEG signals, effective feature extraction must account for both temporal and spectral characteristics. The wavelet transform emerges as a powerful solution, offering distinct advantages over traditional Fourier-based methods. Indeed, the wavelet transform projects signals onto wavelet basis functions that provide localized frequency information. Unlike Short-time Fourier transform (STFT) 's fixed resolution, wavelet transform offers adaptive time-frequency resolution: high frequency resolution at low frequencies and high time resolution at high frequencies (Kehtarnavaz 2008). The wavelet transform's multi-resolution analysis capability makes it particularly suitable for EEG signal processing, where transient features and non-stationary components require careful characterization (Kehtarnavaz 2008). Furthermore, wavelet transform combines precise localization with minimal spectral leakage and Enables clear separation of signal components (Tary et al. 2014). The wavelet transform can be implemented through two primary approaches.

8.4.1.1 Continuous Wavelet Transform (CWT)

Continuous Wavelet Transform uses a wavelet function $\Psi(t)$ and produces a scalogram, similar to a spectrogram for time-frequency analysis. The (CWT) of a signal x(t) is given by:

$$C_x(a, b) = \frac{1}{a} \int_{-\infty}^{+\infty} x(t) \Psi \left(\frac{t-b}{a} \right) dt, \tag{8.1}$$

where a is a positive scale parameter and b is a parameter for the position of the mother wavelet Ψ in time. The main point of the mother wavelet is to provide a source function to generate the daughter wavelets which are simply the translated and scaled versions of the mother wavelet. The proposed features that based on the CWT use two statistical indicators, the mean and the variance as expressed in the following equations:

$$f_i = var((cw_3(i, :)) - var((cw_4(i, :)) \qquad i = 1...24 \tag{8.2}$$

$$f_{i+24} = \frac{var((cw_3(i,:))}{var((cw_4(i,:))} \qquad i = 1...24 \tag{8.3}$$

$$f_{i+48} = \frac{mean((cw_3(i,:))}{var((cw_4(i,:))} \qquad i = 1...24 \tag{8.4}$$

where cw_3 and cw_4 are continuous wavelet transform of signals SC3 and SC4 respectively using the mother wavelet 'Morlet'. The 'Morlet' mother wavelet is chosen in this study according to the results found in our previous work (Mebarkia and Reffad 2019). Hence, 72 features are obtained from continuous wavelet transform.

8.4.1.2 Discrete Wavelet Transform (DWT)

The Discrete Wavelet Transform provides a powerful multi-resolution analysis framework that decomposes signals into hierarchical frequency sub-bands through iterative filtering and down sampling (Subasi 2007). The transformation process involves Low-pass filter ($L[n]$) that generates approximation coefficients (A_1) representing signal trends and High-pass filter ($H[n]$) that produces detail coefficients (D_1) capturing high-frequency components. Each level applies dyadic down sampling (factor of 2), creating critically sampled sub-bands (Fig. 8.5). Hence, features obtained with approximation coefficients provided by the successive decomposition reaching 12 levels are given in Eqs. (8.5) and (8.6).

$$f_{i+96} = \sum_i A_i^2 - \sum_i B_i^2 \qquad i = 1...12 \tag{8.5}$$

$$f_{i+108} = \frac{\sum_i A_i^2}{\sum_i B_i^2} \qquad i = 1...12 \tag{8.6}$$

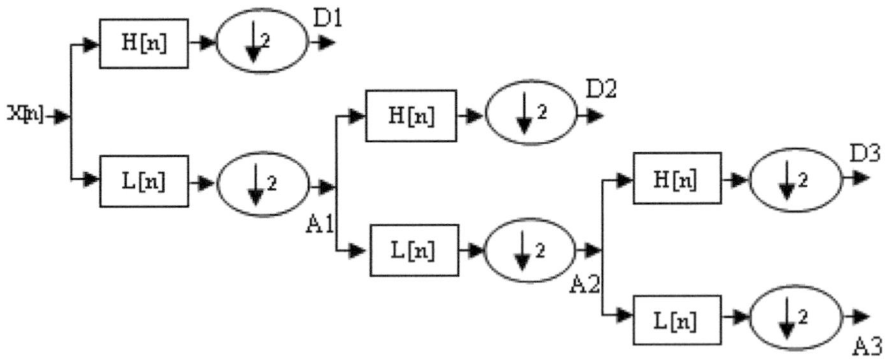

Fig. 8.5 DWT sub-band decomposition process example

where A_i and B_i are approximation coefficients of the decomposed signals in the DWT ith levels for signals SC3 and SC4 respectively. In summary, 24 features are obtained from subtractions and ratios between energies of approximation signals resulting from DWT of SC3 and SC4 in successive decomposition.

8.4.2 Features Based on MFCC

The MFCCs, originally developed for speech recognition (Shanthi and Lingam 2013), were adapted for EEG analysis in this work due to their proven discriminative power. The MFCC extraction process (Fig. 8.6) implements different stages. Prior to signal framing, a high-pass filter is applied (pre-emphasis) to spectrally flatten the signal by boosting high-frequency components.

The pre-emphasis filter is implemented using the first-order difference equation:

$$x(n) = s(n) - as(n-1), \tag{8.7}$$

where

- $x(n)$ represents the output sample i.e. the filtered signal.
- $s(n)$ is the numerical EEG signal.
- a is a constant equals 0.001.

Following high-frequency emphasis, the signals from each segment are partitioned into non-overlapping 0.75 s frames $x_i(n)$. These frames then undergo a windowing process using the specified window function defined as:

$$sh_i(n) = x_i(n)\,w(n), \tag{8.8}$$

where w is the specific window obtained after several trials to get better classification accuracy and is given by:

$$w(n) = 5\,gausswin(M, 0.25) + 3\,kaiser(M), 0 \leq n \leq M-1 \tag{8.9}$$

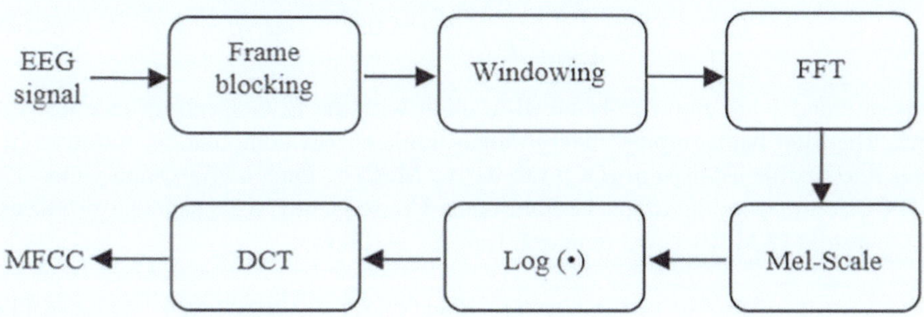

Fig. 8.6 Block diagram of MFCC

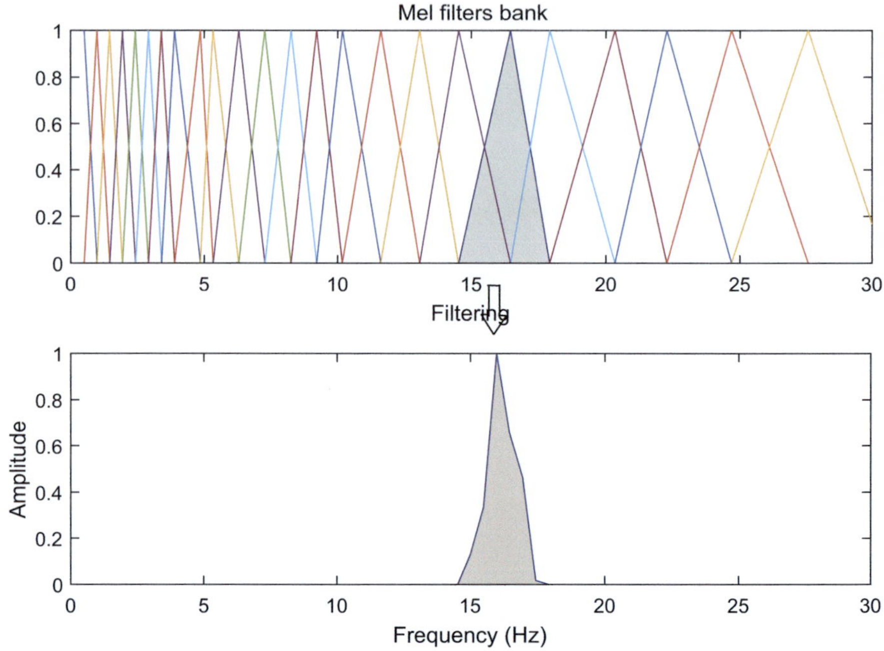

Fig. 8.7 Mel filters bank (up) and an example of a normalized filtered frame in frequency domain in Hz (bottom)

where:

- x_i, sh_i are the ith frame and the ith windowed frame respectively.
- M represents the length of the frame (750 ms) as a number of samples.

For every windowed frame, the spectral representation is obtained through Fourier transformation (Priyadarshani et al. 2012). This spectrum is then processed using a Mel-scale triangular filter bank with equidistant spacing, as illustrated in Fig. 8.7.

Frequencies in Hertz domain and Mel domain are related by the following equation:

$$f_{mel} = a \log_{10}\left(1 + \frac{f}{b}\right), \quad a = 0.23 \quad and \quad b = 3 \tag{8.10}$$

where a and b are empirically tuned to align with the EEG spectrum characteristics. The filter bank outputs undergo logarithmic energy computation, followed by Discrete Cosine Transform (DCT) to derive MFCCs. Each 0.75 s frame yields 12 MFCCs corresponding to the 12-filter bank. Consequently, a 2 s segment produces 24 features (12 MFCCs × 2 temporal frames) as follows:

$$f_{mel(i+72)} = Cmfcc_3(i) - Cmfcc_4(i) \qquad i = 1...24 \tag{8.11}$$

where $Cmfcc_3$ and $Cmfcc_4$ represent the cumulative sum over the two segments of their 24 MFCC rearranged two by two MFCC (the first two MFCC, the second two MFCC, and so on...) of the SC3 and SC4 EEG signals respectively. Finally, a features vector of 120 features was defined as a subtractions or ratios between features from the two channels so the spatial information could be considered.

8.5 Experiment Results

This study evaluates classification accuracy using K-fold cross-validation, a robust method to prevent overfitting by testing on independent folds, to utilize full dataset for both training and validation and to provide variance estimates of model performance. K-fold cross-validation splits data into K equal parts. Then we iteratively train the algorithm on K-1 folds while using the remaining fold as a test set. In this study, we evaluated classification performance using 10-fold cross-validation, where the dataset was first equally divided into 10 subsets (folds). For each iteration, one-fold served as the test set while the remaining nine were used for training, ensuring every data point was tested exactly once. The final reported accuracy represents the mean performance across all 10 test folds, providing a robust estimate of model generalization while mitigating overfitting through this complete rotation of training and test partitions.

8.5.1 EEG Data Set Description

The EEG dataset utilized in this study was obtained from the BCI Lab at the Institute for Human-Computer Interfaces, Graz University of Technology. Specifically, Dataset IIIb includes two-class motor imagery EEG recordings from three participants (originally labeled as O3, S_4, and X11). The data was collected across multiple sessions, with each subject completing three sessions consisting of 4–9 runs per session (Blankertz et al. 2004). After acquisition, the runs were merged and stored in GDF format. The recordings were acquired using a g.tec bipolar EEG amplifier, with signals sampled 125 Hz and bandpass-filtered between 0.5 and 30 Hz, including a Notch filter for noise suppression. During the experiment, visual cues were presented in random order to guide participants in performing either left- or right-hand motor imagery tasks. Electrode placements (C3 and C4) were used to capture signals from the SC3 and SC4 regions, as illustrated in Fig. 8.8. Participants were instructed to initiate imagery movements upon hearing an auditory cue (beep), while real-time feedback was provided via a monitor to help them modulate their brain activity. Each trial corresponded to a labeled EEG signal segment indicating left or right imagery (see Table 8.1). Despite synchronization attempts using auditory cues, EEG signals exhibited temporal variability across trials and subjects. Additionally, the inherent non-stationarity and nonlinearity of EEG data posed challenges for accurate task

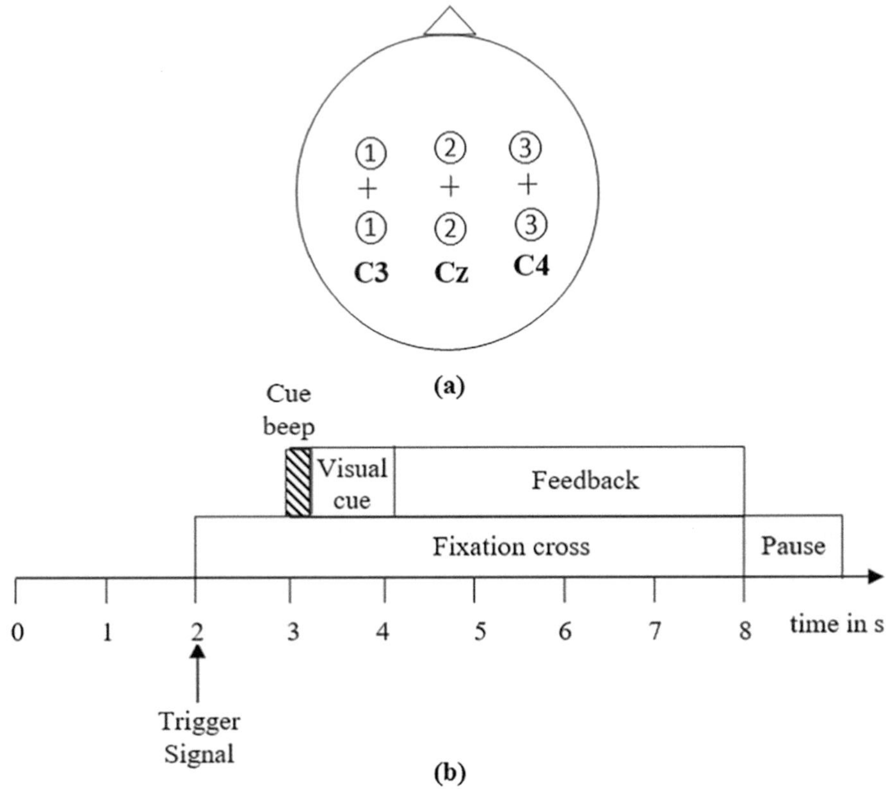

Fig. 8.8 **a** Electrodes position. **b** Timing signal parts (Blankertz et al. 2004)

Table 8.1 Dataset IIIb description (Blankertz et al. 2004)

Subjects	Feedback presentation (s)	Channels	Number of trials	Classes
$O3$	4–8	$C3$, $C4$	640	Left–right
S_4	4–7	$C3$, $C4$	1080	Left–right
$X11$	4–7	$C3$, $C4$	1080	Left–right

classification. For clarity, the subjects O3, S_4, and X11 are hereafter referred to as s_1, s_2, and s_3, respectively.

8.5.2 Classification Evaluation

Identifying specific brain activities from EEG signals is highly challenging due to their inherent complexity, which includes both inter-subject (between individuals) and intra-subject (within the same individual) variability. A key issue is the temporal

Table 8.2 SVM classification accuracies (CA) in percent for all subjects

Time positions in seconds	s_1	s_2	s_3
3.00	**81.23**	72.72	72.96
3.25	75.89	77.17	80.74
3.50	76.46	79.43	**82.22**
3.75	74.46	79.84	78.88
4.00	76.56	**81.46**	79.07
4.25	79.62	78.76	79.44
4.50	75.30	79.73	79.44

inconsistency in task execution–subjects rarely perform actions at precisely the same time points. To assess how timing variations affect classification performance, we evaluated all extracted features using a linear SVM classifier with 10-fold cross-validation. The features were computed over 2 s segments, aligned to different time positions across all subjects. The resulting classification accuracy (CA) is presented in Table 8.2.

The bolded values in the results highlight the highest classification accuracy (CA) achieved by optimizing the time alignment for each subject. These findings demonstrate that peak performance occurs at distinct time points for different individuals, reflecting inherent variability in neural response patterns. To further enhance CA, a Genetic Algorithm (GA) was employed to fine-tune the SVM classifier for each subject. Inspired by biological evolution, GA is a robust metaheuristic for solving complex optimization problems. In this study, GA was applied to feature selection, identifying the most discriminative subset of features while minimizing misclassification errors.

The optimization process is illustrated in Figs. 8.9, 8.10 and 8.11, where GA explores the best feature combinations from a pool of 120 binary-encoded candidates. The results reveal peak CAs of: 95.31% for Subject s_1, 88.51% for Subject s_2 and 86.26% for Subject s_3. This confirms that subject-specific optimization significantly improves BCI performance.

The numbers of selected features are 27, 38 and 40 features for subjects s_1, s_2 and s_3 respectively. The common features between the three subjects are f_4, f_8, f_{34} and f_{92}. This four features among 120 features indicate the diversity between subjects and how it is difficult to find features that satisfy all subjects. In fact, the information about the right and left hand movement is represented differently by the different subjects. Consequently, researchers should take the inter-subjects variability into account in order to adapt features to a specific subject.

Let's now compare our results with the existing results obtained in previous works using the same database "BCI competition IIIb". Table 8.3 outlines the classification accuracies of different works to identify left and right hand movement based on MI- EEG signals. Previous studies in this domain have employed various feature

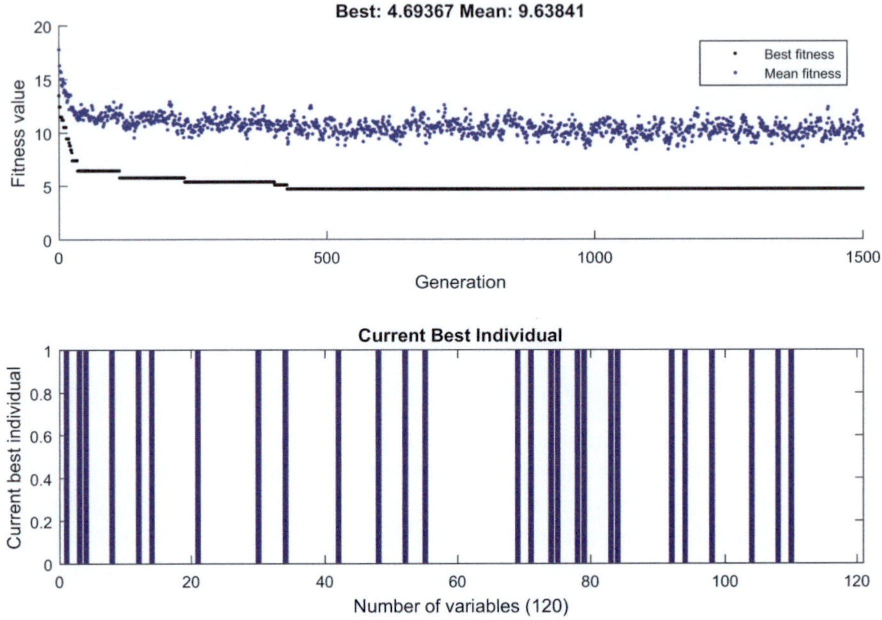

Fig. 8.9 GA optimization using 10-fold cross validation evaluation for all subjects. **a** Misclassification accuracy and selected features for subject s_1

extraction techniques and classification algorithms. The reported results represent the highest mean classification accuracy achieved across the three subjects (s_1, s_2, and s_3). As shown in Tab. 8.3, our optimized feature selection approach from the 120 candidate features significantly enhanced performance, yielding a mean classification accuracy of 90.02%. The superior performance of our method is highlighted in bold within the comparative results.

8.6 Conclusion

In this chapter, the BCI system and its components were explained. Each element from the BCI system is very important especially the features extraction and the classifier modules. Features are a representation of signals that carry the pertinent information.

In this study, the left and right hand movements in MI were identified using SVM classifier fed by selected features. This study highlights the challenges in mental task classification stemming from the non-stationary nature of EEG signals and subject variability in biofeedback synchronization. To address this, we propose a 120-dimensional feature vector combining WT coefficients (capturing time-frequency patterns) and MFCCs (encoding spectral characteristics). The extracted features are

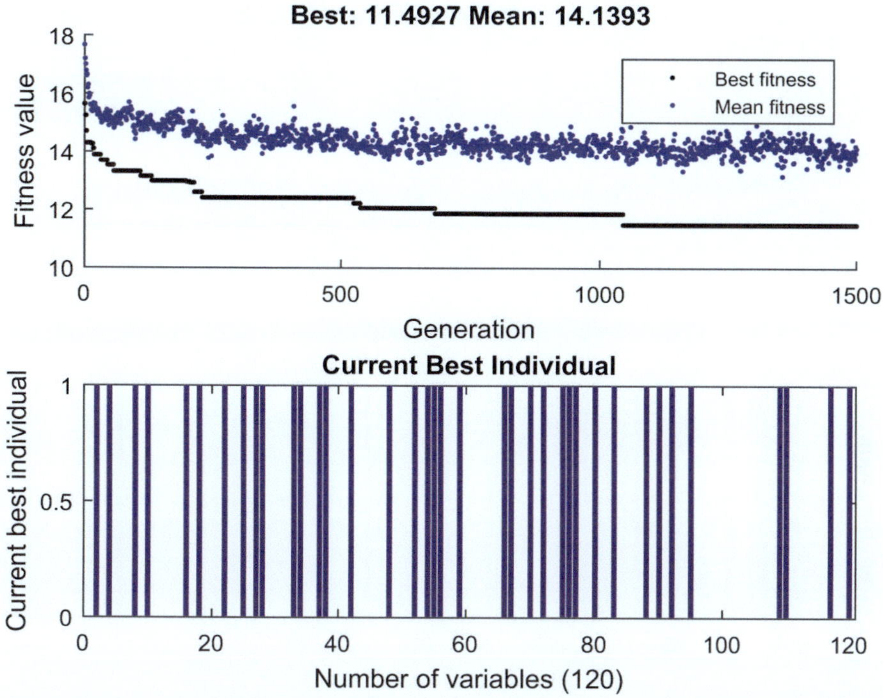

Fig. 8.10 GA optimization using 10-fold cross validation evaluation for all subjects. **b** Misclassification accuracy and selected features for subject s_2

classified using a linear SVM classifier. Without selecting features, the classification accuracy might be weak due to the confusion that might be introduced by features.

To enhance classification accuracy, we implemented a subject-specific optimization of the linear SVM classifier. This optimization utilized a genetic algorithm (GA) to select discriminative features from the 120-D feature space. The GA employed a fitness function combining classification accuracy and feature subset compactness, ensuring robust performance across subjects. The selected features for all subjects found by the GA were distinguished i.e., the common features between them were very few. Researchers should be aware about this result and take intra-subjects diversity more carefully.

Usually, the MFCC are used in the applications related to the speech signal. While wavelet-based features constituted the primary feature set, our analysis revealed that Mel-Frequency Cepstral Coefficients provided complementary discriminative information, yielding a statistically significant improvement in classification accuracy across subjects. The features extraction phase requires more validation and study to improve the accuracy of the classifiers. In addition, it is considered that the combination of the feature vector is essential for a proper classification, thus newer features

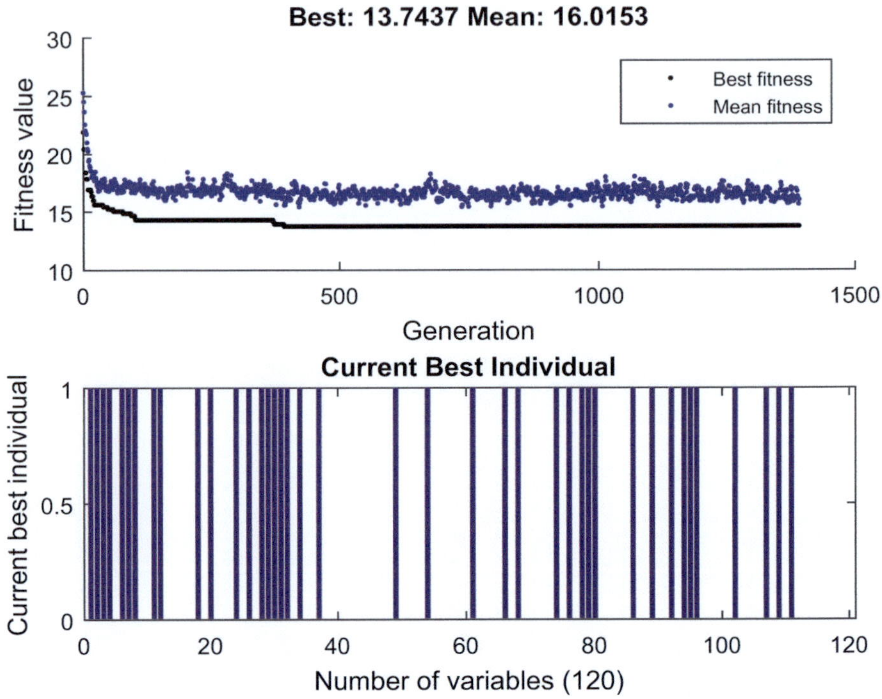

Fig. 8.11 GA optimization using 10-fold cross validation evaluation for all subjects. **c** Misclassification accuracy and selected features for subject s_3

Table 8.3 Summery of average classification accuracies (ACA) in percent for the three subjects in the previous works and present work using the same data base

Works	Features	Classifiers	s_1	s_2	s_3	ACA
Lotte et al. (2007)	Band power	SVM	86.80	75.90	75.40	79.36
Zhong et al. (2008)	Band power	VB	89.30	72.96	75.18	78.96
Brodu et al. (2011)	Wavelet	LDA	–	81.50	80.40	80.95
Brodu et al. (2012)	BP, multifractal	LDA	–	81.70	80.90	81.30
Bashashati et al. (2015)	Wavelet	LR	82.39	83.89	78.15	81.47
Mebarkia and Reffad (2019)	16 diversified	SVM	91.00	89.57	88.58	89.80
This work	MFCC, wavelett	SVM	95.31	88.51	86.26	**90.02**

need to be tested and selected to further improve the classification of MI left-right hand movements with much less computational cost.

References

Al-Qerem, A., Kharbat, F., Nashwan, S., Ashraf, S., Blaou, K.: General model for best feature extraction of EEG using discrete wavelet transform wavelet family and differential evolution. Int. J. Distrib. Sens. Netw. **16**(3), 1–21 (2020)

Bashashati, H., Ward, R.K., Birch, G.E., Bashashati, A.: Comparing different classifiers in sensory motor brain computer interfaces. PLoS ONE **10**(6), e0129435 (2015)

Blankertz, B., Muller, K.-R., Curio, G., Vaughan, T.M., Schalk, G., Wolpaw, J.R., Schlogl, A., Neuper, C., Pfurtscheller, G., Hinterberger, T., et al.: The BCI competition 2003: progress and perspectives in detection and discrimination of EEG single trials. IEEE Trans. Biomed. Eng. **51**(6), 1044–1051 (2004)

Blankertz, B., Tomioka, R., Lemm, S., Kawanabe, M., Muller, K.-R.: Optimizing spatial filters for robust EEG single-trial analysis. IEEE Signal Process. Mag. **25**(1), 41–56 (2008)

Brodu, N., Lotte, F., Lécuyer, A.: Comparative study of band-power extraction techniques for motor imagery classification. In: Proceedings of IEEE Symposium on Computational Intelligence, Cognitive, Algorithms Mind Brain, pp. 1–6 (2011)

Brodu, N., Lotte, F., Lécuyer, A.: Exploring two novel features for EEG-based brain-computer interfaces: multifractal cumulants and predictive complexity. Neurocomputing **79**, 87–94 (2012)

Burges, C.J.: A tutorial on support vector machines for pattern recognition. Data Min. Knowl. Discov. **2**(2), 121–167 (1998)

Chaudhari, R., Galiyawala, H.J.: A review on motor imagery signal classification for BCI. Int. J. Signal Process. **11**(2), 16–34 (2017)

Chaudhary, S., Taran, S., Bajaj, V., Siuly, S.: A flexible analytic wavelet transform based approach for motor-imagery tasks classification in BCI applications. Comput. Methods Programs Biomed. **187**, 105325 (2020)

Coyle, D., Prasad, G., McGinnity, T.M.: A time-frequency approach to feature extraction for a brain-computer interface with a comparative analysis of performance measures. EURASIP J. Adv. Signal Process. **2005**(19), 1–11 (2005)

Duda, R.O., Hart, P.E., Stork, D.G.: Pattern Recognition. Wiley-Interscience (2001)

Golmohammadi, M., Harati Nejad Torbati, A.H., Lopez de Diego, S., Obeid, I., Picone, J.: Automatic analysis of EEGs using big data and hybrid deep learning architectures. Front. Hum. Neurosci **13**, 76 (2019)

Kehtarnavaz, N.: Digital Signal Processing System Design: LabVIEW-Based Hybrid Programming. Elsevier (2008)

Khasnobish, A., Bhattacharyya, S., Konar, A., Tibarewala, D.: K-nearest neighbor classification of left-right limb movement using EEG data. In: Oral Presentation in International Conference on Biomedical Engineering and Assistive Technologies, NIT Jalandhar (2010)

Lotte, F.: Les interfaces cerveau-ordinateur: conception et utilisation en réalité virtuelle. Revue des Sciences et Technologies de l'Information-Série TSI: Technique et Science Informatiques **31**(3), 289–310 (2012)

Lotte, F., Anatole, L., Fabrice, L., Bruno, A.: Studying the use of fuzzy inference systems for motor imagery classification. IEEE Trans. Neural Syst. Rehabil. Eng. **15**(2), 322–324 (2007)

Lotte, F., Bougrain, L., Cichocki, A., Clerc, M., Congedo, M., Rakotomamonjy, A., Yger, F.: A review of classification algorithms for EEG-based brain-computer interfaces: a 10 year update. J. Neural Eng. **15**(3), 031005 (2018)

Lun, X., Yu, Z., Chen, T., Wang, F., Hou, Y.: A simplified CNN classification method for MI-EEG via the electrode pairs signals. Front. Hum. Neurosci. **14**, 338 (2020)

Malmivuo, J., Plonsey, R.: Bioelectromagnetism: Principles and Applications of Bioelectric and Biomagnetic Fields. Oxford University Press, USA (1995)
Mattar, E.A., Al-Junaid, H.J., Al-Seddiqi, H.H.: Biomimetic based EEG learning for robotics complex grasping and dexterous manipulation. Biomimetic Prosthet. (2017)
Mebarkia, K., Reffad, A.: Multi optimized SVM classifiers for motor imagery left and right hand movement identification. Australas. Phys. Eng. Sci. Med. **42**(4), 949–958 (2019)
Ofner, P., Schwarz, A., Pereira, J., Wyss, D., Wildburger, R., Müller-Putz, G.R.: Attempted arm and hand movements can be decoded from low-frequency EEG from persons with spinal cord injury. Sci. Rep. **9**(1), 1–15 (2019)
Pfurtscheller, G., Neuper, C., Schlogl, A., Lugger, K.: Separability of EEG signals recorded during right and left motor imagery using adaptive autoregressive parameters. IEEE Trans. Rehabil. Eng. **6**(3), 316–325 (1998)
Priyadarshani, P., Dias, N., Punchihewa, A.: Dynamic time warping based speech recognition for isolated Sinhala words. In: IEEE 55th International (MWSCAS), pp. 892–895 (2012)
Shanthi, T.S., Lingam, C.: Review of feature extraction techniques in automatic speech recognition. Int. J. Sci. Eng. Technol. **2**(6), 479–484 (2013)
Singh, A., Hussain, A.A., Lal, S., Guesgen, H.W.: A comprehensive review on critical issues and possible solutions of motor imagery based electroencephalography brain-computer interface. Sensors **21**(6), 2173 (2021)
Subasi, A.: EEG signal classification using wavelet feature extraction and a mixture of expert model. Expert Syst. Appl. **32**(4), 1084–1093 (2007)
Tary, J.B., Herrera, R.H., Han, J., van der Baan, M.: Spectral estimation-what is new? what is next? Rev. Geophys. **52**(4), 723–749 (2014)
Teplan, M.: Fundamentals of EEG measurement. Meas. Sci. Rev. **2**(2), 1–11 (2002)
Wang, Y., Jung, T.P.: Improving brain-computer interfaces using independent component analysis. In: Towards Practical Brain-Computer Interfaces, pp. 67–83. Springer, Berlin, Heidelberg (2012)
Wang, Y., Veluvolu, K.C., Lee, M.: Time-frequency analysis of band-limited EEG with BMFLC and Kalman filter for BCI applications. J. Neuroeng. Rehabil. **10**(1), 1–16 (2013)
Wolpaw, J.R., Birbaumer, N., McFarland, D.J., Pfurtscheller, G., Vaughan, T.M.: Brain-computer interfaces for communication and control. Clin. Neurophysiol. **113**(6), 767–791 (2002)
Yu, H., Lu, H., Wang, S., Xia, K., Jiang, Y., Qian, P.: A general common spatial patterns for EEG analysis with applications to vigilance detection. IEEE Access **7**, 111102–111114 (2019)
Yuan, Q., Zhou, W., Xu, F., Leng, Y., Wei, D.: Epileptic EEG identification via LBP operators on wavelet coefficients. Int. J. Neural Syst. **28**(08), 1850010 (2018)
Zhang, R., Xu, P., Guo, L., Zhang, Y., Li, P., Yao, D.: Z-score linear discriminant analysis for EEG based brain-computer interfaces. PLoS ONE **8**(9), 1–7 (2013)
Zhong, M., Lotte, F., Girolami, M., Lécuyer, A.: Classifying EEG for brain computer interfaces using gaussian processes. Pattern Recogn. Lett. **29**(3), 354–359 (2008)

Chapter 9
Robust Hybrid Watermarking Algorithm Based on DCT-PMF-DWT-SVD

Razika Souadek

Abstract This work presents a hybrid algorithm based on DCT-PMF-DWT-SVD. First, the Discrete Cosine Transform (DCT) is applied, followed by the use of the Pixel Movement Function (PMF) to mix the frequencies. Then, three levels of Discrete Wavelet Transform (DWT) are performed before embedding the watermark using the Singular Value Decomposition (SVD). The proposed approach provides excellent robustness against several types of attacks. The use of the Pixel Movement Function enhances security, as the position of the bit selected for watermark embedding remains unpredictable. The system also achieves high imperceptibility. All steps of the algorithm are described in detail, and the experimental results demonstrate its effectiveness. To evaluate performance, the PSNR (Peak Signal-to-Noise Ratio) and NC (Normalized Correlation) metrics were used.

Keywords Discrete Cosine Transform (DCT) · Discrete Wavelet Transform (DWT) · Singular Value Decomposition (SVD) · Pixel Movement Function (PMF)

9.1 Introduction

The rapid development of technology has enabled real-time communication and media exchange over the internet. In response, the research community has increasingly focused on digital watermarking for information authentication. This technique consists of embedding imperceptible and robust information within an original image to ensure its protection (Tirkel et al. 1993; Van Schyndel et al. 1994). The opposite process permits to extract this information from the watermarked image (Hussein 2010; Hernández et al. 1998). The metrics used to evaluate the algorithm effectiveness are: robustness, invisibility and ability. Digital image watermarking is divided into

---

R. Souadek (✉)
The Environmental Engineering and Technologies Laboratory, Mechanical and Electromechanical Engineering Department, Institute of Sciences and Technology, University Center Mila, Mila, Algeria
e-mail: r.souadek@centre-univ-mila.dz

N. Derbel and Q. Zhu (eds.), *Advances in Communication, Signal and Image Processing*,
https://doi.org/10.1007/978-981-95-0401-5_9

two categories: spatial domain methods and frequency domain methods, depending on how the watermark is embedded.

In the spatial domain, the watermark is hidden in the Least Significant Bits (LSB) of the cover image. This method maintains high visual quality but remains fragile against various attacks (Reddy et al. 2009).

The frequency domain are used various reversible transformations like the Discrete Cosine Transform (DCT) (Lin et al. 2010), Discrete Wavelet Transform (DWT) (Dawei et al. 2004), and Discrete Fourier Transform (DFT) (Qi and Qi 2007). The transformation coefficients are used to embed the watermark, followed by applying the inverse transform to reconstruct the watermarked image.

The hybrid system is known to contain a set of transforms in one algorithm. Its aim is obtain a good response of the metrics in the watermarking algorithm. Among them we mention the following algorithms: DCT-DWT (Al-Haj 2007), DWT-SVD (Yin 2007) and DCT-SVD (Ganic and Eskicioglu 2004).

Image watermarking is a critical area of study which has engendered researchers to submit a lot of significant research to enhance the security and copyright authors. A brief overview of some watermarking algorithms is presented below:

Souadek and Boukezzoula (2020) proposed a watermarking method combining the Discrete Wavelet Transform (DWT), Transfer Function (TF), and Pixel Movement Function (PMF). Its aim is to modify the pixel positions within the blocks of the original image. All these techniques and transformations enhance the system's robustness, imperceptibility, and embedding capacity.

Hu and Wiang (2021) presented a new watermarking algorithm based on lower-order Zernike moments for hiding the watermark. This technique aims to limit the distortions introduced by watermarking. The compensation technique is used to compensate for the information loss caused by quantization error, watermark error, and rounding error. The resulting image achieves high quality and is resistant to geometric distortions.

The algorithm proposed in this chapter uses two transforms, DCT and DWT, along with two techniques, PMF and SVD. On top of used the hybrid algorithm, the blend of the transforms allowed us to develop our image watermarking process. Where we used mathematical techniques to increase the security and efficiency. The combine between them achieve a good robustness against various attacks. We use in first step the DCT transform and apply a PMF function to blend the frequencies. After that, we apply three levels of DWT transform then embed the watermark done by SVD technique.

9.2 Transform Domain

9.2.1 Discrete Cosine Transform

The DCT transforms the signal from the spatial domain to the frequency domain by converting pixel values (Lin et al. 2010). The DCT (discrete cosine transform) is a derivative of Fourier transforms. Indeed, the DCT transforms an $N \times N$ matrix into another $N \times N$ matrix, whose amplitudes are arranged according to their frequencies: the low frequencies are located at the top left and the high frequencies at the bottom right. Low frequencies are the most important for watermarking image, while high frequencies are less important. The value of coordinate (0, 0) corresponds to the average value of the elements of the matrix. The intensity of the 2D DCT (Two Dimensional) is shown as follows: (Lu et al. 2006; Jiang et al. 2002).

$$F(u, v) = \frac{2}{N} c(u)c(v) \sum_{x=0}^{N-1} \sum_{y=0}^{N-1} f(x, y) \cos\left[\frac{\pi}{N} u \left(x + \frac{1}{2}\right)\right] \cos\left[\frac{\pi}{N} v \left(y + \frac{1}{2}\right)\right] \tag{9.1}$$

$$c = \left| \begin{matrix} \frac{1}{\sqrt{2}} & \text{if } (u = 0)\ (v = 0) \\ 1 & \text{otherwise} \end{matrix} \right. \tag{9.2}$$

$$F(0, 0) = \frac{1}{8} \sum_{x=0}^{N-1} \sum_{y=0}^{N-1} f(x, y) \tag{9.3}$$

with: u, v, x and y are changed between 0 to $N - 1$ (N: set of natural numbers corresponds to the image size).

In the third formula show, the coefficient $F(0, 0)$ of the DCT transform is measured by the average value of the 8×8 pixels of one block of the image and is used for the extraction of color data (Bajaj and Lay 2007). The rest pixels are show the frequency feature of the image where are called AC (Poursistania et al. 2011). To return to the image, we use the following formula:

$$f(x, y) = \frac{2}{N} \sum_{u=0}^{N-1} \sum_{v=0}^{N-1} c(u)c(v)F(u, v) \cos\left[\frac{\pi}{N} u \left(x + \frac{1}{2}\right)\right] \cos\left[\frac{\pi}{N} v \left(y + \frac{1}{2}\right)\right] \tag{9.4}$$

9.2.2 Discrete Wavelet Transform

This Transform is compacted a high energy and used in various domains well known. The DWT discrete the frequency into four bands. The LL sub-band represents a low-resolution version of the original image, obtained through low-pass filtering, and

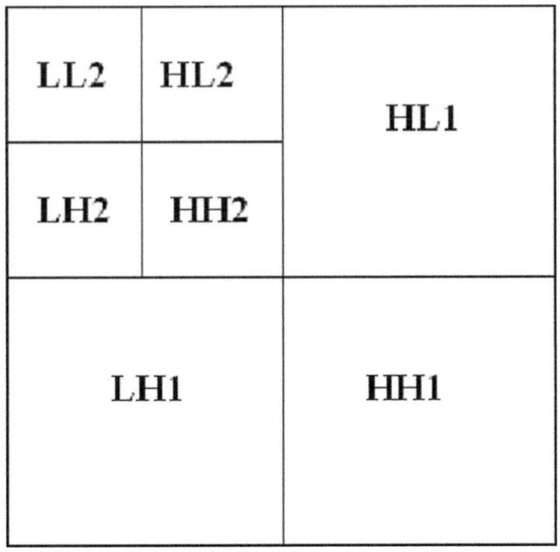

Fig. 9.1 Two levels of Discrete Wavelet Transform

retains most of the important information. The HH component contains the highest frequency details along the diagonal. The HL and LH sub-bands result from the combination of low-pass and high-pass filtering along the vertical and horizontal directions, respectively, where image details are captured. Finally, the LL sub-band can be further subdivided into four additional sub-bands (Fig. 9.1) (Ali et al. 2013).

In the following, we will discuss the transform mechanism used in this work mathematically and practically. A simple DWT transform is used named Haar wavelet (or Daubechies db1 wavelet).

It is the simplest of the wavelets: defined on the interval [0, 1] (or sometimes on $[-\frac{1}{2}, \frac{1}{2}]$). where the function is constant in the same part (Eq. 9.5 and Fig. 9.2). This wavelet is very simple and therefore easy to implement algorithmically (Ali et al. 2013).

$$H(x) = \left| \begin{array}{ll} 1 \text{ if } & x \in [0, \frac{1}{2}[\\ -1 \text{ if } & x \in]0, \frac{1}{2}, 1] \end{array} \right. \tag{9.5}$$

9.3 Mathematical Tools

9.3.1 SVD Technical

Singular Value Decomposition (SVD) was originally developed for real square matrices and later extended to complex matrices. Lately, SVD has been applied in various image processing fields (Liu et al. 2009).

Consider a random matrix A of dimension m_timesn of rank r (where r represents the number of non-zero singular values). There are two orthogonal matrices U and V of sizes m_timesm and m_timesn, respectively, and a "pseudo-diagonal" matrix

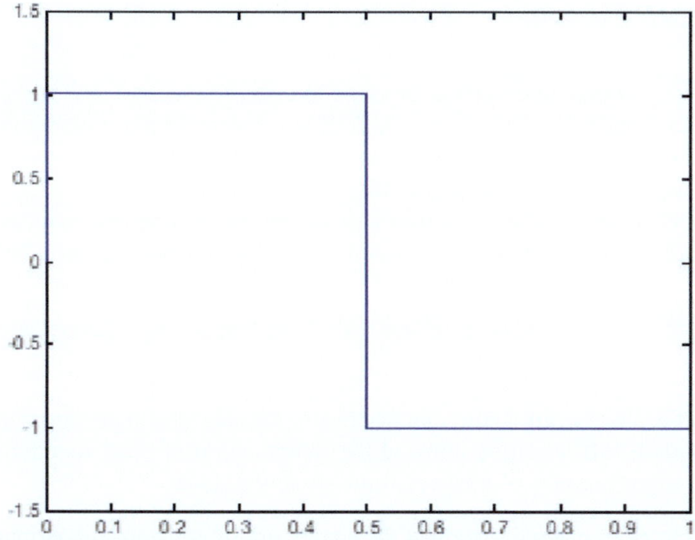

Fig. 9.2 The Haar wavelet

S of size m_timesn, such that:

$$A = USV^T \tag{9.6}$$

with:

$$U\ U^T = I(m)$$
$$V\ V^T = I(n)$$

and:

$$S(m,n) = \begin{bmatrix} S_1 & 0 & \cdots & 0 \\ 0 & S_2 & \ddots & \vdots \\ \vdots & \ddots & \ddots & 0 \\ 0 & \cdots & 0 & S_n \end{bmatrix} \tag{9.7}$$

Numbers $S_1, S_2 \cdots S_n$ denote the singular values of A, with $S_1 > S_2 > \cdots > S_n$.

The aims of the using SVD method in image processing is the singular values represent the energy of the image. It have a very good stability. In addition, the factorization in SVD is unique.

9.3.2 PMF Function

The pixels movement function was invented to increase security, as it was employed in the article (Souadek and Boukezzoula 2020) to change the places and arrangement of the pixels, its gave good results to a large extent. However, in this work we used it on the image. This is after applying the DCT transform.

Mathematically, the PMF function is realizing by shifting one pixel in the odd columns and odd lines for each iteration N (Souadek and Boukezzoula 2020):

$$y_{i+1}(i, j) = \overline{y}_i(i-1, j) \tag{9.8}$$

$$x_{i+1}(i, j) = \overline{x}_i(i, j-1) \tag{9.9}$$

where i and j: change into 1 to n, and where n represents the dimension of the image.

The applied PMF function allowed the modification of pixel locations. In this research, various benefits of this technique are highlighted:

- The bit location selected to embed the watermark bit remains unpredictable.
- Applying the PMF function after the DCT allows for arbitrary selection of bits to hide the watermark across different frequency components.

9.4 Proposed Algorithm

In this approach, we propose a new image watermarking scheme based on a hybrid system. The Discrete Cosine Transform (DCT) including a PMF pixel displacement function that requires a secret key N to determine the iteration counts. Discrete wavelet transform and SVD technical were also used in this algorithm. The aim of this method is to change the site of the pixels by PMF. Then applying the DCT to the original image to ensure safety, so that unauthorized people cannot extract the watermark. Moreover, the addition of the DWT and the SVD makes it possible to evaluate the quality of the image after the insertion steps. All these tools permitted to adopt the best way to strengthen the algorithm against various attacks. Experiments were carried out to demonstrate that the proposed method brings an improvement on the quality of the watermark. It done by: the imperceptibility of the watermark, capacity of insertion and robustness against the various attacks. The algorithm proposed in this paper is presented and evaluated with the $PSNR$ (Eq. 9.10) and NC (Eq. 9.11) parameters. These metrics help assess the method in terms of robustness and invisibility. Like as JPEG compression, the noising addition and geometric attacks.

The proposed watermarking algorithm will present. The watermarking algorithm evaluated by using the following metrics $PSNR$ (Eq. 9.10) and NC (Eq. 9.11). These parameters help us to the evaluation process in term robustness against the attacks and invisibility of watermark.

$$PSNR = 10 \log \frac{I_{max}^2}{\frac{1}{m \times n} \sum_i \sum_j [I(i,j) - I_w(i,j)]^2} \tag{9.10}$$

$$NC(w, w') = \frac{\sum_i \sum_j w(i,j) w'(i,j)}{\sqrt{\sum_i \sum_j w(i,j)^2} \sqrt{\sum_i \sum_j w'(i,j)^2}} \tag{9.11}$$

where:
I and I_w present the host image and the watermarked images respectively.
w and w' present the watermark image and the extracted watermark respectively.

9.4.1 Embedded Process

The embedding algorithm summarized in the following steps (Fig. 9.3):

- Select the cover image I of size 512–512 pixels and the watermark image of size 64 to 64 pixels.
- Apply two dimensions of Discrete Cosine Transform (2D DCT) to the image I.
- the site pixels by Pixel Movement Function (PMF).
- Apply levels of Discrete Wavelet Transform and choose the HH_3 sub-band.
- Apply the Singular Value Decomposition (SVD) to the HH_3 sub-band to obtain the components (U_1, S_1, V_1) and to the watermark image w to obtain the components (U_w, S_w, V_w).
- Insert the component S_w into the component S_1 with a key k.

$$S_E = S_1 + k\, S_w \tag{9.12}$$

- Apply the inverse Singular Value Decomposition then inverse Discrete Wavelet Transform.
- Apply the inverse Pixel Movement Function and inverse Discrete Cosine Transform to obtain a watermarked image I_w.

9.4.2 Extracted Process

The extraction process are highlighting the steps (Fig. 9.4):

- Select the original image I and the watermarked image I_w, each with dimensions 512×512 pixels.
- Apply the Discrete Cosine Transform on both images.
- Change the position pixels by Pixel Movement Function.

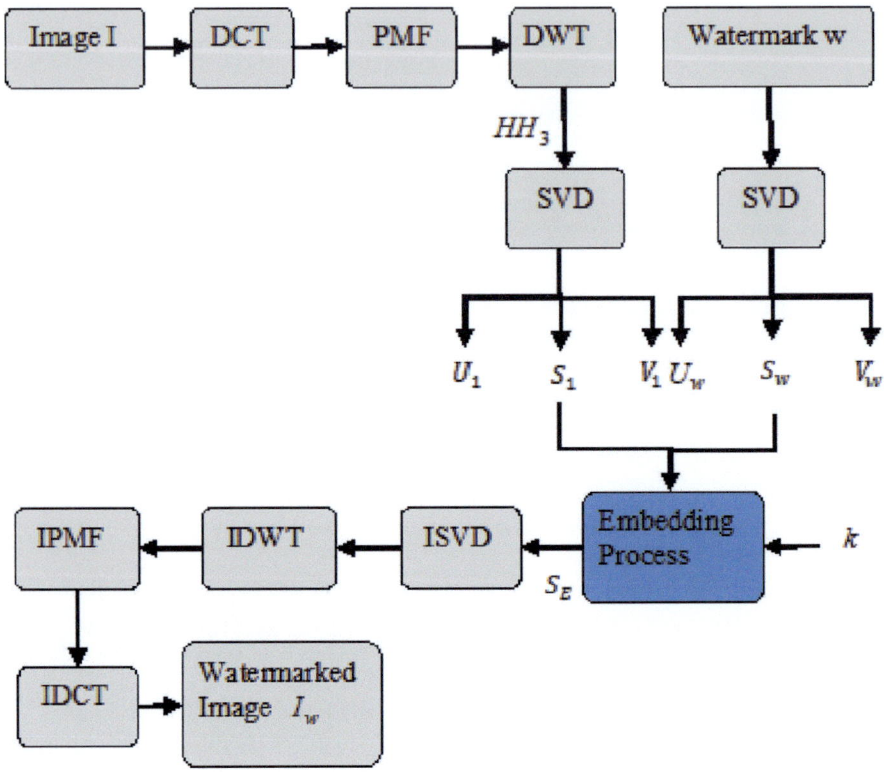

Fig. 9.3 Embedding process diagram

- Apply three levels of Discrete Wavelet Transform.
- Select the sub-bands HH_3 and HH_3^w
- Select the sub-band (U_1, S_1, V_1), and HH_3^w sub-band to obtain the components (U_2, S_2, V_2).
- Extract the watermark image W_E with a key k

$$W_E = U_w \left(\frac{S_2 - S_1}{k} \right) V_w^T \tag{9.13}$$

9.5 Results and Discussion

The proposed algorithm was evaluated using the test images Lena and Airplane, each with dimensions of 512×512 pixels. A watermark image of size 64×64 pixels was used. Figure 9.5 displays the original image and the watermarked image Lena

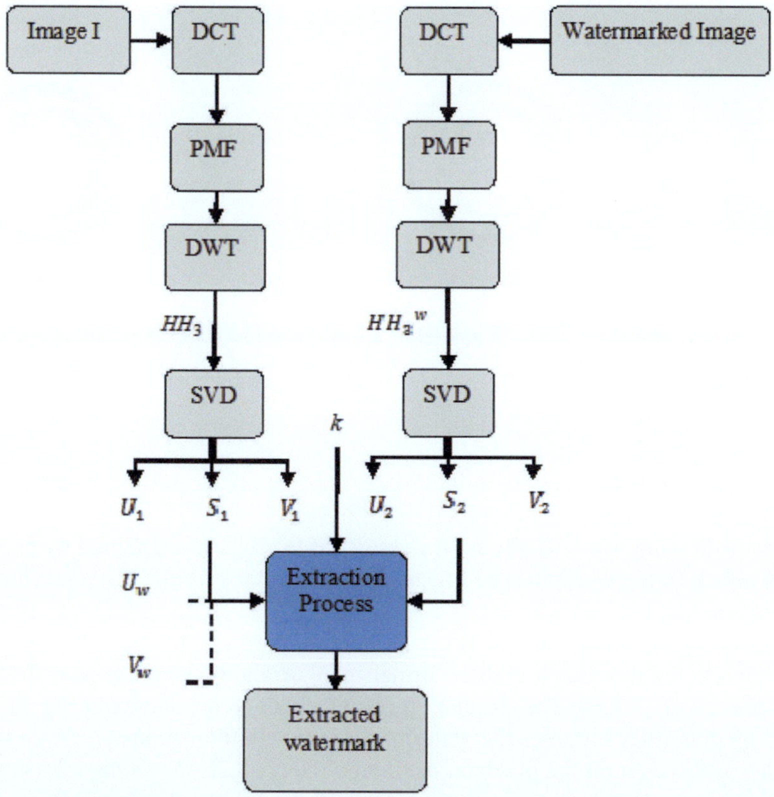

Fig. 9.4 Extracted process diagram

(WI), where the PSNR reached 35.5903 dB and the normalized correlation (NC) was 0.9999.

To evaluate the effectiveness of the proposed algorithm, multiple experiments were conducted to confirm its robustness. Various types of attacks, illustrated in Fig. 9.6, were applied: (SP) Salt and Pepper noise with 0.1% variance, (GN) Gaussian noise with zero mean and 1% variance, (MF) Median filtering based on pixel neighborhoods, (AF) Average filtering using neighboring pixels, (GF) Gaussian low-pass filtering, (SH) sharpening filter, (RS) rescaling from 512 × 512 to 256 × 256 and back to 512 × 512, (RO) rotation by 10°, (TR) translation by 20 × 20 pixels, (GC) gamma correction with a factor of 0.6, (HE) histogram equalization, and (JPEG) JPEG compression with a quality factor of 50.

The proposed algorithm was verified under various types of attacks to evaluate its resistance (Fig. 9.6). Experimental results show that the algorithm ensures both high invisibility and strong robustness. The NC metric remains at a high level, indicating a good similarity between the watermark before and after attacks. To define the limits

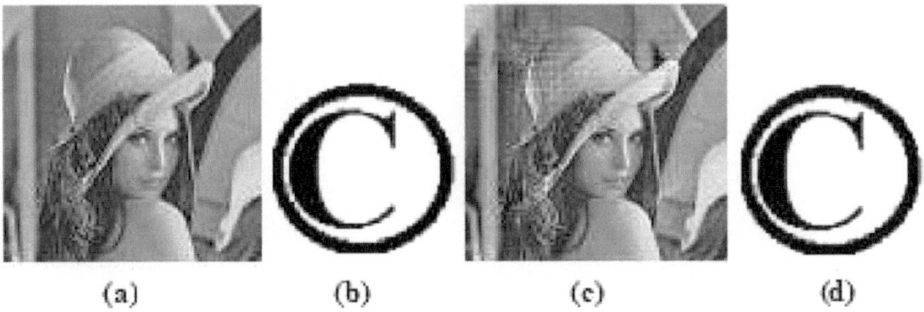

Fig. 9.5 Performance of the proposed algorithm: **a** Cover lena image, **b** Watermark image, **c** Lena image after watermark embedding with a $PSNR$ of 35.5903 dB, **d** Extracted watermark with NC equal to 0.9999

of our approach, different attack intensities were applied, and the corresponding $PSNR$ and NC values were recorded in Table 9.1.

In the following, we will discuss the results of Fig. 9.7. It's obtained by applying several attack to the watermarked image. The NC values results are represented in the Histograms.

Salt and pepper noise able to remove the watermark from the image. The presented approach resisted this attack with a Normalized Correlation factor greater than 0.85 for a variance of 1–30%. The simulation results obtained are shown in Fig. 9.7a.

The median filter removes the watermark by matrix filter of size $(a \times a)$ pixels. The filter slides over all the pixels in the image. Figure 9.7b shows the robustness of the algorithm against the median filter attack with an NC greater than 0.93 up to the size $17\times$ 17 pixels.

The average filter is a low-pass filter can be removing the watermark. Figure 9.7c summarizes the results of the NC parameters compared of the average filter size. The NC values obtained are greater than 0.91 which confirms the algorithm robustness.

The Gaussian filter among known attacks in watermarking image. To verify the robustness of the presented algorithm, we tested it against a Gaussian filter whose its size up to 17×17 pixels. The results of NC are illustrated in Fig. 9.7d is greater than 0.98 which show the efficiency of the algorithm against this filter.

Rescaling is a geometric attack that involves changing the size of the image. This attack affects the watermark since the pixels of image are compacted. The algorithm is tested against this attack up to the 32×32 pixel scale (Fig. 9.7e). We note that the NC factor is greater than 0.89 and the algorithm is still robust.

Rotation is a geometric attack where the image rotates around its central axis. We experiment the algorithm against this attack up to a rotation of 35° (Fig. 9.7f) with results of NC greater than 0.85.

The translation allows to be shifted the pixels of the image by creating black spaces within. This attacks can be to destroy the watermark. In Fig. 9.7g we notice that the NC is greater than 0.90 up to the size 75×75 translation pixels.

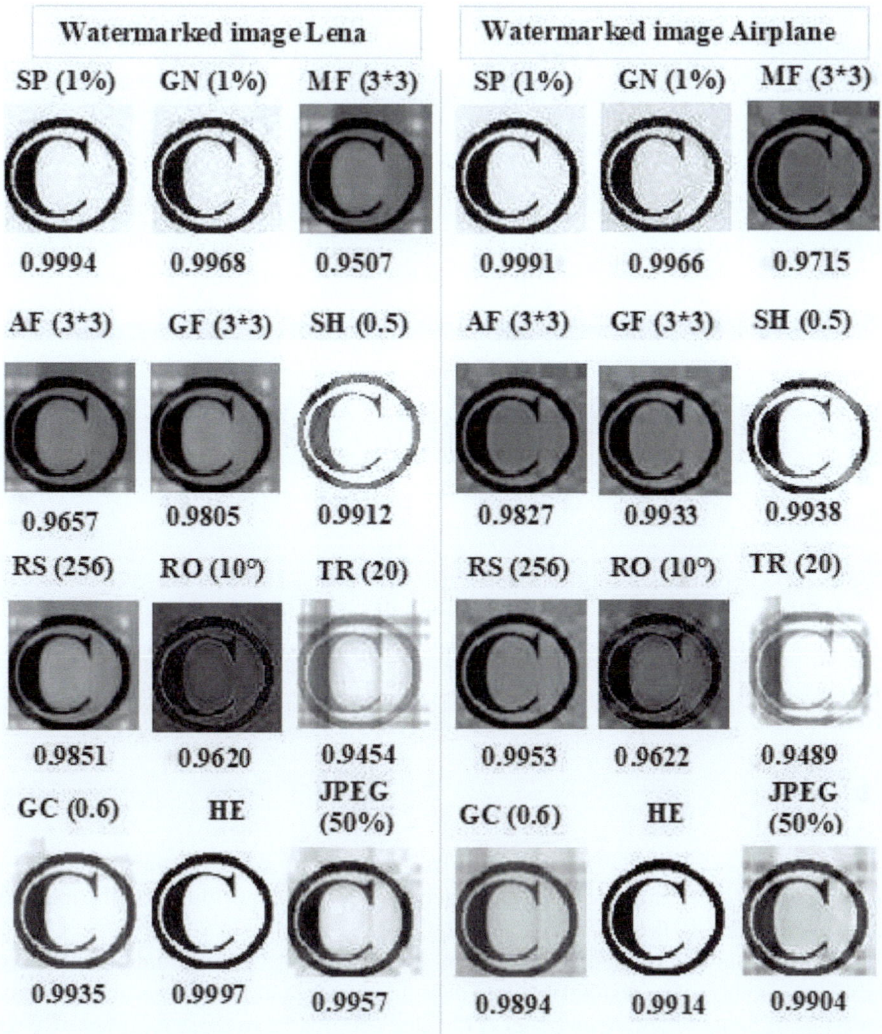

Fig. 9.6 Recovered watermarks from attacked watermarked images, showing the corresponding NC values, (SP) Salt and pepper, (GN) Gaussian noise, (MF) Median filtering, (AF) Average filtering, (GF) Gaussian low-pass filtering, (SH) sharpening filter, (RS) Rescaling , (RO) rotation of 10°, (TR) Translation of 20 × 20 pixels, (GC) gamma correction of 0.6 means, (HE) Histogram Equalization, (JPEG) JPEG compression

Table 9.1 $PSNR$ and NC metrics for test images subjected to various attacks with different intensities

Attacks		Watermarked image		Watermarked image	
		Lena		Airplane	
		$PSNR$	NC	$PSNR$	NC
SP	10%	31.4771	0.9841	31.2437	0.9758
	20%	30.1392	0.9514	29.9032	0.9295
GN	10%	29.6693	0.9412	29.9700	0.9224
	20%	28.8101	0.8815	29.0210	0.8299
MF	5×5	36.9972	0.9796	36.7512	0.9918
	11×11	36.1868	0.9336	35.3238	0.9673
AF	5×5	36.7728	0.9808	36.3555	0.9897
	11×11	35.7719	0.9339	34.9296	0.9610
GF	5×5	37.0581	0.9827	36.9742	0.9930
	11×11	37.0669	0.9821	36.9716	0.9927
RS	128×128	36.6579	0.9681	36.0495	0.9808
	64×64	35.7258	0.9412	34.8496	0.9622
RO	15°	29.7354	0.9413	28.9203	0.9494
	20°	29.6046	0.9663	28.7551	0.9272
TR	30×30	29.3912	0.9445	28.9969	0.9250
	50×50	28.9482	0.9422	28.5158	0.9124
$JPEG$	20%	35.5596	0.9933	35.6360	0.9903
	60%	35.5757	0.9964	35.6705	0.9938

The proposed method robustness was assessed under JPEG compression with a ratio reaching 75%. The NC value remained above 0.99, confirming that the method resists effectively to this type of attack (Fig. 9.7h).

Table 9.2 allows a comparison between our proposed technique and a previously published method. The results obtained demonstrate greater robustness under different types of attacks, achieving better NC values compared to those reported by (Ali and Ahn 2014).

9.6 Conclusion

The proposed algorithm developed in this study relies on a hybrid form using Discrete Cosine Transform, Pixel Movement Function, Discrete Wavelet Transform, and the technique of Singular Value Decomposition. The three aim requests of digital image watermarking robustness, invisibility, and ability of hiding are successfully fulfilled. The $PSNR$ value of 35.5903 dB obtained for test images confirms the high visual quality of the watermarked images. Similarly, the NC value of 0.9999

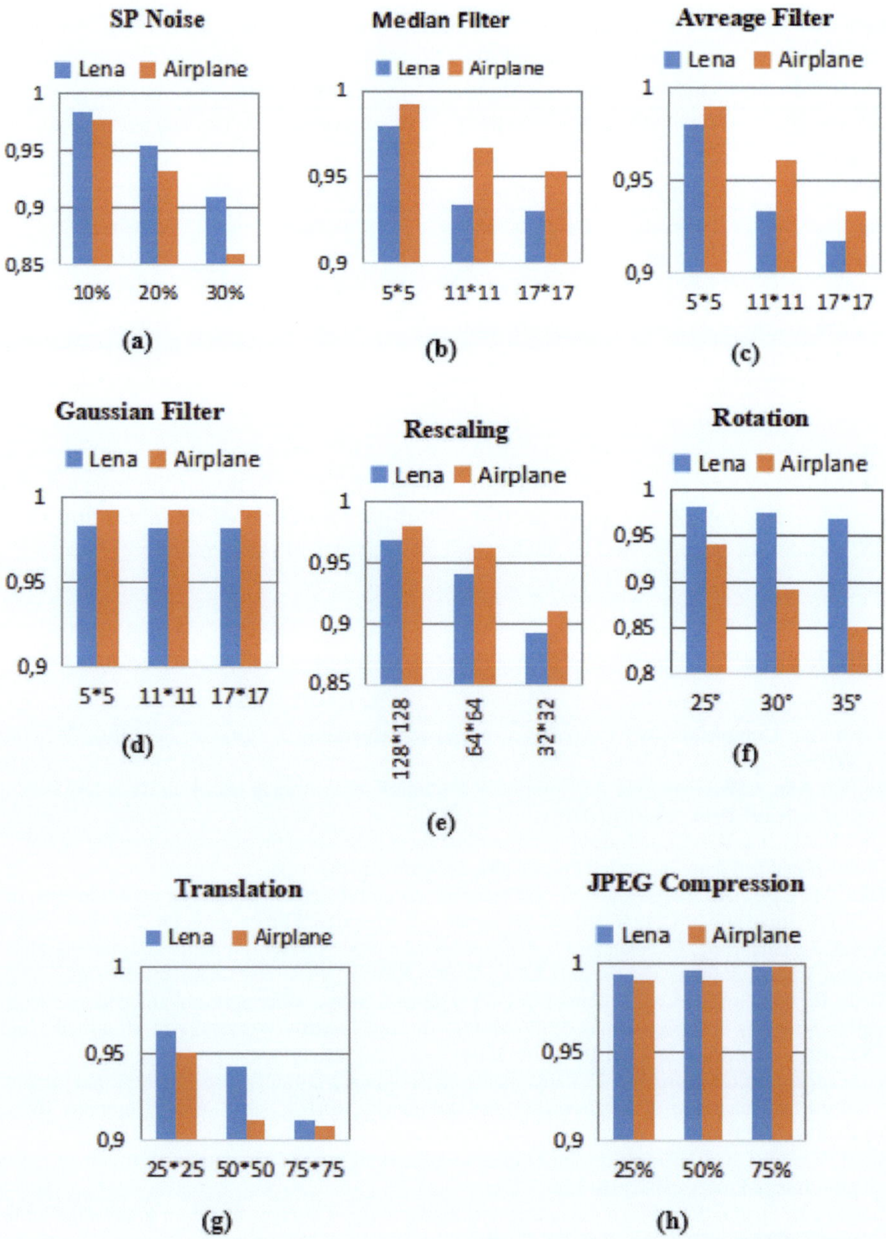

Fig. 9.7 The NC values histograms for test images after various attacks with different intensities

Table 9.2 *NC* values for comparison

Attacks	Algorithm of Ali and Ahn (2014)		Our algorithm	
	Lena	Airplane	Lena	Airplane
SP 1%	0.8322	0 .8212	0.9994	0.9991
GN 1%	0.6391	0.6420	0.9968	0.9966
MF 7×7	0.7619	0.6767	0.9312	0.9577
AF 7×7	0.6694	0.5783	0.9465	0.9676
SH 0.8	0.8085	0.7638	0.9855	0.9916
GF 9 × 9	0.9221	0.8970	0.9821	0.9927
RS 256 × 256	0.9740	0.9658	0.9851	0.9953
C-$JPEG$ 50%	0.9892	0.9889	0.9957	0.9904

for the watermark image highlights the strong similarity between the inserted and extracted watermark. As a perspective, the implementation of the algorithm on a DSP processor is suggested to further assess its practical performance.

References

Al-Haj, A.: Combined DWT-DCT digital image watermarking. J. Comput. Sci. **3**(9), 740–746 (2007)

Ali, M., Ahn, C.W., Pant, M.: An optimized watermarking technique based on de in DWT-SVD domain. IEEE **2013**, 99–104 (2013)

Ali, M., Ahn, C., W.: An optimized watermarking technique based on self-adaptative DE in DWT-SVD transform domain. Signal Process. **94**, 545–556 (2014)

Bajaj, M., Lay, J.A.: Image indexing and retrieval in compressed domain using color clusters. In: IEEE Symposium of Computational Intelligence, Image Signal Processing, pp. 271–274 (2007)

Dawei, Z., Guanrong, C., Wenbo, L.: A chaos-based robust wavelet-domain watermarking algorithm. Chaos, Solitons and Fractals **22**(1), 47–54 (2004)

Ganic, E., Eskicioglu, A.M.: Robust DWT-SVD domain image watermarking: embedding data in all frequencies. In: Proceedings of the Multimedia and Security Workshop on Multimedia and Security—MM&Sec '04, pp. 166–174 (2004)

Hernández, J., Pérez-González, F.: Performance analysis of a 2-D-multipulse amplitude modulation scheme for data hiding and watermarking of still images. IEEE J. Select. Areas Commun. **16**(4), 510–524 (1998)

Hu, R., Xiang, S.: Cover-lossless robust image watermarking against geometric deformations. IEEE Trans. Image Process. **30**, 318–331 (2021)

Hussein, J.: Spatial domain watermarking scheme for colored images based on logaverage luminance. J. Comput. **2**(1), pp. 100–103 (2010)

Jiang, J., Armstrong, A., Feng, G.C.: Direct content access and extraction from JPEG compressed images. Pattern Recogn. **35**, 2511–2519 (2002)

Lin, S., Shie, S., Guo, J.: Improving the robustness of DCT-based image watermarking against JPEG compression. Comput. Stand. Interfaces **32**(1–2), 54–60 (2010)

Liu, F., Han, K., Wang, C.Z.: A novel blind watermark algorithm based On SVD and DCT. In: IEEE International Conference on Intelligent Computing and Intelligent Systems (2009)

Lu, Z., Li, S., Burkhardt, H.: A content-based image retrieval scheme in JPEG compressed domain. Innov. Comput. Inf. Control **2**, 831–839 (2006)

Poursistania, P., Nezamabadi-Poura, H., Moghadamb, R.A., Saeeda, M.: Image Indexing and Retrieval in JPEG Compressed Domain Based on Vector Quantization, Elsevier Ltd, pp. 1005–1017 (2011)

Qi, X., Qi, J.: A robust content-based digital image watermarking scheme. Signal Process. **87**(6), 1264–1280 (2007)

Reddy, P., Prasad, M., Rao, D.: Robust digital watermarking of images using wavelets. Int. J. Comput. Electr. Eng. **1**(2), 111–116 (2009)

Souadek, R., Boukezzoula, N.E.: New image watermarking algorithm based on DWT and pixel movement function PMF. Int. Arab J. Inf. Technol. **17**(1), 1–7 (2020)

Tirkel, A.Z., Rankin, G.A., Van Schyndel, R.M., Ho, W.J., Mee, N.R.A., Osborne, C.F.: Electronic water mark. In: DICTA, pp. 666–672 (1993)

Van Schyndel, R.G., Tirkel, A.Z., Osborne, C.F.: A digital watermark. In: IEEE International Conference: Image Processing, Proceedings. ICIP-94, vol. 2, pp. 86–90 (1994)

Yin, C., Li, L., Lv, A., Li, Q.: Color image watermarking algorithm based on DWT-SVD. In: IEEE International Conference on Automation and Logistics, pp. 2607–2611 (2007)

Chapter 10
Enhancement of SAR Speckle Denoising Using Hybrid Non Local Sigma Filter

Soumaya Fatnassi, Mohamed Yahia, Tarig Ali, and Md. Maruf Mortula

Abstract In order to extract significant information for homogeneous extended targets, it is crucial to use speckle filtering in synthetic aperture radar (SAR) images. In this study,a hybrid filtering technique was proposed, which uses both the Sigma filter and the non-local mean filter. The non-local filtering principle is used to select pixels for this method. Pixels with high patch distances that were not similar in the search area were discarded. To estimate the variability of the pixels, the coefficient variation CV is used witch select homogeneous pixels in the search area. According to the results, the proposed hybrid filter improved the filtering criteria in terms of reducing speckle and conserving spatial detail in comparison to Lee filter and the improved Sigma filter. Airborne (Les-Landes, France acquired by NASA JBL) and Space-borne (Sentinel1 C-band hv SAR image of Sousse, Tunisia) synthetic aperture radar SAR data were used for validation.

Keywords MMSE filter · Non local filter · Sigma filter · SAR · Speckle noise · Sentinel1

S. Fatnassi (✉)
MACS Lab, National Engineering School of Gabes, Gabes, Tunisia
e-mail: soumayafatnassi@yahoo.fr

M. Yahia
American University of Sharjah GIS and Mapping Laboratory Sharjah, Sharjah, UAE
e-mail: myahya@aus.edu

T. Ali · Md. Maruf Mortula
Civil Engineering Department, American University of Sharjah, Sharjah, UAE
e-mail: atarig@aus.edu

Md. Maruf Mortula
e-mail: mmortula@aus.edu

N. Derbel and Q. Zhu (eds.), *Advances in Communication, Signal and Image Processing*,
https://doi.org/10.1007/978-981-95-0401-5_10

10.1 Introduction

The speckle noise appearing in synthetic aperture radar (SAR) images constitutes one problem of the SAR data interpretation (Fatnassi et al. 2019). To maintain spatial details while reducing speckle, several methods have been developed for these reasons.

By averaging the intensities of neighboring pixels, the boxcar filter helps reduce speckle in SAR images (Oliver and Quegan 2004). The smoothing effect that is created around spatial details like texture, points, contours, etc. is a major drawback of this technique.Many other techniques have been developed to avert this deficiency. This filter used different estimation domains such as spatial (Hawwar and Reza 2002) wavelet (Argenti and Alparone 2002) and homomorphic wavelet (Gagnon and Jouan 1997).

Due to their ability to maintain spatial details while reducing speckle, total variation methods (TV) have been widely used to reduce speckle in SAR images (Zhao et al. 2015) the Wavelet (Argenti and Alparone 2002), bandelets (Zhang et al. 2010) and shearlet (Hou et al. 2012) based algorithms constitutes an important brunch of speckle reduction technique. Their main disadvantage is their need to a bias-compensation stage.

One of the patch-based techniques is non-local (NL) filtering, which extends the selection of pixels used for filtering to the global window (Grimson et al. 2015). The main idea is to establish a local window where each pixel value is associated with the similarity between the central and offcentral pixels. The most commonly used measure is the Euclidean distance (Grimson et al. 2015). Zhong et al. (2011), by adapting it to multiplicative noise, the NL Means (NLM) can be used to filter SAR images The NL based filters improved significantly the filtering performance. In SAR speckle filtering, filters that are based on minimum mean square error (MMSE) have been employed and have resulted in a high level of speckle reduction.

The minimum mean square error (MMSE) based filters have been applied in SAR speckle filtering and give a high speckle reduction level.

Since the introduction of Lee sigma filter (Lee 1980, 1983), many versions have been proposed such as Frost et al. (1982), Kuan et al. (1985), the improved Lee sigma (2009) and Feng et al. (2011), etc. These filters are commonly used in Geographical Information Systems (GIS) and remote sensing software due to their efficiency, simplicity in speckle reduction, and low computational demands. An iterative MMSE (IMMSE) filter has been introduced (Yahia et al. 2017, 2020). Usually, a filter that assures a high level of speckle reduction is used to initialize the IMMSE filter. The IMMSE method preserves filtered homogeneous areas and improves spatial details that were initially smoothed after several iterations. The simplicity and ability to preserve spatial details are the primary benefits of using MMSE-based filters. Hence, Compared to NL-based methods, these methods provide less speckle reduction.

In this chapter, the filtering performances of the Lee Sigma filter are enhanced by hybridize it with the NL filtering. The selection of the pixels used for the filtering is extended to NL filtering principle rather than restricting it the sigma range.

10.2 Related Works

The intensity pixel y(i) of a SAR image is affected by a multiplicative speckle noise

$$y\,(i) = x\,(i)\,v\,(i)\,. \tag{10.1}$$

where $x(i)$ is the noise free pixel and $v(i)$ is the speckle noise with unit mean and standard deviation σ_v. The objectives of SAR speckle filtering are:

- Minimize speckle in extend ed homogeneous areas.The ideal situation is to average all pixels. The optimal situation is to preserve the value of the original pixel (i. e $\widehat{x}(i) = \overline{y}(i)$)
- Ensure that the spatial details are maintained in structured areas. The ideal solution is to keep the value of the original pixel (i.e. $\widehat{x}(i) = y(i)$).

Generally, the filtered pixel should verifies $\widehat{x}(i) \in [y(i), \overline{y}(i)]$ or $\widehat{x}(i) \in [\overline{y}(i), y(i)]$

In the MMSE filters (Lee 1980; Lee et al. 2009), the filtered pixel $\widehat{x}(i)$ is declared as:

$$\widehat{x}(i) = \overline{y}(i) + b(i)\big[y(i) - \overline{y}(i)\big] \tag{10.2}$$

where

$$b(i) = \frac{var[x(i)]}{var[y(i)]} \tag{10.3}$$

and

$$\mathrm{var}(x(i)) = \frac{var[y(i)] - \overline{y}^2(i)\sigma_v^2}{1 + \sigma_v^2} \tag{10.4}$$

where $\overline{y}(i)$ and $\mathrm{var}(y(i))$ were the mean and the variance of $y(i)$, respectively. In homogeneous areas, $b(i) = 0$ so $\widehat{x}(i) = \overline{y}(i)$ while in heterogeneous areas, b = 1 and $\widehat{x}(i) = y(i)$. In general, $0 < b(i) < 1$ and $\widehat{x}(i) \in [\overline{y}(i), y(i)]$.

Generally, a square window is applied to determinate the local statistics (i.e. mean and variance) used to calculate the parameter b. But, this solution produced blurring effects around spatial details. To repress this limitation, Lee et al. introduced the sigma range (Lee et al. 2009). When computing the local statistics, pixels outside the sigma range are discarded. A moving window was applied. The computing process of $\widehat{x}(i)$ included pixels within the sigma range $(n_1(y), n_2(y))$ (10.2), where n_1, n_2 define Define the sigma range's limits. The essence of the NLM method is given by

$$\widehat{x}_{NLM}(i) = \sum_{j \in \Omega} w(i, j)\; y(j) \tag{10.5}$$

where Ω is the research window, the weight $w(i, j)$ is the weight witch represent the degree of similarity between the patches Λ_i and Λ_j

$$w(i, j) = \frac{1}{Z(j)} \exp\left(-\frac{d(\Lambda_i, \Lambda_j)}{h}\right) \tag{10.6}$$

where $Z(i)$ is the factor of normalization and h is the smoothing parameter

$$Z(i) = \sum_j \exp\left(-\frac{d(\Lambda_i, \Lambda_j)}{h}\right) \tag{10.7}$$

For single look SAR speckle, $d(\Lambda_i, \Lambda_j)$ can be expressed as (10.8)

$$d(\Lambda_i, \Lambda_j) = \sum_{m \in \sum} G(k) \frac{\left[z_i(m) - z_j(m)\right]^2}{\overline{y}_i} \tag{10.8}$$

where Σ is the set of pixels in the patch, G was the standard Gaussian kernel function and $\overline{y_i}$ is the mean value of y_i in the patch Λ_i.

10.3 Motivation and the Proposed Filter

The main idea of Lee Sigma filter is to estimate the parameter b witch based on the local statistics using a square window. This solution decreases the complexity. In local neighborhood, the statistics are determined using the intensity of the pixels.

However, in the non- local neighborhood, the statistics connected to a pixel are calculated using the patches more than the intensity values.

In this chapter, the authors suggested a hybrid Sigma filter. Therefore, pixels used for the estimation of the parameter b are chosen by analogy to the non-local filtering principle. The weight of each pixel in the search area is determined by comparing its patch to the centered pixel patch. Pixels that are not similar in the search area with high patch distances are not used to compute b in (10.3). Nevertheless, variability in the scene around the processed pixel determines the number of non-similar pixels. All pixels are analogous in homogeneous zone hence in heterogeneous areas we have a high number of non similar pixels. In fact, To pick the number of pixels that are not similar, we must conduct a variability test.

The variability of pixels is estimated using the coefficient variation CV in this research. The local CV determinate the textures, edges and points a better value than the pixels in homogeneous areas. The CV is expressed as bellow:

$$\mathrm{CV}_{\widehat{x}_k} = \frac{std\left[\widehat{x}_k(i)\right]}{\overline{\widehat{x}}_k} \tag{10.9}$$

where $std(\cdot)$ is the standard deviation. The number of selected pixels is given by N_s

$$N_S = N \times \frac{1}{CV^5} \tag{10.10}$$

where N indicates the sum of pixels within the search area.

10.4 Implementation of the Hybrid NL Sigma Filter

The Hybrid NL Sigma Filter described as below

- Identify the search window Ω_1, and the patch $\sum_i$;
- Determine the Coefficient variation;
- Determine the Number of selected pixels in (10.10);
- Discard half of the pixels that have the least similarity coefficient values;
- Determine b using (10.3);
- Compute the filtered pixel using (10.2);

10.5 Experimental Results

To demonstrate the performance of the proposed despeckling technique, SAR data were used. The airborne SAR was the Les-Landes, France acquired by NASA JBL (Fig. 10.1a). The spaceborne data was the Sentinel 1 C-band hv SAR image of Sousse, Tunisia (Fig. 10.1b).

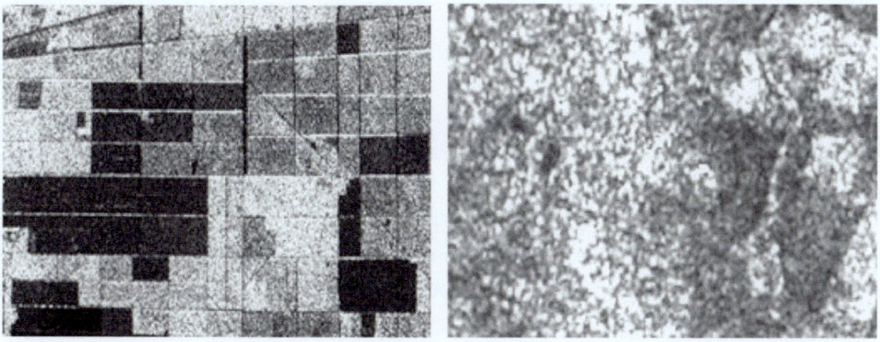

(a) (b)

Fig. 10.1 **a** AirSAR image, **b** Sentinel 1 SAR image

10.6 Implemented Filters

- Lee's filter (1983), a moving window W_1 was used to calculate b in (10.3).
- Improved Sigma filter (Lee et al. 2009), a moving window W_1 was employed. The bounds of the range. Pixels $(n_1(y), n_2(y))$ defined by pixels within the sigma range were included when computing $\widehat{x_i}$ in (10.2), where n_1, n_2 define the bounds of the sigma range.
- Proposed Hybrid Lee's filter. Search window $\Omega_1 = 13 \times 13$, patch $\sum_1 = 3 \times 3$

10.7 Evaluation Criteria

To quantitatively assess the performance of different speckle filtering methods in terms of the speckle reduction level, the equivalent number of looks (ENL) is chosen:

$$ENL(i) = \frac{\left[\widehat{\overline{x}}(i)\right]^2}{var[\widehat{x}(i)]} \tag{10.11}$$

To demonstrate the ability of the algorithm to preserve spatial detail, the authors used the degree based on the ratio of averages (EPD-ROA) (Feng et al. 2011). The $EPD - ROA$ in horizontal direction is:

$$EPD - ROA_H(i) = \frac{\sum\limits_{m,n} \left|\widehat{x}(m,n)_{/\widehat{x}(m,n+1)}\right|}{\sum\limits_{m,n} \left|y(m,n)_{/y(m,n+1)}\right|} \tag{10.12}$$

where m and n are the x, y coordinates of the pixel in the selected zone, respectively. $EPD - ROAV$ is obtained by substituting in (10.12) the indexes $(m, n+1)$ by $(m+1, n)$. For original image, $EPD - ROA$ equal one and if it is closer to one, we have a better ability of spatial detail preservation.

10.8 Comparison Using AirSAR Data

Figure 10.2 presents the Air SAR image. The Lee filter achieved speckle reduction, but spatial details were degraded. In contrast,the improved Sigma filter was able to better preserve spatial details. The proposed hybrid NL-Sigma filter provided the optimal balance between reducing speckles and maintaining spatial details. A rectangular zone is zoomed (see Fig. 10.3) to evaluate more the performances of the studied speckle filters. It is evident that the Lee filter has minimized speckle in the homogeneous areas, but the lines between the rectangular zones have a significant smoothing effect. In fact, the improved Sigma filter better maintained these lines. The Hybrid

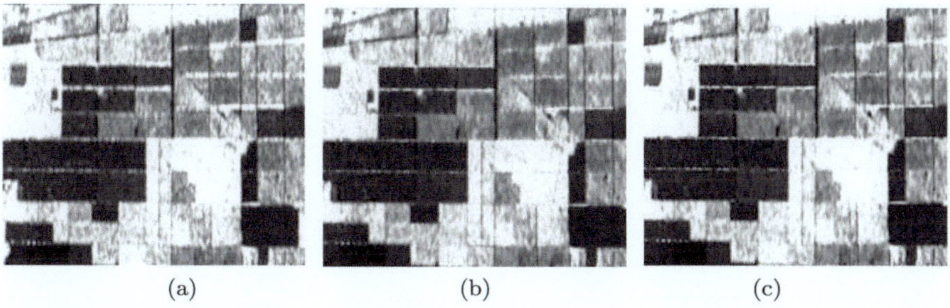

Fig. 10.2 Filtered AiSAR image: **a** Lee's filter, **b** improved Sigma filter, **c** Hybrid NL-Sigma filter

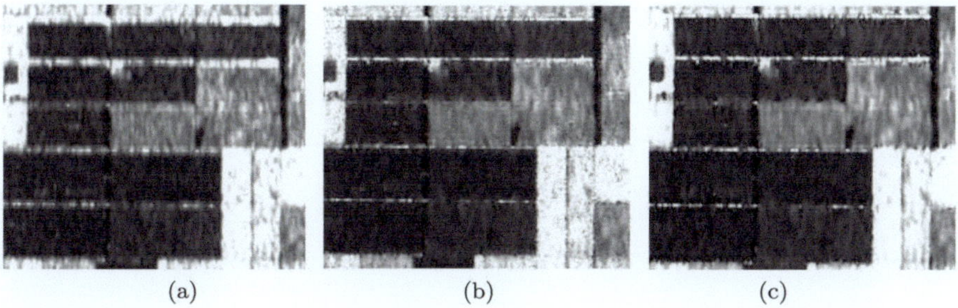

Fig. 10.3 Zoomed Filtered AiSAR image **a** Lee's filter, **b** improved Sigma filter, **c** Hybrid NL-Sigma filter

Table 10.1 Performances of the filters using airborne data

	ENL	EPD_h	EPD_v
Lee's filter	34	0.7407	0.7036
Improved Sigma filter	19	0.8082	0.7836
Hybrid NL-Sigma filter	51	0.8123	0.7899

NL Sigma filter performed better than the other filters in terms of reducing speckle ($ENL = 51$) and special detail preservation ($EPD - ROA = 0.8123, 0.7899$).

Quantitative results are provided in Table 10.1. The proposed hybrid filter was the most efficient solution to balance speckle reduction ($ENL = 51$) and spatial detail preservation ($EDP - ROA = 0.8123, 0.7899$).

10.9 Comparison Using Spaceborne Data

Figure 10.4 display the performance of the studied filter on Sentinel 1 C-band hv SAR image of Sousse, Tunisia. As for AirSAR data, the result showed that Lee's filter gave high speckle reduction; however, the spatial details are considerably blurred.

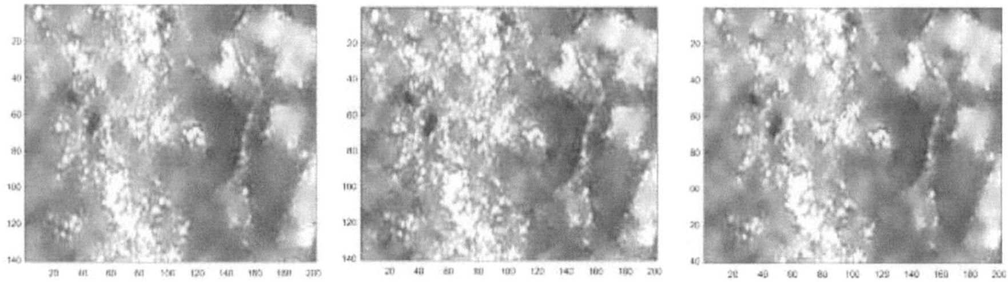

Fig. 10.4 Filtered Sentinel 1 SAR image a Lee's filter, b improved Sigma filter, c Hybrid NL-Sigma filter

Table 10.2 Performances of the filters using spaceborne data

	ENL	EPD_h	EPD_v
Lee's filter	172	0.9687	0.9693
Improved Sigma filter	124	0.9724	0.9729
Hybrid NL-Sigma filter	214	0.9694	0.9701

The improved Sigma filter gave better spatial detail preservation. The Hybrid NL-Sigma filter that was proposed again provided the best solution for both speckle filtering and preserving spatial detail.

The visual interpretation is confirmed by Table 10.2, which provides quantitative results. We once again observe that the suggested hybrid filter was the most suitable compromise between reducing speckle ($ENL = 214$) and spatial detail preservation ($EDP - ROA = 0.9694, 0.9701$).

10.10 Conclusion

In this chapter, the authors proposed a hybrid filtering method that combines the Lee Sigma filter and the non-local mean filter. Therefore, Choosing the percentage of similar patches correctly allowed the proposed filter to outperform the Lee filter and the improved Sigma filter in terms of speckle reduction and spatial detail preservation.

Experimental results demonstrate the effectiveness of this hybrid filter. Future research will revolve around optimizing the Hybrid NL Sigma filter for digital image filtering.

References

Argenti, F., Alparone, L.: Speckle removal from SAR images in the undecimated wavelet domain. IEEE Trans. Geosci. Remote Sens. **40**, 2363–2374 (2002)

Fatnassi, S., Yahia, M., Abdelfattah, R.: Sar speckle filtering using hybrid non local and iterative MMSE filters. Int. Geosci. Remote Sens. Symp. (IGARSS), 1781–1784 (2019)

Feng, H., Bou, H., Gong, M.: SAR image despeckling based on local homogeneous-region segmentation by using pixel-relativity measurement. IEEE Trans. Geosci. Remote Sens. **49**(7), 1756–1760 (2011)

Frost, V.S.,Stiles, J.A., Shanmugan, K.S., Holtzman, J.C.: A model for radar image and its application to adaptive digital filtering for multiplicative noise. IEEE Trans. Pattern Anal. Mach. Intell. PMAI-4:(2), 157–166 (1982)

Gagnon, L., Jouan, A.: Speckle filtering of SAR images: a comparative study between complex-wavelet-based and standard filters. Wavelet Appl. Signal Image Process. Proc. SPIE **3169**, 80–91 (1997)

Grimson, R., Morandeira, N.S., Frery, A.C.: Comparison of nonlocal means despeckling based on stochastic measures. In: IEEE International Symposium on Geoscience and Remote Sensing (IGARSS), pp. 3091–3094 (2015)

Hawwar, Y., Reza, A.: Spatially adaptive multiplicative noise image denoising technique. IEEE Trans. Image Process. **11**(12), 1397–1404 (2002)

Hou, B., Zhang, X., Bu, X., Feng, H.: SAR image despeckling based on on subsampled Shearlet transform. IEEE J. Select. Top. Appl. Earth Observ. Remote Sens. **5**, 809–823 (2012)

Kuan, D.T., Sawchuk, A.A., Strand, T.C., Chavel, P.: Adaptive noise filtering for images with signal dependent noise. IEEE Trans. Pattern Anal. Mach. Intell. PAMI-7, 165–177 (1985)

Lee, J.S.: Digital image enhancement and noise filtering by use of local statistics. IEEE Trans. Pattern Anal. Mach. Intell. PAMI-2(2), 165–168 (1980)

Lee, J.S.: Digital image smoothing and the sigma filter. Comput. Vis. Graph. Image Process. **24**(2), 255–269 (1983)

Lee, J.S., Wen, J.-H., Ainsworth, T.L., Chen, K.S., Chen, A.J.: Improved sigma filter for speckle filtering of SAR imagery. IEEE Trans. Geosci. Remote Sens. **47**(1), 202–213 (2009)

Oliver, C., Quegan, S.: Understanding Synthetic Aperture Radar Images. SciTech Publishing, Amazon France (2004)

Yahia, M., Hamrouni, T.-A., Abdelfattah, R.: SAR speckle denoising using iterative filter. In: International Geoscience and Remote Sensing Symposium IGARSS, pp. 1756–1760 (2017)

Yahia, M., Ali, T., Mortula, M.M., Abdelfettah, R., El Mahdy, S., Arampola, N.S.: Enhancement of SAR speckle denoising using the improved immse filter. IEEE J. Select. Top. Appl. Earth Observ. Remote Sens. **13**, 859–871 (2020)

Zhang, W., Liu, F., Jiao, L., Hou, B., Wang, S., Shang, R.: SAR image despeckling using edge detection and feature clustering in bandelet domain. IEEE Geosci. Remote Sens. Lett. **7**(1), 131–135 (2010)

Zhao, Y., Liu, J.G., Zhang, B., Hong, W., Wu, Y.-R., Wu, Y.R.: Adaptive total variation regularization based SAR image despeckling and despeckling evaluation index. IEEE Trans. Geosci. Remote Sens. **53**(5), 2765–2774 (2015)

Zhong, H., Lu, L., Jiao, L.: Fast non local lee filter for SAR image despeckling using directional projection. In: International Conference on Radar, pp. 1600–1603 (2011)

Chapter 11
Advancing JPEG Compression: Integer DTT Enhancement with an Effective Quantisation Matrix

Nabila Brahimi, Toufik Bouden, and Tahar Brahimi

Abstract For image compression, the Discrete Tchebichef Transform (DTT) has recently been proposed as a competitive alternative to the Discrete Cosine Transform (DCT). This chapter presents a compression framework based on an integer approximation of the DTT. For a set of standard test images with varying statistical characteristics, an adopted quantisation matrix is derived from the energy distribution of the integer DTT coefficients. Due to the non-normality of most integer DTT matrices, their diagonal components are integrated into the quantisation process. Experimental results demonstrate that employing the JPEG standard with integer DTT and the proposed quantisation matrix yields improved compression efficiency relative to the conventional JPEG luminance quantisation matrix, while maintaining high visual quality in the reconstructed grayscale images.

Keywords DTT · JPEG standard · Image compression · Quantisation matrix · Tchebychev transform

11.1 Introduction

Most current compression standards for both still and moving images rely on block-based transforms as a fundamental step in the compression process, typically fol-

N. Brahimi (✉)
NDT Laboratory, Electronic Department, Mohammed Seddik Benyahia-University of Jijel, Jijel, Algeria
e-mail: nabila.brahimi@univ-jijel.dz

T. Bouden
NDT Laboratory, Automatic Department, Mohammed Seddik Benyahia-University of Jijel, Jijel, Algeria
e-mail: bouden_toufik@univ-jijel.dz

T. Brahimi
L2EI Laboratory, Electronic Department, Mohammed Seddik Benyahia-University of Jijel, Jijel, Algeria
e-mail: t.brahimi@univ-jijel.dz

N. Derbel and Q. Zhu (eds.), *Advances in Communication, Signal and Image Processing*,
https://doi.org/10.1007/978-981-95-0401-5_11

lowed by scalar quantisation. The widely adopted JPEG (Joint Photographic Experts Group) (Wallace 1991), which uses Discrete Cosine Transform (DCT), remains the most prominent international standard for still image compression. In recent years, the Discrete Tchebichef Transform (DTT), originally introduced by Mukundan et al. (2001), has been investigated for its effectiveness in lossy image compression. Studies have demonstrated that DTT can outperform traditional methods at various compression rates. Furthermore, its applicability has also been explored in lossless image compression techniques (Xiao et al. 2016).

The DTT exhibits energy compaction and decorrelation properties that are highly comparable to those of the DCT. In particular, the decorrelation efficiency of both transforms is nearly equivalent, with recorded values of 97.34 for DTT and 97.82 for DCT when applied to 8×8 image blocks with a correlation coefficient of 0.9 (Ishwar 2008; Ishwar et al. 2008). Although floating-point operations ensure high computational precision, they significantly increase the demand on system resources in terms of processing time and hardware complexity.

However, integer operations result in faster computation, reduced complexity, and lower costs, while also minimizing power consumption (Cai et al. 2021; Deepsita et al. 2021; Brahimi et al. 2020, 2022a, b).

Due to its effectiveness in image compression, the integer form of the DTT has attracted growing attention in recent research (Yuting et al. 2020) where both forward and inverse transform operations have been extensively investigated (Yuting et al. 2020; Chan and Im 2021; Prattipati et al. 2013).

The effectiveness of image compression, particularly the compression ratio achieved at a given image quality, is closely linked to the quantisation matrix. In the JPEG standard, quantisation matrices for luminance and chrominance components were developed by taking advantage of the perceptual characteristics of the Human Visual System (HVS). Although originally tailored for the exact Discrete Cosine Transform (DCT), these matrices have also been proven to be compatible with DTT (Ji et al. 2013). With the increasing application of integer DTTs in next-generation wireless and satellite communication systems, there is a growing demand for a quantisation matrix specifically adapted to integer DTT approximations. In response to this demand, we introduce a quantisation matrix designed to be compatible with the JPEG standard when used in conjunction with integer DTT. This proposed matrix is constructed by examining the energy distribution of DTT coefficients obtained from a set of representative test images, ensuring an improved compression efficiency suited to the characteristics of the transform.

Although a quantisation matrix for the exact DTT in image compression has been developed (Xiao et al. 2016), to our knowledge, no existing work has proposed a version of the JPEG quantisation matrix adapted for approximate or integer forms of the DTT.

Recent studies in image and video compression have led to the development of integer DTT approximations, as proposed in Farsiani and Sodagar (2020), Oliveira et al. (2016). The compression efficiency achieved by these approximations is comparable to that of the DCT in lossy image coding scenarios. Despite their performance, existing studies have not explored the design of a quantisation matrix that is specifically adapted to the structural and statistical properties of the integer DTT.

This chapter presents an improved version of the baseline JPEG compression standard by incorporating integer DTT approximations.

With regard to the energy distribution of integer DTT coefficients an appropriate quantisation matrix has been proposed.

In this work, a novel quantisation matrix is proposed for the JPEG standard based on integer DTT transform. The development work is performed as follows. After presenting the exact discrete Tchebychev transform and its applications in Sect. 11.2, the related work and motivation are introduced in Sect. 11.3. Then, the energy distribution of integer DTT approximations is presented and illustrated in Sect. 11.4. In Sects. 11.5 and 11.6, the quantisation matrix design based on integer DTT with application to the JPEG standard. The results and implementation discussions are given in Sect. 11.7. The final section of this Sect. 11.8 presents some concluding remarks and highlights key takeaways.

11.2 Exact Discrete Tchebychev Transform and Its Applications

The one-dimensional N-point (DTT) of a discrete input signal $x(n)$ is mathematically defined in Ishwar et al. (2008).

Eight-point transforms are widely employed in image and video processing applications, particularly within various compression standards. Among these, the Discrete Tchebichef Transform (DTT) has garnered increasing attention in the literature due to its advantageous properties, including high coding gain, efficient energy compaction, and computational simplicity. These characteristics make the DTT a strong candidate for resource-constrained environments and real-time multimedia systems.

The transform presented in Prattipati et al. (2013), known as the Discrete Tchebichef Transform (DTT), is an orthogonal transformation derived from the Discrete Tchebichef Polynomials (DTP) framework introduced in Mukundan et al. (2001). Its implementation is highly efficient, requiring only 44 additions and 29 bit-shift operations. As a result, it avoids multiplication operations entirely, which significantly enhances its suitability for implementation low-power hardware applications.

The 8-points DTT matrix is given by:

$$C = D \begin{bmatrix} 1 & 1 & 1 & 1 & 1 & 1 & 1 & 1 \\ -7 & -5 & -3 & -1 & 1 & 3 & 5 & 7 \\ 7 & 1 & -3 & -5 & -5 & -3 & 1 & 7 \\ -7 & 5 & 7 & 3 & -3 & -7 & -5 & 7 \\ 7 & -13 & -3 & 9 & 9 & -3 & -13 & 7 \\ -7 & 23 & -17 & -15 & 15 & 17 & -23 & 7 \\ 1 & -5 & 9 & -5 & -5 & 9 & -5 & 1 \\ -1 & 7 & -21 & 35 & -35 & 21 & -7 & 1 \end{bmatrix} \tag{11.1}$$

$$C = D\ T \tag{11.2}$$

where in (11.2), T is the Tchebychev transform matrix and D is a diagonal scaling matrix which equals:

$$D = \text{diag}\left(\frac{1}{\sqrt{2}}, \frac{1}{\sqrt{42}}, \frac{1}{\sqrt{42}}, \frac{1}{\sqrt{66}}, \frac{1}{\sqrt{142}}, \frac{1}{\sqrt{546}}, \frac{1}{\sqrt{66}}, \frac{1}{\sqrt{858}}\right) \tag{11.3}$$

In image compression application, the input image is divided in 8×8 blocks. If we consider X an input 8×8 block and Y the corresponding block in the transformed domain,The forward and inverse 2-D DTT transforms for all 8×8 blocks can be expressed respectively as:

$$Y = C\ X\ C^T \tag{11.4}$$

$$X = C^T\ Y\ C \tag{11.5}$$

Thus, no computational overhead was added by the matrix D. Therefore, the 8-point matrix T, is the only source of computational complexity. An Integer Tchebychev Transform (ITT) requiring additions and bit shift operation (44 addition and 29 bit-shift operations) is proposed in Prattipati et al. (2013).

The authors in Oliveira et al. (2016), proposed an extremely low complexity DTT approximation. Their methodology consists of generating a class of parametric approximate matrices and then identify the optimal class member in terms of coding performance. A fast algorithms using the sparse matrix factorisation for their DTT approximations is also introduced.

For 1-D approximate DTT, such algorithm reduces the arithmetic cost of C_{a1} to only 24 additions and 06 bit-shifting operations, no multiplication operation is required. The amount of calculation is significantly reduced.

The integer DTT transform introduced in Oliveira et al. (2016), noted as C_{a1}, can be expressed as follows:

$$C_{a1} = D_{a1} \begin{bmatrix} 2 & 2 & 2 & 2 & 2 & 2 & 2 & 2 \\ -2 & -1 & -1 & 0 & 0 & 1 & 1 & 2 \\ 2 & 0 & -1 & -1 & -1 & -1 & 0 & 2 \\ -2 & 1 & 2 & 1 & -1 & -2 & -1 & 2 \\ 1 & -2 & 0 & 1 & 1 & 0 & -2 & 1 \\ 1 & 2 & -1 & -1 & 1 & 1 & -2 & 1 \\ 0 & -1 & 2 & -1 & -1 & 2 & -1 & 0 \\ 0 & 0 & -1 & 2 & -2 & 1 & 0 & 0 \end{bmatrix} \tag{11.6}$$

$$C_{a1} = D_{a1}\, T_{a1} \tag{11.7}$$

where,

$$D_{a1} = \text{diag}\left(\frac{1}{\sqrt{32}}, \frac{1}{\sqrt{12}}, \frac{1}{\sqrt{12}}, \frac{1}{\sqrt{20}}, \frac{1}{\sqrt{12}}, \frac{1}{\sqrt{14}}, \frac{1}{\sqrt{12}}, \frac{1}{\sqrt{10}}\right) \tag{11.8}$$

The second integer transform reported in Farsiani and Sodagar (2020) is given as follows:

$$C_{a2} = D_{a2} \begin{bmatrix} 1 & 1 & 1 & 1 & 1 & 1 & 1 & 1 \\ -2 & -2 & -1 & 0 & 0 & 1 & 2 & 2 \\ 3 & 0 & -1 & -2 & -2 & -1 & 0 & 3 \\ -2 & 1 & 2 & 0 & 0 & -2 & -1 & 2 \\ 1 & -2 & -1 & 2 & 2 & -1 & -2 & 1 \\ -1 & 2 & -2 & -1 & 1 & 2 & -2 & 1 \\ 0 & -1 & 2 & -1 & -1 & 2 & -1 & 0 \\ 0 & 1 & -2 & 0 & 0 & 2 & -1 & 0 \end{bmatrix} \tag{11.9}$$

$$C_2 = D_{a2}\, T_{a2} \tag{11.10}$$

where:

$$D_{a2} = \text{diag}\left(\frac{1}{\sqrt{8}}, \frac{1}{\sqrt{18}}, \frac{1}{\sqrt{28}}, \frac{1}{\sqrt{18}}, \frac{1}{\sqrt{20}}, \frac{1}{\sqrt{20}}, \frac{1}{\sqrt{12}}, \frac{1}{\sqrt{10}}\right) \tag{11.11}$$

11.3 Related Work and Motivation

The baseline JPEG encoder operates as a transform-based compression system, utilizing an 8 × 8 DCT followed by quantisation and entropy coding, typically combining run-length encoding with Huffman coding. Among these steps, quantisation step plays a pivotal role in achieving lossy compression. By selectively reducing the precision of frequency-domain coefficients, quantisation effectively eliminates redundant high-frequency components, which contribute less to perceived image quality. This strategy aligns with the characteristics of the human visual system, which is significantly more sensitive to low-frequency content than to high-frequency details.

The standard JPEG quantisation matrix, designed to exploit this perceptual property, is defined as follows:

$$Q_{JPEG} = \begin{bmatrix} 16 & 11 & 10 & 16 & 24 & 40 & 51 & 61 \\ 12 & 12 & 14 & 19 & 26 & 58 & 60 & 55 \\ 14 & 13 & 16 & 24 & 40 & 57 & 69 & 56 \\ 14 & 17 & 22 & 29 & 51 & 87 & 80 & 62 \\ 18 & 22 & 37 & 56 & 68 & 109 & 103 & 77 \\ 24 & 35 & 55 & 64 & 81 & 104 & 113 & 92 \\ 49 & 64 & 78 & 87 & 103 & 121 & 120 & 101 \\ 72 & 92 & 95 & 98 & 112 & 100 & 103 & 99 \end{bmatrix} \tag{11.12}$$

DTT-based JPEG standard consists of replacing the DCT transform by the DTT. It is mentioned that this quantisation matrix can be used as an exact luminance quantisation matrix of the DTT (Ji et al. 2013). In Xiao et al. (2016), an optimal quantisation matrix for the DTT-based JPEG baseline is proposed. In reference, an optimal quantisation matrix, Q_{opt}, for DTT-based JPEG baseline is proposed such as the following:

$$Q_{opt} = \begin{bmatrix} 16 & 16 & 16 & 16 & 16 & 16 & 16 & 16 \\ 16 & 16 & 16 & 16 & 16 & 16 & 16 & 16 \\ 16 & 16 & 16 & 16 & 16 & 16 & 16 & 16 \\ 16 & 16 & 16 & 16 & 16 & 16 & 16 & 16 \\ 16 & 16 & 16 & 16 & 16 & 16 & 17 & 19 \\ 16 & 16 & 16 & 16 & 16 & 17 & 19 & 26 \\ 16 & 16 & 16 & 16 & 16 & 16 & 16 & 16 \\ 16 & 16 & 16 & 16 & 16 & 16 & 16 & 16 \end{bmatrix} \tag{11.13}$$

11.4 Energy Distribution of Integer DTT Approximations

When an image is transferred to a transform domain, its energy is redistributed over the coefficients in the new domain. Table 11.1 shows the energy distribution of image Lena coefficients for different DTT approximations.

The proposed approach generates a quantisation matrix by exploiting the energy distribution of integer DTT approximations suitable for the JPEG standard based on the integer DTT. Due to their efficiency in approximating the DTT and providing the best compromise between coding performance and computational cost, the two integer approximations in Oliveira et al. (2016) and Farsiani and Sodagar (2020) were considered. Tables 11.1, 11.2 and 11.3 show the energy distribution of the Lena test image coefficients, using the exact DTT and the DTT approximations reported in Farsiani and Sodagar (2020), Oliveira et al. (2016).

As shown in Tables 11.1, 11.2 and 11.3, the energy distribution of the coefficients differs from one transform to another.

The selection of quantisation factors should account not only for the spatial location of each coefficient within the transform block but also for its associated energy. Coefficients that exhibit higher energy levels typically carry more significant image information and should therefore be quantised using smaller step sizes to preserve their content. Conversely, coefficients with lower energy contribute less to the overall image quality and can be quantised more aggressively using larger step sizes.

Table 11.1 Energy distribution of different DTT approximations for image Lena: (a) Exact DTT

(a)

1 72.1 dB	9 –5.9 dB	32 –28.3 dB	50 –41.1 dB	59 –59.2 dB	63 –53.4 dB	60 –61.3 dB	64 –∞dB
2 25.4 dB	14 –8.7 dB	17 –12.6 dB	28 –25.3 dB	47 –37.8 dB	52 –46.9 dB	43 –34.1 dB	58 –57.8 dB
7 2.2 dB	27 –24.8 dB	31 –27.6 dB	37 –31.6 dB	49 –39.9 dB	56 –53.5 dB	40 –33.3 dB	62 –61.6 dB
6 2.7 dB	24 –24.1 dB	26 –24.8 dB	34 –29.9 dB	46 –37.7 dB	54 –51.8 dB	36 –30.7 dB	57 –57.4 dB
5 9.5 dB	21 –10.8 dB	22 –19.8 dB	25 –24.6 dB	38 –32.4 dB	51 –44.7 dB	29 –25.8 dB	53 –48.8 dB
4 17.6 dB	11 –7.9 dB	13 –8.3 dB	19 –15.1 dB	23 –22.6 dB	41 –33.4 dB	18 –14.6 dB	45 –37.1 dB
8 0.7 dB	30 –26.9 dB	33 –29.1 dB	42 –33.9 dB	48 –39.7 dB	55 –53.0 dB	39 –32.5 dB	61 –61.3 dB
3 20.8 dB	10 –6.6 dB	12 –8.2 dB	15 –10.8 dB	20 –18.6 dB	35 –30.2 dB	16 –12.3 dB	44 –35.4 dB

Table 11.2 Energy distribution of different DTT approximations for image Lena: (b) DTT approximation in Oliveira et al. (2016)

(b)

1 72.1 dB	2 29.8 dB	3 7.3 dB	5 –3.8 dB	7 –6.2 dB	6 –4.8 dB	11 –13.1 dB	4 0.8 dB
9 –10.0 dB	8 –8.3 dB	10 –12.5 dB	15 –24.4 dB	12 –22.6 dB	18 –25.2 dB	29 –29.5 dB	14 –24.3 dB
49 –33.1 dB	17 –24.8 dB	22 –26.9 dB	44 –32.0 dB	23 –28.0 dB	34 –30.3 dB	27 –29.0 dB	30 –29.6 dB
40 –31.7 dB	13 –23.4 dB	16 –24.6 dB	32 –29.9 dB	19 –25.2 dB	26 –28.8 dB	20 –25.9 dB	21 –26.2 dB
59 –37.8 dB	31 –29.8 dB	35 –30.8 dB	57 –36.7 dB	41 –31.8 dB	51 –33.7 dB	42 –31.9 dB	37 –31.3 dB
63 –38.6 dB	24 –28.3 dB	25 –28.8 dB	56 –36.0 dB	46 –32.4 dB	48 –32.9 dB	38 –31.5 dB	28 –29.4 dB
62 –38.1 dB	33 –30.3 dB	43 –32.0 dB	60 –37.9 dB	36 –31.1 dB	50 –33.4 dB	39 –31.6 dB	45 –32.4 dB
64 –40.8 dB	47 –32.8 dB	54 –35.1 dB	61 –38.1 dB	52 –34.2 dB	58 –36.7 dB	55 –35.9 dB	53 –34.5 dB

Table 11.3 Energy distribution of different DTT approximations for image Lena: (c) DTT approximation in Farsiani and Sodagar (2020)

(c)

1 72.1 dB	2 13.3 dB	20 –13.0 dB	15 –10.8 dB	36 –24.7 dB	6 –0.1 dB	50 –29.6 dB	4 3.0 dB
3 5.7 dB	13 –9.2 dB	24 –17.1 dB	14 –10.5 dB	38 –25.0 dB	8 –3.7 dB	35 –23.5 dB	5 2.2 dB
21 –13.4 dB	31 –20.8 dB	44 –27.7 dB	23 –15.2 dB	37 –25.0 dB	12 –8.5 dB	34 –21.8 dB	7 –0.8 dB
29 –20.2 dB	49 –28.8 dB	57 –34.0 dB	30 –20.3 dB	51 –30.9 dB	22 –13.7 dB	41 –26.8 dB	11 –5.9 dB
27 –19.2 dB	47 –27.9 dB	58 –34.0 dB	32 –20.8 dB	52 –31.6 dB	18 –12.7 dB	48 –28.3 dB	9 –4.8 dB
28 –19.7 dB	43 –27.3 dB	54 –32.2 dB	33 –21.4 dB	55 –32.9 dB	19 –12.7 dB	45 –27.8 dB	10 –5.3 dB
39 –26.1 dB	59 –34.4 dB	64 –40.2 dB	42 –27.1 dB	61 –35.8 dB	26 –19.0 dB	53 –31.9 dB	17 –11.4 dB
46 –27.9 dB	60 –34.5 dB	63 –39.4 dB	40 –26.3 dB	62 –37.6 dB	25 –18.5 dB	56 –33.4 dB	16 –11.3 dB

11.5 Construction of Quantisation Matrixs Using Integer Implementations of DTT

In the JPEG standard, the quantisation process consists of dividing each Discrete Cosine Transform (DCT) coefficient by a quantisation factor. This step is designed to eliminate low-magnitude coefficients-typically associated with perceptually insignificant image details-while simultaneously reducing the dynamic range of the remaining coefficients. The visual degradation resulting from this process is generally imperceptible to the human eye. However, it is crucial to recognize that not all transform coefficients carry the same amount of information: low-frequency components usually encapsulate the most significant visual data, whereas high-frequency components contain finer, less perceptually relevant details.

Image compression schemes based on the Discrete Tchebichef Transform (DTT) adopt a similar principle, employing a linear uniform quantisation approach to compress the transformed coefficients. In this context, low frequency coefficients are quantised using smaller step sizes to preserve crucial image content, while high-frequency coefficients are quantised with larger step sizes to enhance compression by discarding less relevant information. Consequently, the quantisation factor should be position-dependent within the transform block-larger for coefficients that convey minimal perceptual information. To this end, we propose a quantisation matrix specifically tailored to perceptual properties of integer-based DTT coefficients, thereby optimizing both compression efficiency and visual quality.

$$Q_p = \begin{bmatrix} 16 & 16 & 18 & 19 & 24 & 36 & 36 & 31 \\ 19 & 16 & 20 & 22 & 24 & 40 & 23 & 40 \\ 20 & 22 & 24 & 28 & 37 & 53 & 50 & 54 \\ 24 & 20 & 25 & 35 & 45 & 73 & 73 & 58 \\ 22 & 21 & 37 & 74 & 70 & 92 & 101 & 103 \\ 24 & 43 & 50 & 64 & 100 & 104 & 120 & 92 \\ 45 & 100 & 62 & 79 & 100 & 70 & 70 & 101 \\ 41 & 41 & 74 & 59 & 70 & 90 & 100 & 99 \end{bmatrix} \tag{11.14}$$

11.6 Construction of Quantisation and Inverse Quantisation Matrix Through the Integration of Diagonal Matrix *D*

From a hardware implementation standpoint, the multiplication introduced by the diagonal matrices in (11.3), (11.8) and (11.11) can be incorporated into the quantisation step.

11.6.1 Equivalent Quantisation Matrix

The DTT quasi-orthogonal transform of 2-dimentional data X is T:

$$\begin{cases} Y = C\ X\ C^T \\ \quad = (D\ T)\ X\ (D\ T)^T \\ \quad = D\ T\ X\ T^T\ D \end{cases} \tag{11.15}$$

where: X is the entry block, Y is the transformed block by the transform C, T^T denotes the transpose of the transform matrix T and $C = D\ T$.

Let:

$$Y^* = T\ X\ T^T \tag{11.16}$$

so

$$Y = D\ Y^*\ D \tag{11.17}$$

Let d is the element of the diagonal matrix, D. The entry of Y at position (i, j) is:

$$y(i, j) = d_i\ d_j\ y^*(i, j) \tag{11.18}$$

Assuming that the elements of the original quantisation matrix is $Q = [q(i, j)]$, the quantised coefficient $y^Q(i, j)$ is:

$$\begin{cases} y^Q(i, j) = \dfrac{y(i, j)}{q(i, j)} \\ \qquad\quad = \dfrac{d_i\ d_j\ y^*(i, j)}{q(i, j)} \\ \qquad\quad = \dfrac{y^*(i, j)}{\dfrac{q(i, j)}{(d_i\ d_j)}} \end{cases} \tag{11.19}$$

Hence, the quantisation matrix of integer approximate DTT can be developed from the conventional quantisation matrix element $q(i, j)$ dividing it by $d_i\ d_j$ that is $\dfrac{q(i, j)}{(d_i\ d_j)}$. Accordingly, the equivalent quantisation matrix is $D^{-1}\ Q\ D^{-1}$.

11.6.2 Equivalent to Inverse Quantisation Matrix

In the decoding process, the inverse quantisation matrix and the inverse transform are combined as follows:

$$\begin{cases} C^T\,(Y^Q\;Q)\;C = (D\;T)^T(Y^Q \otimes Q)(D\;T) \\ \qquad\qquad\qquad = T^T\;D\,[Y^Q \otimes Q]\;D\;T \\ \qquad\qquad\qquad = T^T\;(D\;Y^Q\;D) \otimes Q\;T \\ \qquad\qquad\qquad = T^T\;[Y^Q \otimes D\;Q\;D]\;T \end{cases} \tag{11.20}$$

The symbol $\otimes$ represents the matrix dot where $y(i, j)^Q$ denotes the quantisation coefficient of $y(i, j)$. This implies that $d_i\ d_j\ y(i, j)^Q$ must be calculated before performing the inverse transform. This step can be integrated with the inverse quantisation process to make the inverse quantisation step as $q(i, j)\ d_i\ d_j$. Therefore, when we use the integer approximate DTT transform matrix T to perform the forward and inverse transform, the quantisation and inverse quantisation of the matrices are different.

11.7 Experimental Results and Analysis

For a fair assessment of our proposed quantisation matrix of greyscale image compression, the JPEG-like standard is considered. The PSNR in our case is calculated from the Mean Squared Error (MSE) as follows:

$$PSNR = 10\ \log_{10} \frac{255^2}{MSE} \tag{11.21}$$

$$MSE = \frac{1}{N^2} \sum_{n=1}^{n} \sum_{m=1}^{n} [X(m, n) - Y(m, n)]^2 \tag{11.22}$$

The image compression/decompression system follows the framework of the DTT-based JPEG standard, with three important steps: transform (exact DTT or integer approximate DTT transforms), quantisation and entropy coding.

Table 11.4 Performance evaluation of the proposed quantisation matrix. ($PSNR$ expressed in dB)

Transforms	Quantisation matrices	Lena		Boat		Bridge	
		CR	*PSNR*	*CR*	*PSNR*	*CR*	*PSNR*
Exact DTT	Q_{JPEG}	13.01	29.80	11.71	28.31	08.29	24.49
	Q_{opt}	13.06	29.77	11.75	28.41	08.29	24.51
	Q_p	13.10	29.67	11.75	28.41	08.29	24.45
DTT approx. in Oliveira et al. (2016)	Q_{JPEG}	13.04	31.09	11.74	29.59	08.32	25.70
	Q_{opt}	13.23	31.10	11.92	29.61	08.41	25.71
	Q_p	13.33	31.77	11.86	30.01	08.37	26.02
DTT approx. in Farsiani and Sodagar (2020)	Q_{JPEG}	13.11	31.94	11.63	30.55	08.22	26.22
	Q_{opt}	12.84	32.31	11.78	30.59	08.30	26.23
	Q_p	12.86	33.40	11.70	31.19	08.28	26.59

For each transform, at the same compression ratio, we compute the PSNR using the standard quantisation matrix JPEG, Q_{JPEG}, the optimal quantisation matrix, Q_{opt} in (Xiao et al. 2016) for exact DTT-based JPEG standard and the proposed quantisation matrix, Q_P. The three standard greyscale test images of size 512 × 512: Lena, Boat and Bridge were used for performance comparison purpose. We choose These images for their different content and the amount of detail they contain, which is indicated by their parameters: spatial frequency measure (SFM) (Eskicioglu and Fisher 1995) and spectral activity measure (SAM) (Grgic et al. 2001). Table 11.4 shows the PSNR values obtained using similar compression ratio values for the various cases considered.

When using the exact DTT-based JPEG baseline, the PSNR values of our proposed quantisation matrix are slightly lower than those of other existing quantisation matrices. That is because our proposed quantisation matrix is involved to integer Techebichef transform approximations, which we can see an improvement of PSNR values when we used the DTT approximations. For the approximate DTT-based JPEG, the PSNR values appear the more important using the proposed quantisation

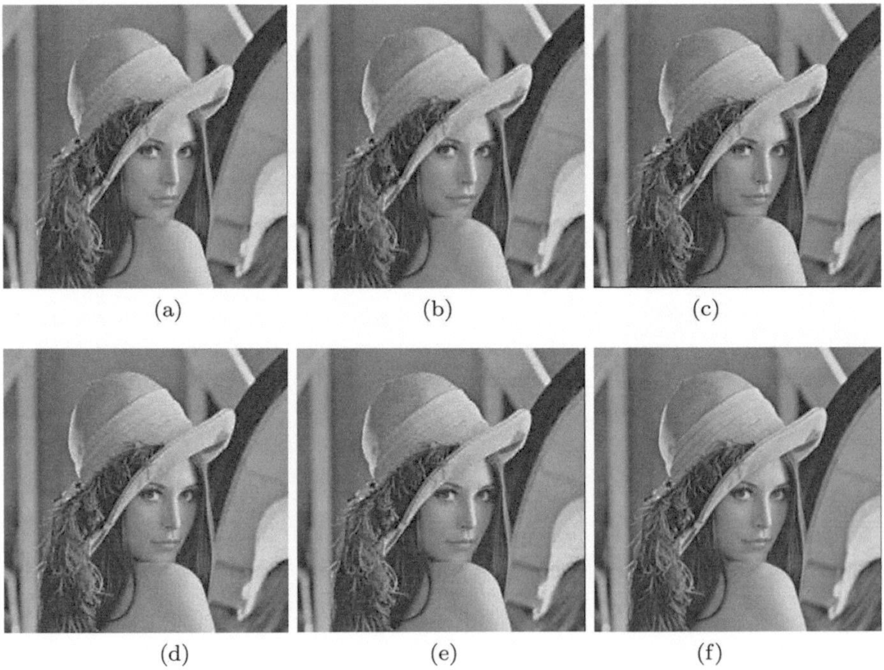

Fig. 11.1 The reconstructed image by different quantisation matrices: **a** DTT, Q_{JPEG}: CR = 17.84, PSNR = 29.19 dB, **b** DTT, Q_{opt}: CR = 17.35, PSNR = 29.42 dB, **c** DTT, Q_p: CR = 17.85, PSNR = 29.42 dB, **d** approximate DTT reported in Farsiani and Sodagar (2020), Q_{JPEG}: CR = 17.87, PSNR = 29.64 dB, **e** approximate DTT reported in Farsiani and Sodagar (2020), Q_{opt} : CR = 17.39, PSNR = 30.56 dB, **f** approximate DTT reported in Farsiani and Sodagar (2020), Q_p: CR = 17.92, PSNR = 30.65 dB

for all considered test images, and this is for the two considered approximate DTT. For instance, an improvement of about 0.6 dB is obtained when we compare our proposed quantisation matrix with the optimal quantisation matrix Q_{opt} proposed in Xiao et al. (2016).

Figure 11.1 illustrates the reconstructed images obtained using both the exact DTT and its approximate version, each evaluated with different quantisation matrices.

As it can be seen from Fig. 11.1a–c, the quality of the reconstructed images using the quantisation matrices Q_{opt} or Q_p is slightly better than that of Q_{JPEG} when employing the exact DTT transform in the JPEG standard. Morever, by Using the approximate DTT (Farsiani and Sodagar 2020) in JPEG standard, it can be shown that the visual quality of the images reconstructed with the proposed quantisation matrix is superior to that of the two alternative quantisation matrices. Note that the transform in Farsiani and Sodagar (2020) requires less number of arithmetic operations compared to the exact DTT transform.

11.8 Conclution

In this chapter, a new and efficient approach based on integer Tchebychev approximations for an efficient image compression technique has been proposed. Recently, integer Tchebychev approximations appear to be the most recent transforms that still attract the interest among the data compression researchers.

The JPEG quantisation matrix is destined for the Discrete Cosine Transform, for the purpose of designing a new quantisation matrix suitable for integer DTT-based JPEG standard, we investigated the energy distribution of some integer DTT approximations.

The experimental results show that the Tchebychev transform appears the same as the discrete cosine transform with the same compression rate, but integer DTT approximations combined with the proposed quantisation matrix outperforms the DTT combined with the optimal quantisation matrix by an average about 0.6 dB in terms of PSNR. Furthermore, compared to the JPEG quantisation matrix, our quantisation matrix offers an improvement of more than 0.43 dB is achieved.

References

Brahimi, N., Bouden, T., Brahimi, T., Boubchir, L.: A novel and efficient 8-point dct approximation for image compression. Multimedia Tools Appl. **79**(11), 7615–7631 (2020)

Brahimi, N., Bouden, T., Brahimi, T., Boubchir, L.: Efficient multiplier-less parametric integer approximate transform based on 16-points dct for image compression. Multimedia Tools Appl. **81**(26), 37723–37746 (2022)

Brahimi, N., Bouden, T., Brahimi, T., Boubchir, L.: Lossy image compression based on efficient multiplier-less 8-points dct. Multimedia Syst. **28**(1), 171–182 (2022)

Cai, L., Qian, Y., He, Y., Feng, W.: Design of approximate multiplierless dct with csd encoding for image processing. In: 2021 IEEE International Symposium on Circuits and Systems (ISCAS), pp. 1–4. IEEE (2021)

Chan, K.-H. Im, S.-K.: Discrete tchebichef transform for versatile video coding. In: Proceedings of the 2021 International Conference on Multimedia Retrieval, pp. 623–626 (2021)

Deepsita, S.S., Divya, K., Mahammad, S.N.: Energy efficient and multiplierless approximate integer DCT implementation for HEVC. In: 2021 IFIP/IEEE 29th International Conference on Very Large Scale Integration (VLSI-SoC), pp. 1–6. IEEE (2021)

Eskicioglu, A.M., Fisher, P.S.: Image quality measures and their performance. IEEE Trans. Commun. **43**(12), 2959–2965 (1995)

Farsiani, S., Sodagar, A.M.: Hardware and power-efficient compression technique based on discrete tchebichef transform for neural recording microsystems. In: 2020 42nd Annual International Conference of the IEEE Engineering in Medicine & Biology Society (EMBC), pp. 3489–3492. IEEE (2020)

Grgic, S., Mrak, M., Grgic, M.: Comparison of jpeg Image Coders, p. 3. University of Zagreb, Faculty of Electrical Engineering and Computing Unska (2001)

Ishwar, S.: Discrete tchebichef transform and its application to image/video compression. Ph.D. thesis, Concordia University (2008)

Ishwar, S., Meher, P.K., Swamy, M.: Discrete tchebichef transform-a fast 4 × 4 algorithm and its application in image/video compression. In: 2008 IEEE International Symposium on Circuits and Systems, pp. 260–263. IEEE (2008)

Ji, L., Zhi-Guo, Z., Bin, X., Ze-Lin, Y., Dun, W.: Based on discrete orthogonalt chebichef transform for image compression. Classification No. of Chinese Library Classification: TP391 [A], pp. 12–4261 (2013)

Mukundan, R., Ong, S.H., Lee, P.A.: Image analysis by tchebichef moments. IEEE Trans. Image Process. **10**, 1357–1364 (2001)

Oliveira, P.A., Cintra, R.J., Bayer, F.M., Kulasekera, S., Madanayake, A.: Low-complexity image and video coding based on an approximate discrete tchebichef transform. IEEE Trans. Circuits Syst. Video Technol. **27**(5), 1066–1076 (2016)

Prattipati, S., Ishwar, S., Swamy, M., Meher, P.K.: A fast 8 × 8 integer tchebichef transform and comparison with integer cosine transform for image compression. In: 2013 IEEE 56th International Midwest Symposium on Circuits and Systems (MWSCAS), pp. 1294–1297. IEEE (2013)

Wallace, G.K.: The jpeg still picture compression standard. Commun. ACM **34**(4), 30–44 (1991)

Xiao, B., Lu, G., Zhang, Y., Li, W., Wang, G.: Lossless image compression based on integer discrete tchebichef transform. Neurocomputing **214**, 587–593 (2016)

Yuting, W., Xuedong, L., Zaijun, H.: Implementation of an image compression algorithm based on low complexity integer approximate discrete tchebichef transform. In: 2020 IEEE 4th Information Technology, Networking, Electronic and Automation Control Conference (ITNEC), vol. 1, pp. 2183–2187. IEEE (2020)

Chapter 12
Ant Colony Optimization Algorithm for Multiple Sclerosis Lesion Segmentation Based on MR Brain Image

Dalenda Bouzidi, Fahmi Ghozzi, and Ahmed Fakhfakh

Abstract Recently, Medical imaging is the process by which a physician can examine the inside of a patient's body without operating it through the development of devices for the diagnosis and treatment of patients. One intriguing medical imaging method that is thought to be a highly helpful tool for identifying the tumor growth of multiple sclerosis is magnetic resonance imaging (MRI). An efficient method for separating brain tumors from MRI pictures is segmentation. For the automatic identification of MS outliers, a number of recent methods for the segmentation and classification of MRI sequences have been put forth. The "Ant Colonies Optimization ACO" meta-heuristic technique is presented in this paper for MRI image segmentation. In order to optimize their overall rendering and compare with consensual segmentation, we suggest using the ant colony technique to estimate the segmentations of brain MRI images from a public MR dataset of 30 MS patients that were imaged with a 3T MR scanner using conventional sequences. We used MATLAB GUI program to evaluate the performance of this algorithm optimized by maximizing the intra-class variance criterion MVar.

Keywords Multiple sclerosis · Outliers · Segmentation · ACO · Consensus · MVar

D. Bouzidi (✉)
SM@RTS, Digital Research Centre of Sfax, University of Sfax, Sfax, Tunisia
e-mail: dalendabouzidi91@gmail.com

F. Ghozzi
Signals, Systems, Artificial Intelligence and Networks (SM@RTS), National School of Electronics and Telecommunications of Sfax, Sakiet Ezzit, Tunisia
e-mail: Fahmi.ghozzi@enetcom.usf.tn

A. Fakhfakh
Laboratory of Signals, Systems, Artificial Intelligence and Networks (SM@RTS), National School of Electronics and Telecommunications of Sfax, Sakiet Ezzit, Tunisia
e-mail: ahmed.fakhfakh@enetcom.usf.tn

N. Derbel and Q. Zhu (eds.), *Advances in Communication, Signal and Image Processing*,
https://doi.org/10.1007/978-981-95-0401-5_12

12.1 Introduction

Multiple sclerosis is a chronic neurological disease that primarily affects women and affects immature adults (those aged 20–40). Inflammatory damage and demyelination of the central nervous system (CNS) are hallmarks of multiple sclerosis (MS). This leads to the loss and destruction of the axons' myelin sheaths, along with a glial scar and the formation of sclerosis plaques throughout the brain and spinal cord.

The progression of the disease into 4 progressive forms which are as follows (Camara et al. 2008): Relapsing Remitting Multiple Sclerosis (RR-MS): also known as cyclic remitting. This form is characterized by the succession of distinguished and interspersed outbreaks, each followed by a more or less complete duration of remission. It is also marked by the lack of progression between pushes. Secondary Progressive Multiple Sclerosis (SP-MS): This form appears after a first relapsing-remitting form. It is marked by the onset of a phase that is gradually accentuated with the appearance or not of new outbreaks. The handicap gradually worsens without any clear remissions. Usually people with relapsing-remitting disease will develop progressive disease after 15 years of being diagnosed with the first form.

Primary Progressive Multiple Sclerosis (PP-MS): also referred to as the primary progressive form. This aspect is distinguished by a slow and constant accumulation of disability, sometimes accompanied by periods of ephemeral improvement.

Progressive Relapsing Multiple Sclerosis (PR-MS) from the start: This phase begins with the early stages of MS. It is marked by the appearance of additional outbreaks and more or less complete recovery phases. The progression of the handicap persists between two attacks. The diagnosis and analysis of multiple sclerosis are significantly impacted by MRI. Additionally, MRI can be utilized to conduct retrospective studies that improve our understanding of the disease and test the efficacy of a treatment. However, this does not preclude the existence of alternative systems that offer improved performance, robustness, and dependability in identifying pathological lesions. This work's main objective is to demonstrate the efficacy of a metaheuristic known as "Optimization by Ant Colonies". To identify multiple sclerosis outliers, we suggest using ant colony meta-heuristics to estimate brain image segmentations from MRI scans and compare them with the consensus segmentation that is obtained. Our contribution primarily focuses on optimizing the various parameters employed in this method to get satisfactory consensus segmentation results. The structure of this work is as follows: Following the presentation of relevant literature in Sect. 12.2, the metaheuristics algorithm employing the database and the ant colony optimization technique employed are introduced in Sect. 12.3, and then in Sect. 12.4, the obtained results are represented. Finally, we would end up with a conclusion and suggest some perspectives.

12.2 Related Works

The literature has a number of modern segmentation techniques. The most well-known classification techniques include fuzzy partitioning, also known as FCM (Dunn 1974), genetic algorithms (GA) (Maulik and Bandyopadhay 2000), and the mobile center algorithm, also known as K-Means (Selvakumar et al. 2012). A computer-assisted technique for cerebral MRI segmentation was proposed by Senthil et al. (2014). It uses a Kmeans classification algorithm and morphological filtering to detect tumor localization. This method avoids poorly grouped areas and allows the tumor to be separated from various cerebral MRI images. However, it yields varied outcomes for varying numbers of groups. We may also note the work of (Carass et al. 2017) which used thresholding approaches with Lesion-TOADS and voxel connectivity analysis. In another study Souplet et al. (2008) an automatic threshold for the FLAIR image was used to pick tumors that corresponded to acute areas. Schmidt et al. (2012) employed a region growth algorithm to revise a threshold on a probabilistic lesion map that was computed based on brain segmentation. Furthermore, the spatial, statistical, and textural features which were retrieved using the Volumetric Grayscale Co-occurrence Matrix (GLCM) and the Grayscale Path Length Matrix (GLRL) are taken into consideration in the submitted method. This optimization was carried out by combining an SVM tool with a genetic algorithm (Schmidtet al. 2012). Many studies have previously been conducted, including (Sahnoun et al. 2018), to enhance MRI image contrast so that healthy tissues and diseased tissues impacted by MS pathology can be seen more clearly. The performance of white matter-based lesion segmentation in brain magnetic resonance images with multiple sclerosis is evaluated using yet another method under different circumstances, including varying degrees of pretreatment, image augmentation, and training data set composition (Ravnik 2018). Additionally, the authors of the research study by Devers and Fauvel (2017) recommend that researchers use the ACO meta-heuristic optimization algorithm. In order to improve an image segmentation criterion and increase the readability of an MRI picture, the goal is to find and use the ant colony algorithm. Once more, the objective will be to identify the gray levels in a given image that will make it as "sharp" as possible and allow the various items to be most easily distinguished from one another. This is especially helpful when it comes to MRI images; in fact, it might help visualize the areas of the brain with lesions (or "holes") more clearly, which would speed up the process of identifying any problems. In order to confirm and validate our findings based on the consensus segmentation, this motivates us to use the ACO metaheuristic algorithm on this new database.

12.3 Methods and Materials

12.3.1 Database

In this study, the evaluation of Ant Colony Optimization algorithm was carried out using one real and novel database, including 30 MS patients created by Lesjak et al. (2017) and it is available in the site web (http://lit.fe.uni-lj.si/tool), each patient's MR scans consisted of 2D T1-weighted, 2D T2-weighted, and a 3D FLAIR where each type of MRI image has specific parameters are present in the Table 12.1. Figure 12.1 represents the 30 MS patients of the database images.

Table 12.1 MR image of dataset setting

	T1	T2	FLAIR
Dimension	2D	2D	3D
Repetition time TR (ms)	2000	5000	6000
Echo time TE (ms)	20	120	392
Inversion time TI	800 ms		1800 ms
Flip angle FA	120	120	20
Sampling	0.42 × 0.42 × 3.30 mm	0.57 × 0.57 × 3.00–3.33 mm	0.47 × 0.47 × 0.88 mm

12.3.2 ACO Algorithm

In order to accomplish this task, we have drawn inspiration from a few studies that discuss the optimization of ant colonies for picture segmentation in the literature. This approach uses a colony of artificial ants as a heuristic to solve discrete combinatorial problems. At gray level 0, we have the starting place of the ants, which will not be considered because it might be fascinating. As a result, the ants will travel on the gray levels. Then, depending on a number of factors, including the quantity of pheromones, the ants move with a specific potential from one gray level to another. The ants will offer a tuple of multiple thresholds after their journey is complete (Fig. 12.2).

According to the following studies Devers and Fauvel (2017), Nakib (2008) whereby researchers are encouraged to use the ACO meta-heuristic optimization technique by the authors Dorigo and Blum (2005) was the first to propose this optimization algorithm. The addition of a parameter that changes ant visibility is the only change made to the original method. Nevertheless, the obtained results are not particularly satisfactor.

Then we suggest the metaheuristic ACO algorithm in our article (Bouzidi et al. 2019) to improve the image segmentation method by maximizing the intraclass variance. This allows us to get as close as possible to the optimal solution. We make modifications using the intra-class variance criterior MVar (Nakib et al. 2005). A group of ants serves as the foundation for RunACO's overall operation. Each ant will propose segmentation thresholds. Each ant's image result from these thresholds will be assessed, and the best image will be retained. The various pitches of the method proposed in Fig. 12.3 will be explained next: The implementation is carried out in multiple stages, with each ant proposing a threshold and being tested in each round. The "get" method is used to retrieve the graphical interface's parameters, which are then arranged in RunACO.m in a comprehensible manner. We obtain the image data, the number of thresholds, the pheromone evaporation rate, the number of ants, the stopping circumstances, and the alpha and beta, which we will talk about later. The number of iterations, or the number of iterations without a change in the optimal solution, is used to determine the halting conditions. The 256 × 256 matrix

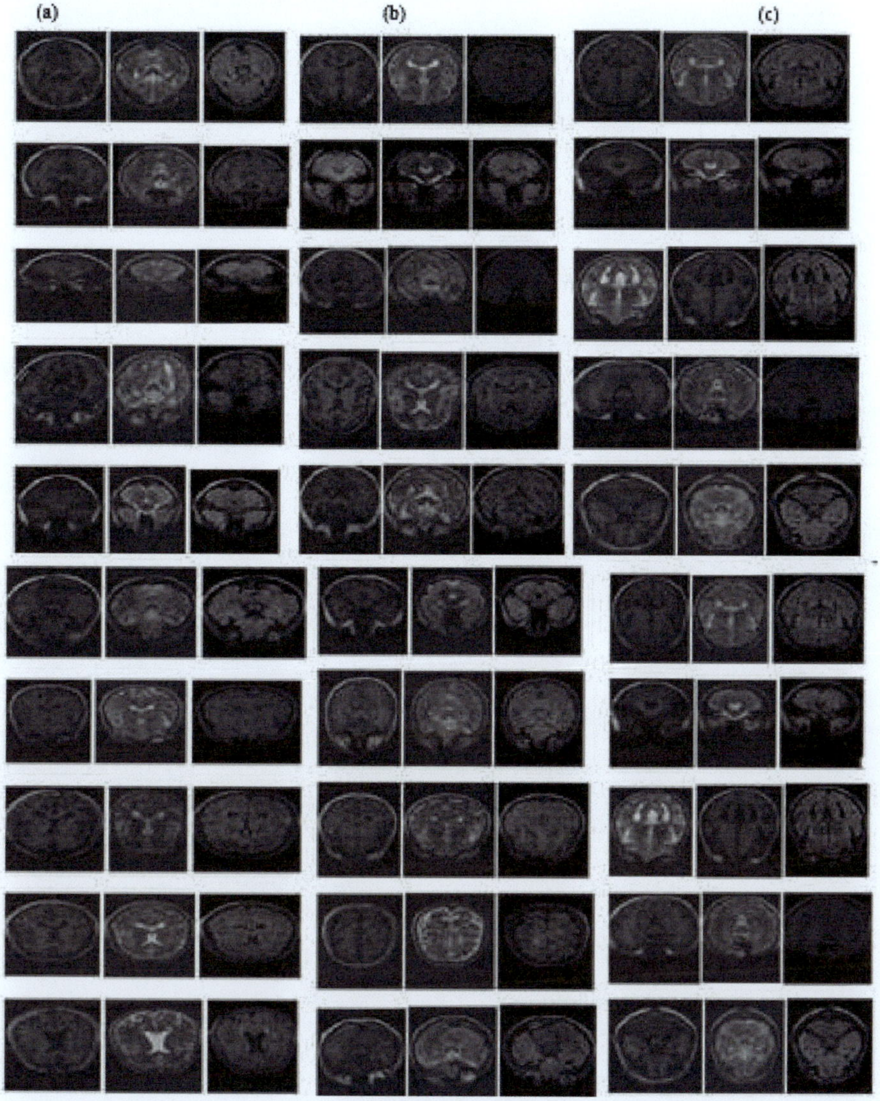

Fig. 12.1 The database of 30 MS patients (vertical): **a** patient 1 to patient 10, **b** patient 11 to patient 21, **c** patient 22 to patient 30, with 3 MR images T1-w, T2-w and FLAIR (horizontal)

of pheromones serves as the foundation for the entire program. The code refers to it as "transi". The number of pheromones on the jump from threshold i to threshold j is taken for each i, j. Recall that a solution is constructed as a collection of thresholds, and that the solution is expanded by each threshold until the required number of thresholds is attained. Therefore, even though these values are weights rather than probabilities, we may say that the transi matrix is in a sense an incidence matrix. The

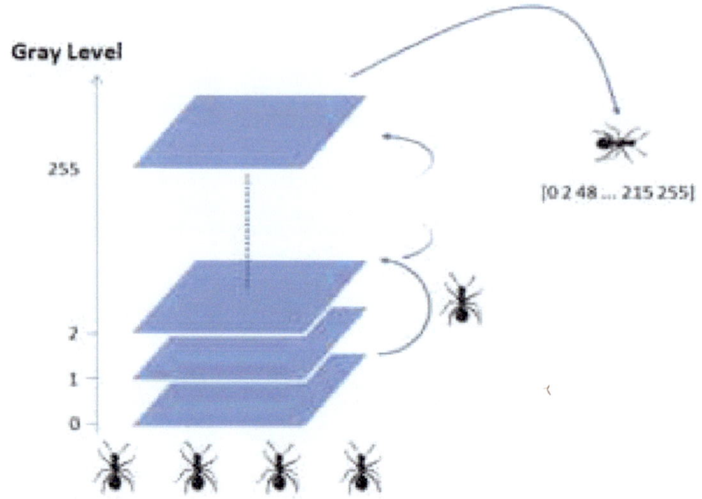

Fig. 12.2 Diagram illustrating the movement of ants on the grayscale

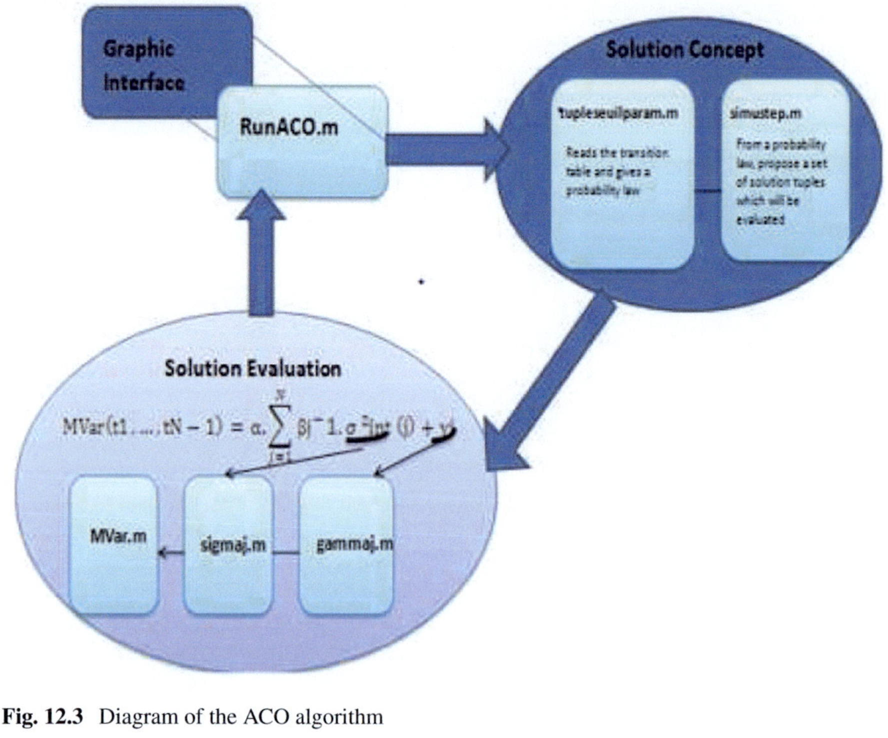

Fig. 12.3 Diagram of the ACO algorithm

program "tupleseuilparam.m" is used to construct these tuples, but it also requires two additional parameters.

These parameters are taken into account in the design of the solution. "tupleseuil param.m" selects a row of the transition matrix (pheromones) and applies the α to it. We then multiply each element of the list by the visibility to the power β. We then divide the list by the sum of its terms, we then obtain a probability law (The sum of the terms is equal to 1 and each of the positive or zero values gives a probability).

This probability law is used in a "simustep.m" simulation which will generate a random threshold. This process is repeated several times per ant, in order to constitute tuples of thresholds. The program will keep the best one according to the optimization of the best value of the intraclass variance MVar (1.3). This operation is repeated a certain number of times for each ant.

Initially, k ants will move randomly on gray levels. Then, other ants are allowed to move on the gray levels n times according to a given probability, and deposit a quantity $\Delta\ \tau i, j(t)$ of pheromones at each displacement. $\tau i, j(t)$ represents the probability of jumping from a threshold i to a threshold j. So we have the formula:

$$\Delta\tau_{i,j}(t) = \begin{cases} \frac{1}{L^k(t)} & \text{if} \quad (i, j) \in T_K(t) \\ 0 & \text{if} \quad (i, j) \notin T_K(t) \end{cases} \tag{12.1}$$

where $T_K(t)$ is the solution provided by the ant k at iteration t, and $L^k(t)$ is the sum of the segmentation thresholds contained in the solution. This means that on its way, the ant will drop at each of its jumps the inverse of the sum of the thresholds it will have traveled. At each loop turn, the pheromones evaporate according to the following expression:

$$\tau_{i,j}(t+1) = (1-\rho)\tau_{i,j}(t) + \sum_{k=1}^{m} \Delta\tau_{i,j}(t) \tag{12.2}$$

where $\Delta\ \tau i, j(t)$ is the quantity of pheromones deposited by the ant, m the number of ants used at the iteration t and ρ the evaporation rate which is fixed. Initially, pheromones are initialized by a small amount $\tau_0 \geq 0$ Moreover, at each iteration, the best thresholds will be selected thanks to the MVar criterion:

$$MVar(t1, ..., tN-1) = \alpha \cdot \sum_{j=1}^{N} \beta_j^{-1}\, \sigma^2\, \text{int}(j) + \gamma j \tag{12.3}$$

When the stop criterion is exceeded the algorithm then stops the number of iterations given for which the $MVar$ criterion does not allow to find better segmentation thresholds and does not allow to execute: where N represents the number classes to segment (assumed to be known a priori), $\sigma^2\text{int}(j)$ the intraclass variance of the class j, t_1 to t_{N-1} the selected segmentation thresholds, and βj, γ_j and α the variance correction parameters intraclass used in the Otsu method (Otsu 1979). α is $(1000 \times M) - 1 \times \sqrt{NR}$ where M is the total number of pixels in the image and $N(R)$ is the number of regions, that is, the number of related components in the segmented image. This term is used to normalize and penalize over-segmented images. The term $\beta j = (1 + \log(N(j)))$ is chosen so that the term $\beta j^{-1} \sigma^2 \text{int}(j)$ is weak for large classes. Nj is the number of pixels in class j. The term $\gamma_j = C(Nj)/Nj$ is high when the segmented image has many regions of the same size, especially if they are small. $C(Nj)$ is the number of regions whose cardinal is equal to Nj for large regions, $C(Nj)$ is, in most cases, equal to 1, whereas for small regions it becomes greater than 1.

So, according to the research work of Devers and Fauvel (2017), the optimization problem comes from parameter β which is very important. When β is so elevated, we can notice from Fig. 1.5 that it prevented us from having relevant images because they were simply unreadable with very small thresholds and we cannot detect any lesion (Fig. 12.4).

Next we evaluated the variation of parameter β as a function of DSC coefficient which is detailed in Sect. 12.4 with respect to the consensus (Fig. 12.5); we obtain the following graph where when we increase the β the coefficient of similarity compared to the reference will directly decrease so we can not detect existing MS lesions. In our case, we take ($\beta = 0.5$) in order to maximize the DSC.

Once more, the number of ants, k, helps determine the implementation time in order to display the threshold tuples selected by each ant, which facilitates image segmentation (Fig. 12.6). Since more repetitions will be required for each ant to reach the top three segmentation thresholds, the time will grow proportionately when we increase the number of ants, k. The following values were selected for the various parameters of the applied method for all simulations following the values validation test: the initialization probability $\tau_0 = 0.01$, the evaporation rate ($\rho = 0.1$), and the number of ants $k = 20$. The other settings are defined in Table 12.2.

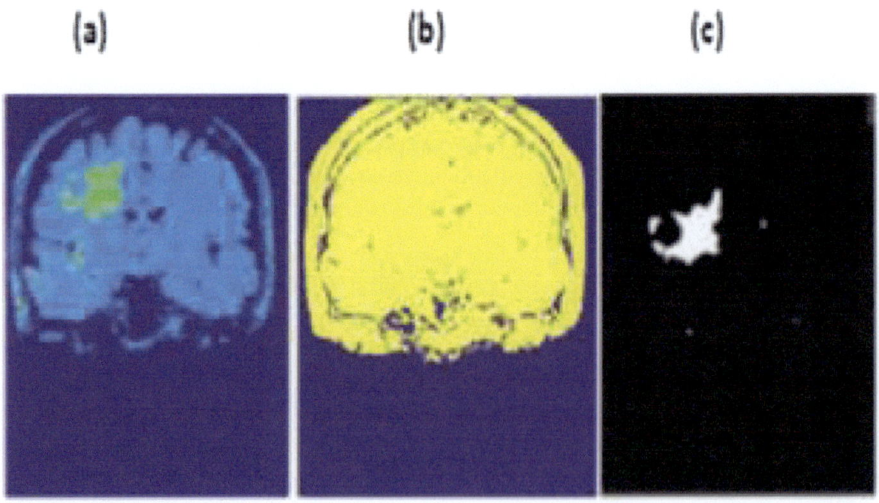

Fig. 12.4 Coronal Flair image of patient5: **a** segmented image with $\beta = 0.5$, **b** segmented image with $\beta = 0.5$, **c** consensus

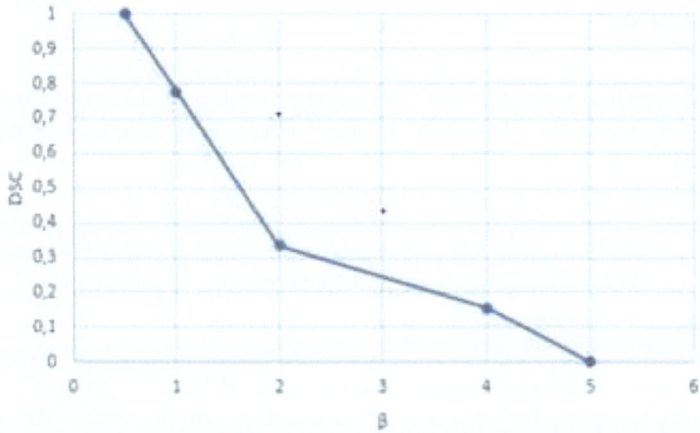

Fig. 12.5 Graphic of DSC in terms of β

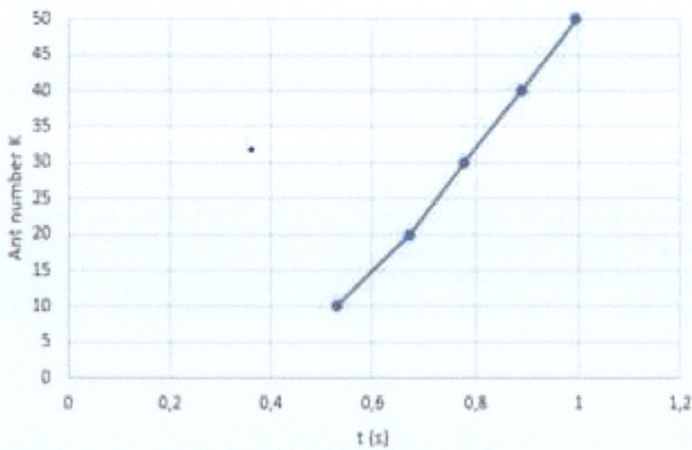

Fig. 12.6 Illustration of ant number k depending on time t

Table 12.2 ACO algorithm setting

Setting	Values
Ants number k	$20 \times N$
Evaporation coefficient ρ	0.1
α	1
β	0.5

We employ the implement MATLAB program, a computer-friendly language with an ergonomic interface, to apply this approach. Here, MATLAB's ergonomic features make it much easier to handle the material that was provided to us. MATLAB can be used for a variety of tasks, including as segmenting an image, displaying it, or even implementing a scalable method.

12.4 Results

The following values were chosen for the different settings of the applied method: the number of ants k = 20,the evaporation rate($\rho = 0.1$) and the initialization probability $\tau_0 = 0.01$. The other settings are defined in Table 12.2.

The generated segmentation and the associated consensus segmentation for each test image are compared in order to validate our method. The segmentation results on a diseased MRI picture in several classes are shown in Fig. 12.7. The outcome of segmenting the abnormal MRI picture of Fig. 12.7a is shown in Fig. 12.7b. In this instance, the system identified three classifications, enabling us to measure the cerebrospinal fluid's volume.

The results of segmentation on a pathological MR image in several classes are shown in this picture. The outcome of segmenting the abnormal MR image of Fig. 12.7a is displayed in Fig. 12.7b. In this instance, the algorithm detected the presence of three classes, which allows us to test the volume of the cerebrospinal fluid. We can also notice that the cortex was also well detected. The segmentation results for classes four and five are shown in Fig. 12.7c, d. Comparing this segmented image to the one in Fig. 12.7b, we can observe that the volume of the cerebrospinal fluid is well segmented, but the cortex is not well detected, whereas the lesions are well detected. Therefore, we see a white matter congestion and nearly no gray matter when there are three classes. There is a clear separation between the cerebrospinal fluid (CSF) and the rest of the structure. Additionally, a lightcolored patch that is nearly in the center of the white matter is visible. When there are five clusters, the component interpretation is entirely enhanced. Good gray matter, white matter, and CSF detection; the tumors light color in the brain's core makes it nearly impossible to miss. Figure 12.7b displays the segmentation results using the ACO method;

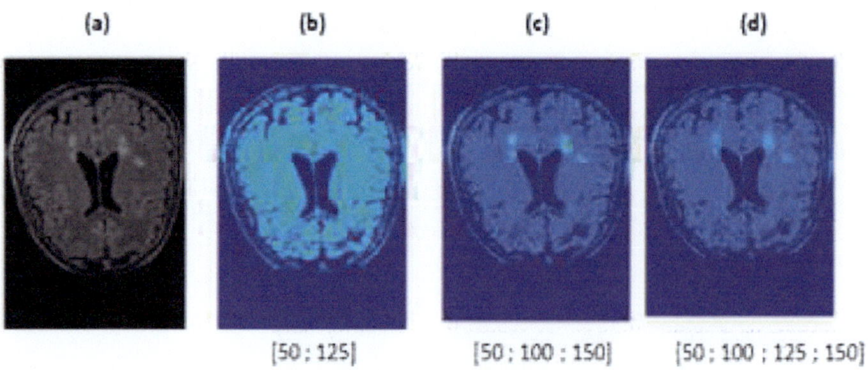

Fig. 12.7 Segmentation of pathological MRI images. **a** Original pathological image, **b** segmentation of the image into 3 classes (50; 128), **c** segmentation of the image into 4 classes (50; 100; 150), **d** segmentation of the image in 5 classes t = (50; 100; 125; 150)

the approach yields good results, with all lesions appearing (yellow in color) even though some lesions were mistakenly recognized. We evaluated the quality of the segmentation obtained in respect to a number of estimators frequently employed in the field of Richard et al. (2004):

Sensitivity: It corresponds to the True Positive Fraction (TPF).

$$Sensitivity(\%) = \frac{TP}{TP + FN} \tag{12.4}$$

Precision: It corresponds to False Positive Fraction (FPF) that demonstrates the ability to select all existing lesions corresponds to the value:

$$Precision(\%) = \frac{TP}{TP + FP} \tag{12.5}$$

Similarity: It also corresponds to the Kappa Index (or Dice coefficient) and measures the overlap between the segmentation obtained and the ground truth. The Dice Similarity Coefficient (DSC) (Senthil et al. 2014) can therefore be calculated with respect to consensus segmentation:

$$DSC(\%) = \frac{2 \times TP}{2 \times TP + FP + FN} \tag{12.6}$$

Our primary objective is to identify every outlier. A discovered region is referred to as a true positive (TP) if at least one voxel overlaps the reference; if not, it is referred to as a false positive (FP). A false negative (FN) is any labeling that our approach is unable to detect (Zijdenbos et al. 2002). Whose values vary between 0 and 1, and a value greater than 0.6 signifies a high degree of ground truth conformance. Our test is in compliance with the reference, as indicated by the Dice similarity coefficient value of 0.8 that was achieved for this test. As a result, we found more lesions than the ground truth (Fig. 12.8c).

We compute the sensitivity, the coefficient DSC, and the precision using the ACO algorithm on the novel database, as shown in Table 12.3.

Once more, the DSC, accuracy, and sensitivity values for the two techniques are extremely near to one another; the ACO algorithm's DSC values fall between 0.698 and 1.

Nonetheless, the following Formula (12.7) is used to determine the average DSC value for the ACO algorithm:

$$DSC_{mean} = \frac{1}{30}\sum_{i=1}^{30} DSC_i \tag{12.7}$$

We then find that, in comparison to the consensus segmentation, we achieved an 86% similarity. This indicates that the lesion segmentations produced by the

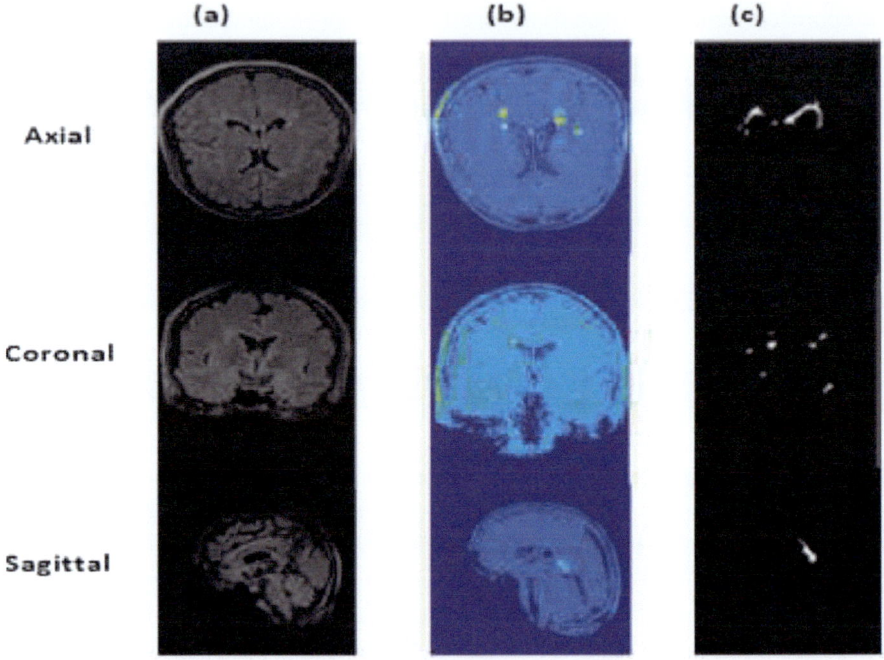

Fig. 12.8 Example of the result of ACO. **a** Original abnormal image (axial, coronal and sagittal FLAIR image slice of patient10), **b** our test, **c** reference

employed tumor segmentation algorithm were highly consistent and repeatable (Fig. 12.9).

These improvements are obtained thanks to the optimization of the parameters related to the intra-class variance. We can therefore estimate that we have succeeded in finding the optimal thresholds that produce an acceptable segmentation, by optimizing the MVar criterion using the ACO algorithm. We observe in the Fig. 12.10 a set of solution tuples of the pathological image of patient 10 (Fig. 12.10a) from the database seen previously.

The color's tint indicates the solutions' MVar; a terrible MVar is blue, and an excellent MVar is yellow. The segmentation with a greater Mvar value has very narrow thresholds, as can be shown. The image will be oversegmented by these criteria, and the visual rendering will be subpar. As a result, we can talk about the value of MVar in this image to explain the segmentation's relevance. Optimal thresholds were empirically determined in the prior analysis of this image (Fig. 12.10). Since these thresholds lack pictorial interest, they are always worse than solutions composed of small classes if they display a comparatively big MVar. Because the technique was created to address combinatorial optimization issues, this approach produces results that are acceptable. It is superior to other metaheuristics in terms of complexity and parameterization, nevertheless. The use of segmentation is still relevant in a variety

Table 12.3 Values of DSC, precision and sensitivity

Patient	DSC	Precision	Sensitivity
1	0.835	0.93	0.76
2	0.698	0.71	0.55
3	0.8	066	1
4	0.698	0.71	0.55
5	0.815	0.8	0.8
6	0.864	0.88	0.84
7	0.727	0.66	0.8
8	0698	0.71	0.55
9	0.863	0.86	0.86
10	0.8	0.57	0.8
11	0.822	0.68	0.86
12	1	1	1
13	0.769	1	0.53
14	0.888	0.8	1
15	0.836	0.97	0.73
16	0.857	0.85	0.85
17	1	1	1
18	0.8	1	0.8
19	0.909	1	0.83
20	0.8	0.92	0.89
21	0.914	0.91	0.94
22	0.938	0.92	0.95
23	0.665	0.76	0.67
24	1	1	1
25	0.792	0.62	0.8
26	0.827	0.8	0.76
27	0.7	1	0.7
28	0.988	1	0.9
29	0.657	0.8	0.65
30	1	1	1

of contexts, including the quantity of classes and their makeup. Since there isn't a single effective universal technique for MRI picture segmentation, the uniqueness of each solution presents a challenge. This type of issue can be difficult to resolve and may result in incorrect interpretations.

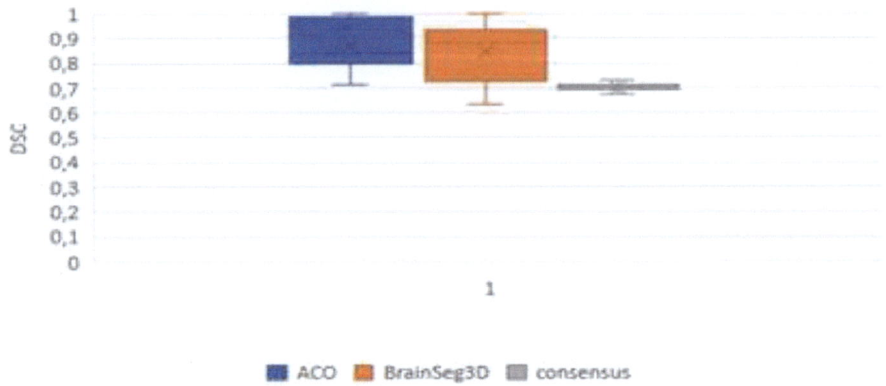

Fig. 12.9 DSC Graph for ACO, BrainSeg3D and Consensus of inter- and intra-rater variabilities

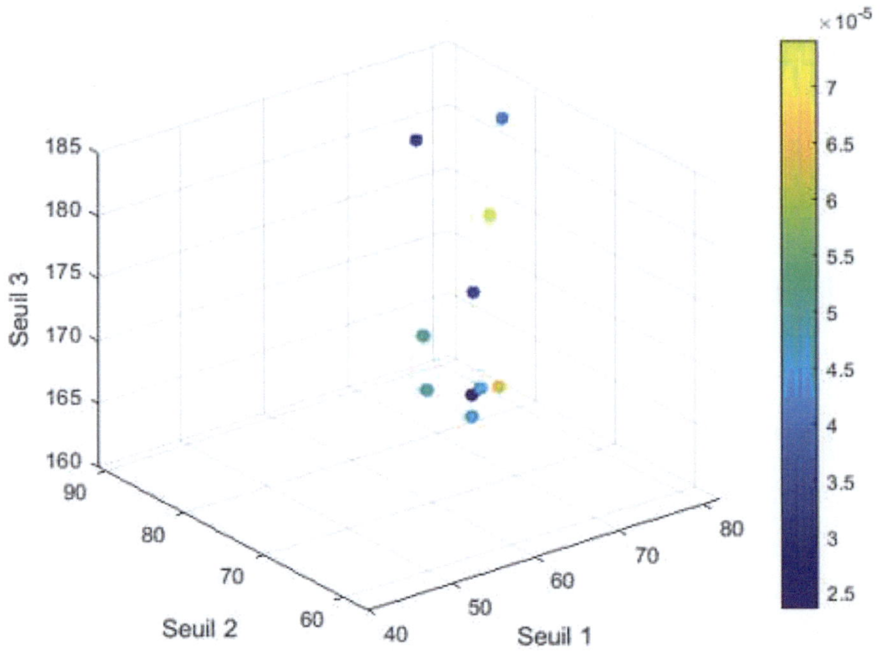

Fig. 12.10 Distribution of tuples threshold solutions

12.5 Conclusion

In this work, we have detailed the results of an optimized metaheuristic ACO algorithm applied on Matlab dedicated primarily to brain segmentation of MRI images. We compared the obtained results from our contribution applied to a new database by ground truth and other techniques used in the literature according to different criteria. Our optimized method has two fundamental advantages: flexibility, in the sense that it allows with the help of operators provided by the ACO algorithm, the generation of more adequate solutions for the segmentation of a given image and the optimization of different criteria. In addition, the method detects the correct clusters; this means that the extraction of the regions of the image is well done. The goal of brain image segmentation is to identify correctly the major structural tissues. Based on the visual analysis and the index defined by the Dice performance evaluation formula, our main objectives are achieved by obtaining satisfied results from the segmentation of medical images as well as setting better parameters. Since we were able to detect approximately all of the existing lesions for most of the patients in the database except for some falsely detected lesions compared to the consensus. In any case, we find that very small lesions are difficult to detect. Further improvements in the proposed treatment algorithms are needed if the goal is to detect MS disease early.

References

Bouzidi, D., Ghozzi, F., Fakhfakh, A.: Ant colony optimisation ACO to detect Multiple sclerosis lesions using magnetic resonance imaging. In: Bouktir, T., Neji, T. (eds.) Proceedings of the 4th International Conference on Recent Advances in Electrical Systems, pp. 73–77, Tunisia (2019)

Bouzidi, D., Ghozzi, F., Taouil, K., Fakhfakh, A.: Brain Seg 3D to detect Multiple sclerosis lesions using magnetic resonance imaging. In: IEEE International Conference on Design Test of Integrated Micro Nano-Systems, Tunisia (2020)

Camara, O., Schnabel, J.A., Ridgway, G.R., Crum, W.R., Douiri, A., Scahill, R.I., Hill, D.L.G., Fox, N.C.: Accuracy assessment of global and local atrophy measurement techniques with realistic simulated longitudinal Alzheimer's disease images. NeuroImage **42**, 696–670 (2008)

Carass, A., Roy, S., Jog, A., Cuzzocreo, J.L., Magrath, E., Gherman, A., Button, J., Nguyen, J., Prados, F., Sudre, C.H., Cardoso, M.J., Cawley, N., Ciccarelli, O., Wheeler-Kingshott, C.A.M., Ourselin, S., Catanese, L., Deshpande, H., Maurel, P., Commowick, O., Barillot, C., Tomas-Fernandez, X., Warfield, S.K., Vaidya, S., Chunduru, A., Muthuganapathy, R., Krishnamurthi, G., Jesson, A., Arbel, T., Maier, O., Handels, H., Iheme, L.O., Unay, D., Jain, S., Sima, D.M., Smeets, D., Ghafoorian, M., Platel, B., Birenbaum, A., Greenspan, H., Bazin, P.-L., Calabresi, P.A., Crainiceanu, C.M., Ellingsen, L.M., Reich, D.S., Prince, J.L., Pham, D.L.: Longitudinal multiple sclerosis lesion segmentation: resource and challenge. NeuroImage **148**(C), 77–102 (2017)

Devers, L., Fauvel, L.: Ter Miashs. Université Bordeaux (2017)

Dorigo, M., Blum, C.: Ant colony optimization theory: a survey. Theor. Comput. Sci. **344**, 243–278 (2005)

Dunn, J.: A fuzzy relative of the ISODATA process and its use in detecting compact well separated clusters. Int. J. Cybern. **V**(3:3), 32–57 (1974)

Ghribi, O., Sellami, L., Ben Slima, M., Mhiri, C., Dammak, M., Ben Hamida, A.: Multiple sclerosis exploration based on automatic mri modalities segmentation approach with advanced volumetric evaluations for essential feature extraction. Biomed. Signal Process. Control **40**, 473–487 (2018)

Lesjak, Ž., Galimzianova, A., Koren, A., Lukin, M., Pernuš, F., Likar, B., Spiclin, Ž.: A novel public MR image dataset of multiple sclerosis patients with lesion segmentations based on multi-rater consensus. Springer Science+Business Media, LLC (2017)

Maulik, U., Bandyopadhay, S.: Genetic algorithm-based clustering technique. Pattern Recogn. **V**(33), 1455–1465 (2000)

Nakib, A., Oulhadj, H., Siarry, P.: Modified intraclass variance minimization for segmentation of retinal angiographic images. In: Conférence de la Société Française de Génie Biologique et Médical (SFGBM), pp. 134–135, Nancy, Février (2005)

Nakib, A.: Conception de métaheuristiques d'optimisation pour la segmentation d'images. Application à des images biomédicales. In: Thèse de doctorat en traitement d'images. University of Paris 12 (2008)

Otsu, N.A.: Threshold selection method from gray-level histograms. IEEE Trans. Syst. Man Cybern. **9**(1), 62–66 (1979)

Ravnik, D.: Ataset variability leverages white-matter lesion segmentation performance with convolutional neural network (2018)

Richard, N., Dojat, M., Garbay, C.: Automated segmentation of human brain MRimages using a multi-agent approach. Artif. Intell. Med. **30**, 153–175 (2004)

Sahnoun, M., Kallel, F., Dammak, M., Mhiri, C., Ben Mahfoudh, K., Ben Hamida, A.: A comparative study of MRI contrast enhancement techniques based on Traditional Gamma Correction and Adaptive Gamma Correction: case of multiple sclerosis pathology. In: 4th International Conference on Advanced Technologies for Signal and Image Processing (ATSIP) (2018)

Schmidt, P., Gaser, C., Arsic, M., Buck, D., Förschler, A., Berthele, A., Hoshi, M., Ilg, R., Schmid, V.J., Zimmer, C., Hemmer, B., Mühlau, M.: An automated tool for detection of flair hyperintense white matter lesions in multiple sclerosis. Neuroimage **59**(4), 3774–3783 (2012)

Selvakumar, J., Lakshmi, A., Arivoli, T.: Brain tumor segmentation and its area calculation in brain MRimages using K-means clustering and Fuzzy C-menas algorithm. In: International conference on Advances in Engineering, Science,and Management(ICAESM) (2012)

Singh, S., Joseph, R.P., Manikandan, M.: Brain tumor MRI image segmentation and detection in image processing. Int. J. Res. Eng. Technol. **03**(01) (NC-WiCOMET-2014) (2014)

Souplet, J. C., Lebrun, C., Ayache, N., Malandain, G.: An automatic segmentation of T2-FLAIR multiple sclerosis lesions. In: Proceedings of the Multiple Sclerosis Lesion Segmentation Challenge Workshop (MICCAI '08), Proceedings in The MIDAS Journal, New York, NY, USA (2008)

Zijdenbos, A.P., Forghani, R., Evans, A.C.: Automatic pipeline analysis of 3-d mri data for clinical trials: application to multiple sclerosis. IEEE Trans. Med. Imaging **21**, 1280–91 (2002)

Chapter 13
Boosting Cancer Disease Detection Based on SVM and TwinSVM Classifications of MIRNA Expression

Ines Slimene, Imen Messaoudi, Afef Elloumi Oueslati, and Zied Lachiri

Abstract Nowadays people are suffering from different diseases caused by pollution, stress and bad nutrition. Doctors and biologists have to diagnose, detect and find corresponding treatment. Even most of the diseases are treated; some type of disease like cancer needs early detection and preventive treatment. In this paper, we provide a solution for early cancer disease detection. Our method consists of analyzing miRNAs expression for many cancer types leveraging a novel machine learning approach grounded in the Support Vector Machine framework: Twin-Support Vector Machine (twinSVM). The predicted area under the curve (AUC) yielded up to 92.44% over a cancer disease-miRNA association dataset with the polynomial kernel. The accuracy, precision and recall were 94.73%, 98.16%, and 96%, respectively. With our model, the achieved AUC outperformed those obtained by methods existing in the state of the art. Further, a comparison with SVM showed that TWSVM outperformed SVM in terms of accuracy, and AUC. Indeed, SVM with linear kernel reached 93.55% as accuracy and 91.84% as AUC value.

Keywords miRNA · TwinSVM · SVM · Disease detection

I. Slimene (✉) · Z. Lachiri
SITI Laboratory, Electrical Engineering Department, National School of Engineers of Tunis, University of Tunis El Manar, Tunis, Tunisia
e-mail: ines.slimene@enit.utm.tn

Z. Lachiri
e-mail: zied.lachiri@enit.rnu.tn

I. Messaoudi
Industrial Computing Department, Higher Institute of Information Technologies and Communications, University of Carthage, Tunis, Tunisia
e-mail: imen.messaoudi@enit.rnu.tn

A. E. Oueslati
Electrical Engineering Department, National School of Engineers of Carthage, University of Carthage, Tunis, Tunisia
e-mail: afef.elloumi@enit.utm.tn

N. Derbel and Q. Zhu (eds.), *Advances in Communication, Signal and Image Processing*,
https://doi.org/10.1007/978-981-95-0401-5_13

13.1 Introduction

MicroRNAs (miRNAs) are short, non-protein-coding RNAs that function as regulators of gene expression in various biological processes. Since their discovery in 1993 by Lee et al. (1993), studies have revealed the presence of more than 500 miRNAs that can target nearly 30% of human messenger ribonucleic acid (mRNA) (Lewis et al. 2005).

A miRNA interacts with one or more precise sequences in the 3'-UTR region of an mRNA (a single-stranded RNA molecule that is complementary to a gene DNA strand). If the complementarity between strands is imperfect, a repression of the translation of the target mRNA is generally observed.

The discovery of these small regulatory RNAs has, therefore, redefined the biological notions relating to organisms and to a spectrum of health problems including cancers. Indeed, until then, studies mainly focused on genes encoding proteins and are considered as the unique cause of DNA transformation. Since then, many studies have demonstrated a link between the deregulation of miRNA expression and the development of cancers.

MiRNAs participate in organism development; however, their deregulation goes hand in hand with certain cancers' appearance and development. In that case, studies destined to measure miRNA expression have revealed alterations in clinical samples of cancer patients (Calin and Croce 2006). The mapping of genes encoding these miRNAs confirmed these observations. They also revealed that miRNAs are present in chromosomal regions amplified or dehydrated in tumor cells (Calin et al. 2004).

The search for the miRNAs biological function and role in tumorigenesis opens a new investigation field in biomedical research because miRNA expression has been shown to be closely related to the onset and progression of specific cancers.

In the patient cells, miRNAs can be over or under-expressed. On the other hand, miRNAs have been identified as key regulators with tumor suppressor functions. MiR-15 and mir-16 are the first discovered tumor suppressors. Mir-21 is also a well-known miRNA. whose up-regulation is implicated in the pathogenesis of most cancer diseases such as glioma, lung, and liver cancers (Wu et al. 2015). Mir-200 is another example of miRNA which is involved in conjunctival MALT Lymphoma (Cai et al. 2012).

Hence, the microRNA expression became an invaluable tool in the study of multiple aspects of diseases from diagnosis to treatment. Nevertheless, traditional biological methods are money and time consumer and need a lot of sophisticated materials. Computational techniques are then proposed to promote miRNA-disease association studies while offering a minimum of both costs and treatment time.

Within this framework, many machine learning algorithms have been developed in recent years to predict cancer disease based on miRNA expression.

A large number of miRNAs are implicated in a disease appearance; which represents a high dimensional features dataset. Therefore, finding the most representative miRNA for each disease is another focus for the detection of the diseases. Thus, a features selection technique will be presented and used in this study to define the most important features for our datasets.

Chen et al. (2018) provided an effective model for miRNA-disease association detection using miRNA similarity and common diseases characteristics. The model was tested on five different cancer types' datasets (downloaded from HMDD v2.0 database). Results were, 90%, 98%, 98%, 98% and 100% for Colon Neoplasms, Lymphoma, Prostate Neoplasms, Breast Neoplasms, and Esophageal Neoplasms, respectively.

Another model was presented by Chen et al. (2017) to detect the association between miRNA and disease using a Ranking-based KNN method. The Obtained AUC was 82.21%.

Ang Li and al. have used dual random walk with restart and network projection technology for miRNA-disease association prediction. Results have shown an AUC value of about 0.9029 (Li et al. 2021).

Aside from the efficiency of cited approaches, a new machine learning model that outperformed the existing models appeared in recent researches of diseases detection. This model is Twin Support Vector Machines (Twin SVM). The main idea of TwinSVM is to determine two non-parallel hyperplanes by solving two SVM problems. Each hyperplane is closer to one of the two classes. As a consequence, the computational cost becomes 1/4 of the SVM one.

For example, the TwinSVM model was used by Sadewo et al. (2020) to detect pancreatic cancer. In this work, pancreatic data as CA, hemoglobin, leukocytes, hematocrit, and platelets were considered. These biological data present patient and normal samples. The best results were obtained with an RBF kernel producing an accuracy of 98%, a sensitivity of 97%, and a specificity of 100%.

Alam et al. (2017) based their work on twinSVM for Alzheimer's disease detection and LDA for the features extraction. Dataset considered here were obtained from the Alzheimer's Disease Neuroimaging Initiative (ADNI) database. This model yielded an accuracy of $92.65 \pm 1.18\%$.

The TwinSVM model was used also for heart disease detection. Dataset was obtained from he publicly available UCI machine learning repository. The considered attributes were: age, gender, Maximum Heart Rate Achieved, Serum cholesterol, etc. .... This model was compared with other classifiers such as K-Nearest Neighbors (KNN), Random Forest, Decision Trees, and Logistic Regression. In this work, an accuracy value of 90.72% was reached (Brik et al. 2021).

Note that all these works have used medical images or other clinical data, but they don't have dealt with miRNAs. On this basis, we have thought to develop a new disease prediction algorithm based on the RPM measure of miRNAs in features selection and different machine learning techniques (Slimene et al. 2021). The main idea of this work was to distinguish between healthy persons and persons with breast invasive carcinoma disease. In terms of accuracy, we reached 0.99 with Support vector machine (SVM) against 0.98 with both Logistic regression (LR) and K-nearest neighbors (KNN). These fruitful results constitute the principal motivation of this work. To fulfill optimized predictions of miRNA-disease associations, we opted for using SVM and we have sought one of its most successful variants to compare with. This variant is the twinSVM.

In conclusion, the aim of this paper is to present a new disease detection method using miRNA expression, SVM and twinSVM models. As for our dataset, we consider 1064 miRNA that can intervene in different disease types. Therefore, we need to minimize the number of features with the objective of reducing complexity and defining the most involved miRNA in a cancer disease. For this, we compare two different features selection techniques which are Elastic Net, Ridge, and Lasso regularization.

This chapter is composed of 5 sections. Section 13.2 presents our work's methodology, where we start by data collection and preprocessing. Then we apply a features selection method to choose the most representative features. Finally, we apply a SVM and TwinSVM models to our data which aims to classify them in normal and patient tissue. In Sect. 13.3 we show and discuss our results.

13.2 Methodology

The procedure for the cancer disease detection we followed here is provided in Fig. 13.1. It consists of using miRNA expression in the feature extraction, SVM and Twin Support Vector Machines (TWSVM) as classifiers.

The methodology consists of five steps. We start with data collection, followed by a preprocessing step. Then, we test and compare the performance of different features extraction models. The final steps are SVM and TWSVM models' train and validation. Different kernel types will be compared to find the representative model that best suits our data.

In the next subsections, we discuss in detail each component.

13.2.1 Data Collection

In this study, we are using miRNA expression derived from sequencing and collected from the Firehose Broad GDAC database (https://gdac.broadinstitute.org/).

Here, we consider 276 samples belonging to healthy subjects and 2475 other samples belonging to patients with four cancer diseases:

Fig. 13.1 Our model for the cancer disease detection

- **Breast invasive carcinoma**: This form of cancer originates in the epithelial cells of the milk ducts or lobules. It grows, then, into the breast tissue and can also spread to other body parts.
- **Glioma**: It is a brain tumors originating from glial cells, the supportive tissue surrounding neurons. It presents 33% of brain tumors.
- **Stomach and Esophageal carcinoma**: This cancer begins from abnormal cells that line the stomach. It can appear in the upper part of the stomach close to the esophagus or in the body of the stomach
- **Pan-kidney cohort**: Abbreviated as PAN, this cancer is the result of blood vessel inflammation that causes injury to organ systems. Pan can affect different organs such as nerves, heart, and kidneys.

Figure 13.2a provides the samples number by cancer type. Figure 13.2b presents the number of healthy and patient samples in total.

Our goal is to predict patient individuals independently from previous knowledge about the relationship between one of these diseases and miRNAs. For this, data samples are represented by the Read-Per-Millions expression value (RPM). RPM is an expression unit that offers a digital measure of gene or miRNA abundance (Wagner et al. 2012).

13.2.2 Data Pre-processing

Figure 13.2b shows that our data are unbalanced with 276 healthy samples against 2475 patients ones. If the skew in the class distributions is acute, accuracy and specificity measures can become inaccurate. To balance data, we use Synthetic Minority Oversampling Technique (SMOTE) (Lv et al. 2018). The goal of this technique is to create synthetic samples for the minority class which is in our case healthy samples class.

Figure 13.3 presents data repartition before and after applying the SMOTE technique. Figure 13.3a shows that we have more tumor samples than normal ones. After applying SMOTE technique, we have almost the same number of tumor and normal samples.

In the next step, we tend to normalize our data using a standardization method to put data on the same scale. The purpose is to prevent large-range features from affecting the metrics performance. In our work, data were scaled with standard Scalar and were spun around zero with a standard deviation value equal to one (Raju et al. 2020). Equation (13.1) gives the standard scalar formula:

$$Z_{scaled} = \frac{(X - \mu)}{\sigma} \tag{13.1}$$

with μ = is the data mean value, and σ = is the standard deviation.

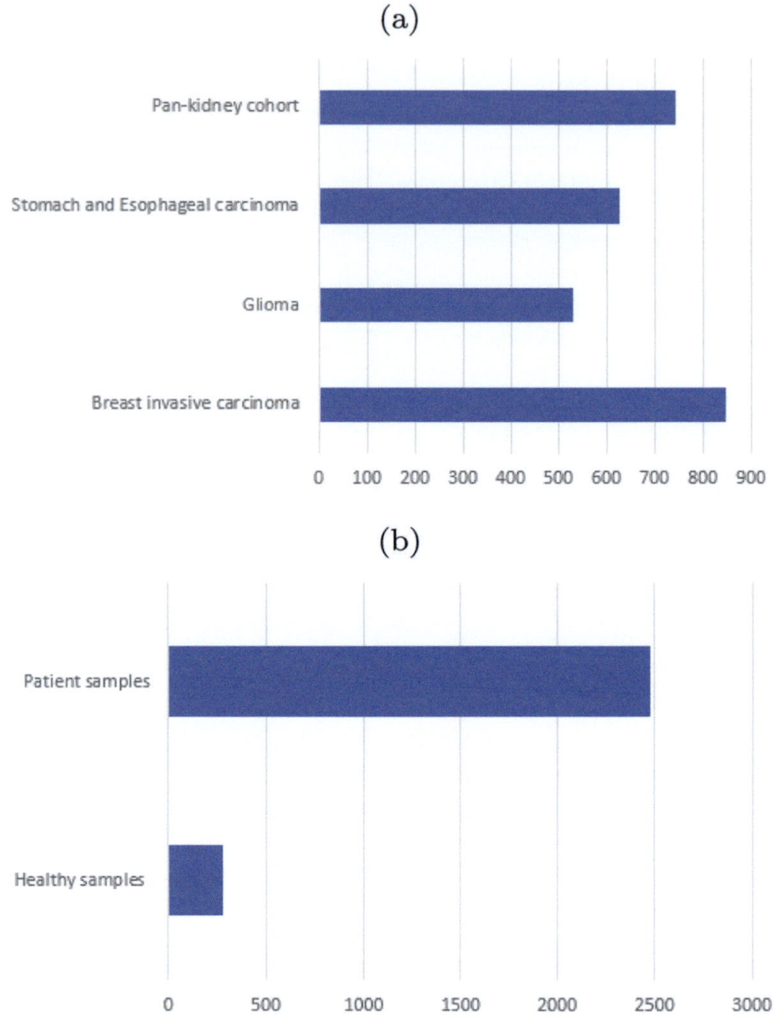

Fig. 13.2 Data repartition by cancer type (**a**) and samples type (**b**)

13.2.3 Features Selection

Dealing with a big number of features could give a poor validation results especially if the features number is close to the samples one. Thus, we can face an overfitting problem. Also, it is computationally expensive. Features Selection is an automatic or manual process of choosing a features subset which most contributes most in prediction. For this reason, many features selection methods have been proposed.

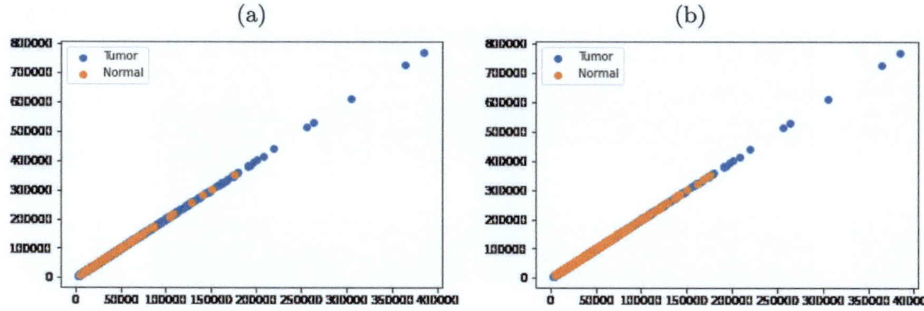

Fig. 13.3 Data repartition before oversampling (**a**) and after oversampling (**b**)

Four main types have been identified:

- Filter methods: features are selected based on the performance measures used independently of the model.
- Wrapper methods: features selection quality is evaluated by a classifier model.
- Hybrid methods: are a combination of the best performances of filters and wrappers.
- Embedded methods: features choice is conducted during modeling execution.

Herein, we have 1064 miRNAs for 2076 samples.

Figure 13.4 shows that some miRNAs are not involved in any disease, for example, hsa-let-7d, hsa-let-7f-1, and hsa-let-7i, are represented by a black color wich means that those miRNAs are not involved in diseases presented in this work. However, for hsa-let-7b, hsa-let-7f-2, hsa-mir-100 we can see lighter colors because they are involved in our diseases.

Therefore, we have to select the most representative miRNAs for our prediction system. For features selection we use an embedded method since the kind of this method works better with linear models like SVM. In the following, we present the features selection models which are Ridge (Buteneers et al. 2013), Lasso (Fonti and Belitser 2017), and ElasticNET (De Mol et al. 2009).

- Ridge: The ridge regression introduces an l2 penalty term to the cost function presented by Eq. (13.2). The purpose is to minimize useless feature coefficients by tuning hyper-parameter value and reduce overfitting risk.

$$\sum_{i=1}^{N_{training}} \left(y^{(i)}_{real)} - y^{(i)}_{pred)}\right)^2 + \alpha \sum_{j=1}^{n} a_j^2 \tag{13.2}$$

a_j denotes the coefficient corresponding to the the jth feature and α is the hyper-parameter that controls the strength of the l_2 regularization term which is the final term.

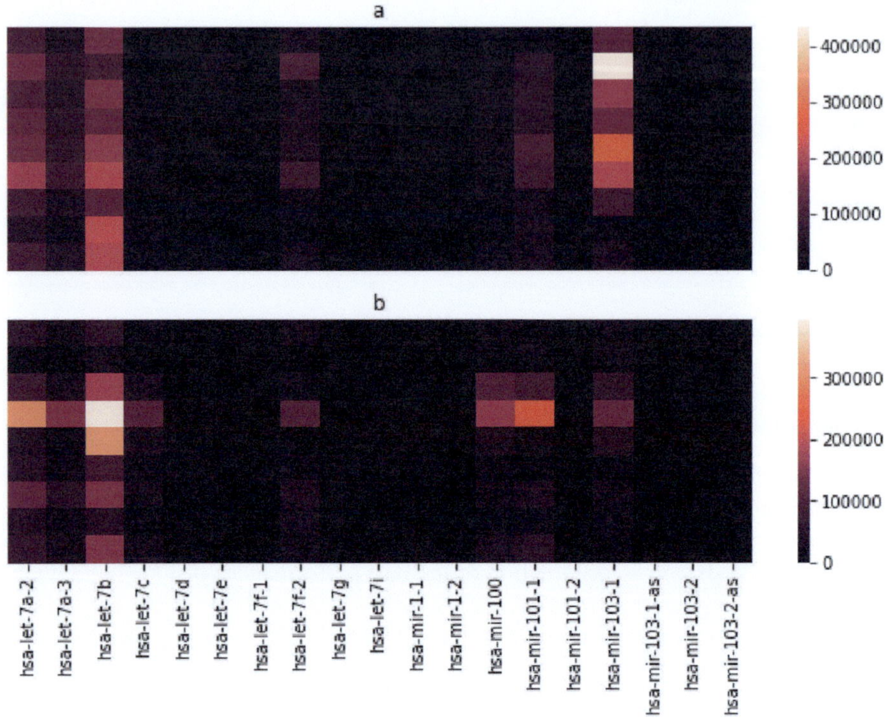

Fig. 13.4 MIRNAs involved health (**a**) and disease (**b**)

- Lasso: The least Absolute Shrinkage and Selection Operator (Lasso) has the same idea a the ridge regression, but it adds l1 penalty, rather than l2 penalty, to the cost function as seen in Eq. (13.3). The role of this penalty is to detect correlated features and put the coefficient of the less important ones to zero.

$$\frac{1}{2N_{training}} \sum_{i=1}^{N_{training}} \left(y_{real)}^{(i)} - y_{pred)}^{(i)}\right)^2 + \alpha \sum_{j=1}^{n} |a_j| \tag{13.3}$$

- Elastic Net: The elastic Net regression is a mix of the Ridge and Lasso models. Another hyperparameter r is added to tune the intensity of l1 penalty. The cost function then becomes:

$$\frac{1}{2N_{training}} \sum_{i=1}^{N_{training}} \left(y_{real)}^{(i)} - y_{pred)}^{(i)}\right)^2 + \alpha r \sum_{j=1}^{n} |a_j| + \frac{1}{2}\alpha(1-r) \sum_{j=1}^{n} a_j^2 \tag{13.4}$$

We have to correctly tune hyper-parameters α and r to obtain the lowest overfitting risk value.

13.2.4 SVM and TWSVM Classifiers

The support vector machines (SVM) is a learning method that was introduced by Vapnik (1999). It is usually used for solving classification or regression problems. The idea is to find an hyperplane that differentiates two classes in an n-dimension space by maximizing the margin distance. The Fig. 13.5 shows an illustration of the SVM classifier.

Since SVM gives good classification results with reduced computational cost, many SVM model variants have been proposed to achieve a better classification performance. One of these variants is TwinSVM model as proposed by Khemchandani et al. (2007). The main idea of the latter model is to find two non-parallel hyperplanes. Each is positioned near one class while maintaining maximum separation from the other. The Fig. 13.6 Illustrates TWSVM classifier.

For the linear case, Eqs. (13.5) and (13.6) present the two nonparallel hyperplanes formulations.

$$x^T\omega_1 + b_1 = 0 \tag{13.5}$$

$$x^T\omega_2 + b_2 = 0 \tag{13.6}$$

where ω_1 and ω_2 are weight parameters, b_1 and b_2 are bias terms, and x is the data vector.

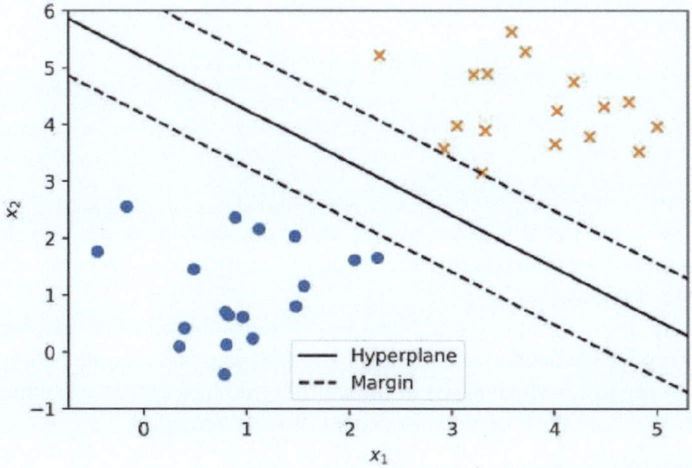

Fig. 13.5 SVM classifier illustration (Mir and Nasiri 2019)

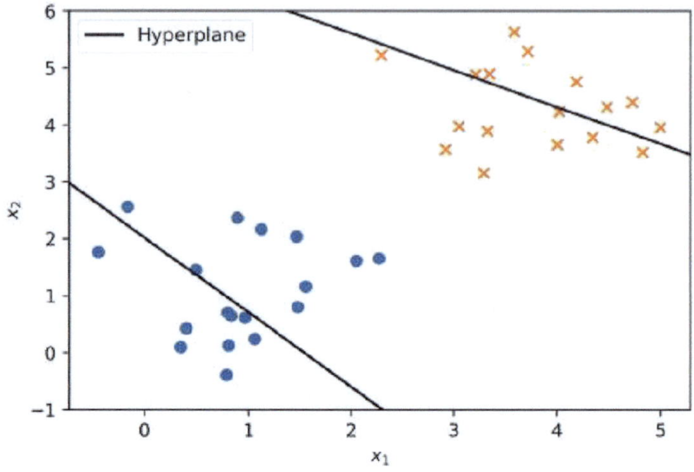

Fig. 13.6 TwinSVM classifier illustration (Mir and Nasiri 2019)

TWSVM model can be obtained by solving the following problems:

TWSVM1:

$$\min_{\omega_1, b_1, \zeta} \frac{1}{2} \|A\omega_1 + e_1 b_1\|^2 + c_1 e_2^T \zeta \tag{13.7}$$

Subject to

$$-(B\omega_1 + e_2 b_1) + \zeta \geq e_2 \tag{13.8}$$

And TWSVM2:

$$\min_{\omega_2, b_2, \eta} \frac{1}{2} \|B\omega_2 + e_2 b_2\|^2 + c_2 e_1^T \eta \tag{13.9}$$

Subject to

$$-(A\omega_2 + e_1 b_2) + \eta \geq e_1 \tag{13.10}$$

where c_1 and c_2 are penalty parameters, ζ and η are slack positive variables, e_1 and e_2 are vectors of one dimension. Here, A and B represent respectively the training points of the two classes.

As we said, these equations find linear hyperplanes; but to work on a high dimension feature space we need to use kernels like SVM to avoid complex computations. This work compares performances of three different kernel functions which are linear, polynomial, and RBF kernels for cancer disease detection.

Table 13.1 SVM and TWSVM kernel functions

Kernel name	Mathematical function
Linear	$K(x_i, x_j) = x(i)x(j)$
Polynomial	$K(x_i, x_j) = (\gamma x(i)x(j) + L)^D$
RBF	$K(x_i, x_j) = exp(-\gamma \lvert x(i) - x(j) \rvert^2)$

Here we have the two hyperplane formulas with kernel:

$$K(x^T, C^T)u_1 + b_1 = 0 \tag{13.11}$$

$$K(x^T, C^T)u_2 + b_2 = 0 \tag{13.12}$$

TWSVM model can be obtained by solving the following quadratic programming problems:

TWSVM1:

$$\min_{\omega_1, b_1, \zeta} \frac{1}{2} \left\| K(A, C^T)u_1 + e_1 b_1 \right\|^2 + c_1 e_2^T \zeta \tag{13.13}$$

Subject to

$$-(K(B, C^T)u_1 + e_2 b_1) + \zeta \geq e_2 \tag{13.14}$$

and TWSVM2:

$$\min_{\omega_2, b_2, \eta} \frac{1}{2} \left\| K(B, C^T)u_2 + e_2 b_2 \right\|^2 + c_2 e_1^T \eta \tag{13.15}$$

Subject to

$$-(K(A, C^T)u_2 + e_1 b_2) + \eta \geq e_1 \tag{13.16}$$

where $C^T = [A, B]^T$, $u_1, u_2 \in R^d$, and K is the kernel function.

We present in Table 13.1 the mathematical function of each of the SVM and TWSVM kernels used in this work.

13.2.5 Performance Evaluation

The purpose of this work is to evaluate the twinSVM classifier based on the features selection method for cancer disease detection from miRNA expression.

To evaluate the features selection method, we use R^2-score which is the proportion of total variance of the predicted model by total variance. This score is described by Eq. (13.17).

Table 13.2 Formulas of the evaluation metrics

Metrics	Formula	Description
Accuracy	$\frac{TP+TN}{TP+TN+FP+FN}$	Used to compare between correctly identified patient and healthy samples
Precision	$\frac{TP}{TP+FP}$	Gives the proportion of correctly positive identification
Recall	$\frac{TP}{TP+FN}$	Gives the proportion of positive samples correctly identified

$$R^2 = 1 - \frac{\sum \left(Y_{actual} - Y_{predicted}\right)^2}{\sum \left(Y_{actual} - Y_{mean}\right)^2} \tag{13.17}$$

Then, we compare the performances of the different kernel functions. Metrics used to evaluate models' performance are presented in Table 13.2.

TP, FN, FP and TN respectively indicate the correctly classified positive samples, incorrectly classified negative samples, incorrectly classified positive samples, and correctly classified negative samples for the experimented model.

13.3 Results and Discussions

In this section, we present the results of our studies that validate the usefulness of our method for different cancer diseases detection.

13.3.1 Features Selection

In the first part of this work, we compare three embedded features selection methods which are: Ridge, Elastic Net, and Lasso models. We start by dividing data into train and test datasets using cross validation with $k = 5$ to evaluate our models. Then we use the gridSearchCV python library to find the best parameters of our chosen model (Pedregosa et al. 2011). For the evaluation, we use the R^2-score. Table 13.3 shows

Table 13.3 Evaluation of the features selection methods

Method	R^2-score
Ridge	0.137
Lasso	0.261
Elastic net	0.270

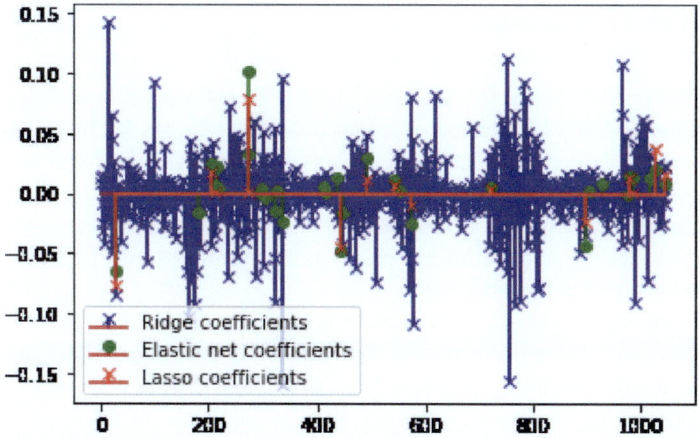

Fig. 13.7 Comparison between ridge, lasso and elastic net

that the R^2-score best value (which is the smallest) is obtained with the Elastic Net model. We notice that Ridge kept almost all the features.

Figure 13.7 shows that Elastic Net gives the more representative features for our model. With the Elastic Net method, we obtained 36 features which are defined with their respective positions in chromosomes in Table 13.4 (Kozomara et al. 2019).

Table 13.4 The selected miRNAs for classification (Chr.: Chromosome, h-m: hsa-mir)

MiRNA	Chr.	Start	End	MiRNA	Chr.	Start	End
h-m-10b	chr2	176150303	176150412	h-m-3199-2	chr22	27920526	27920611
h-m-145	chr5	149430646	149430733	h-m-3613	chr13	49996415	49996501
h-m-15a	chr13	50049119	50049201	h-m-3682	chr2	53849122	53849205
h-m-181b-2	chr9	124693710	124693798	h-m-378c	chr10	130962588	130962668
h-m-182	chr7	129770383	129770492	h-m-421	chrX	74218377	74218461
h-m-183	chr7	129774905	129775014	h-m-454	chr17	59137758	59137872
h-m-21	chr17	59841266	59841337	h-m-585	chr5	169263601	169263694
h-m-210	chr11	568089	568198	h-m-589	chr7	5495819	5495917
h-m-222	chrX	45747015	45747124	h-m-592	chr7	127058088	127058184
h-m-2277	chr5	93620696	93620788	h-m-618	chr12	80935736	80935833
h-m-24-1	chr9	95086021	95086088	h-m-665	chr14	100875033	100875104
h-m-29c	chr1	207801852	207801939	h-m-671	chr7	151238421	151238538
h-m-301b	chr22	21652981	21653058	h-m-708	chr11	79402022	79402109
h-m-30c-2	chr6	71376960	71377031	h-m-760	chr1	93846832	93846911
h-m-3174	chr15	90006755	90006841	h-m-9-1	chr1	156420341	156420429
h-m-3176	chr16	543277	543366	h-m-9-2	chr5	88666853	88666939
h-m-3191	chr19	47226944	47227019	h-m-93	chr7	100093768	100093847
h-m-3199-1	chr22	27920525	27920612	h-m-96	chr7	129774692	129774769

13.3.2 TWSVM Results

In the second part of this work, we build TwinSVM classification with linear, polynomial, and RBF kernel for cancer diseases detection. Tables 13.5 and 13.6 present the comparison between the kernel functions performance with and without features selection (with the Elastic Net method), respectively. We notice that measured metrics values (accuracy, precision, recall, and AUC) are better when we use a features selection technique. Therefore, eliminating redundant and irrelevant features optimizes classifiers performances.

It can be further observed from Table 13.5 that accuracy and precision obtained by the polynomial kernel are greater than those of linear and RBF kernels with respectively a value of 0.9473 and 0.9816. however, the TwinSVM classifier with an RBF kernel gives the best recall value which is 1. The results of Tables 13.5 and 13.6 are illustrated in bar plots in Figs. 13.8 and 13.9.

Figure 13.9 shows that the twinSVM classifier with linear or polynomial kernel gives almost the same precision value with or without applying a features selection method. This means that in our case features selection technique does not influence the correctness of positive cases detection proportion. Whereas, the twinSVM classifier with RBF kernel gives the same recall value with or without features selection.

13.3.3 Evaluation and Comparisons

In the final part of our tests, we first compare our proposed method with certain state of the art methodologies which we previously presented in the introduction of this paper. After that, we test the SVM method and we compare the results of SVM and TWSVM.

Table 13.5 Comparative results for different kernels with features selection

Kernel function	Accuracy	Precision	Recall	AUC
Linear	0.7929	**0.9923**	0.7784	0.8554
Polynomial	**0.9473**	0.9816	**0.9600**	**0.9244**
RBF	0.9100	0.9100	1	0.5

Table 13.6 Comparative results for different kernel without features selection

Kernel function	Accuracy	Precision	Recall	AUC
Linear	0.7811	0.9847	0.7622	0.8489
Polynomial	**0.9300**	**0.9952**	0.9368	**0.8849**
RBF	0.9055	0.9055	**1**	0.5

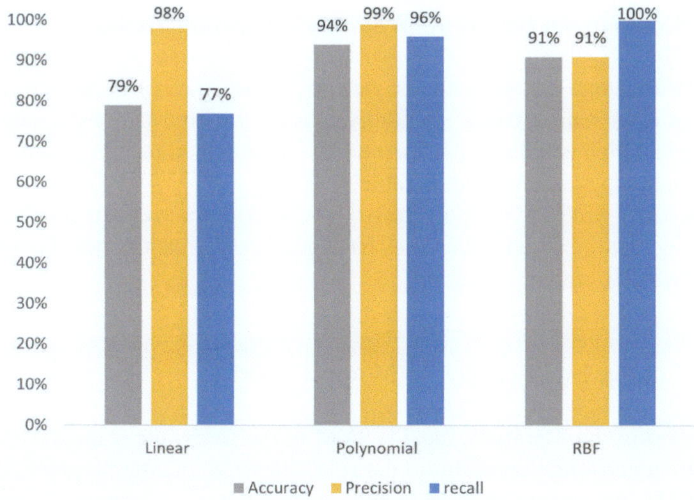

Fig. 13.8 Comparison between different kernel functions performances with features selection

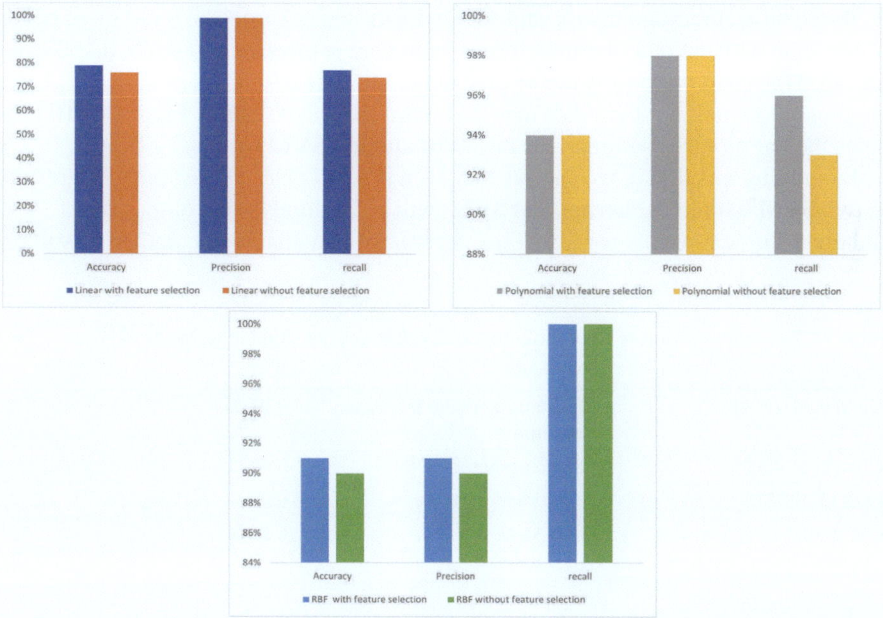

Fig. 13.9 Comparison of classifiers with and without features selection

13.3.3.1 Comparison Between TWSM and the SOTA Methods

Among all the state of the art methods, we chose those in research related to miRNA involvement in human diseases. As for evaluation, we considered the same performance metric which is Area under Curve (AUC). Obtained results for our approach and other methods are presented in Table 13.7.

We note that twinSVM outperforms other algorithms used for miRNA-disease association detection. Other techniques have good performances but stay lower than the twinSVM AUC value.

13.3.3.2 Comparison Between SVM and TWSM

AS mentioned in the literature, the TWSVM outperforms the standard SVM technique in solving different problems of data classification. In order to prove the advantage of TWSM, we tested the SVM technique with the same kernels and we considered the same metrics for evaluation. Table 13.8 illustrates the SVM results with a features selection.

Based on accuracy, precision, and AUC with the linear kernel the SVM gives better results than with the other kernels. These values are respectively: 0.9355, 0.9597 and 0.9184. However, the best value of precision, which is 0.9885, is obtained with the polynomial kernel; but the other metrics record the worst results. Figure 13.10 provides the bar plot that summarizes the SVM and TWSVM results.

Examining this figure, we can see that even if SVM gives almost the same results as twinSVM with linear kernel, TWSVM with polinomial kernel outperforms SVM and gives the best results regarding the accuracy, precision, recall, and AUC values.

Table 13.7 Comparative results of different cancer diseases detection in terms of AUC

Papers	Methods	AUC
Chen et al. (2018)	Extreme gradient boosting machine	0.9123
Chen et al. (2017)	Ranking-based KNN	0.8221
Li et al. (2021)	Dual random walk	0.9029
Our model	twinSVM	**0.9244**

Table 13.8 SVM classification results, based on features selection

Kernel function	Accuracy	Precision	Recall	AUC
Linear	**0.9355**	0.9737	**0.9597**	**0.9184**
Polynomial	0.7186	**0.9885**	0.6949	0.8117
RBF	0.9274	0.9871	0.9313	0.9120

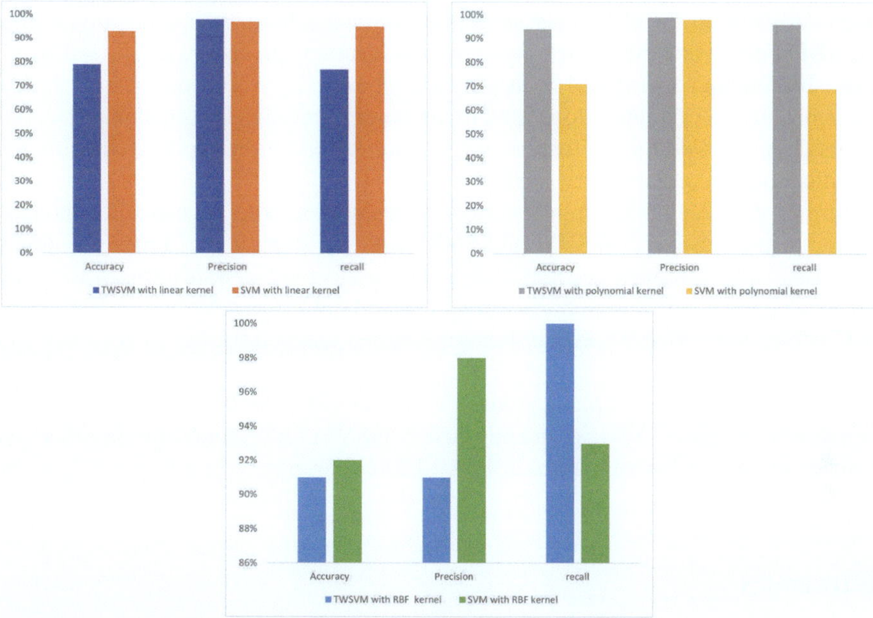

Fig. 13.10 Comparison of the SVM and TWSVM classifiers with features selection

In conclusion, all our experiments show that the twinSVM technique with features selection gives very good performances in terms of accuracy, precision, recall, and AUC. These results are a target of improvement by tuning and optimizing the hyperparameter values.

13.4 Conclusion

This study introduces an innovative and efficient method for cancer diseases detection from miRNA RPM expression. The proposed model is based on twinSVM classification with Elastic Net: an embedded features selection technique that improves evaluation metrics values. The considered datasets are clinical data composed of healthy and patient samples. w dealt here with four types of cancer disease which are: Pan-kidney cohort, Stomach and Esophageal carcinoma, Glioma, and Breast invasive carcinoma.

In our tests, three embedded features selection methods were firstly tested: Ridge, Elastic Net, and Lasso. For evaluation, the R^2-score was used and the best value was obtained with the Elastic Net model. Only 36 features were kept from 1064 miRNAs in different 2076 samples.

In the literature, the TwinSVM classifier showed its efficiency even with complex data by using two hyper-planes rather than one for the SVM method. For this, we based our work on TwinSVM classifier three kernels, which are: linear, polynomial, and RBF. Four evaluation metrics were used: Accuracy, precision, recall, and AUC. Tests showed that TwinSVM with polynomial kernel gave us the best accuracy, precision, and AUC values, compared to the other state-of-the-art models. Also, our results demonstrated the usefulness of the features reduction step in enhancing the classification evaluation metrics values.

Further, to prove that our model performances surpass other classification models, a comparison between TWSVM and SVM with the same kernel types was done. Obtained results demonstrated that TWSVM outperformed SVM in terms of accuracy and AUC. Indeed, SVM with linear kernel reached 0.9355 as accuracy and 0.9184 as AUC value; and TWSVM with polynomial kernel reached 0.9473 as accuracy and 0.9244 as AUC value.

In our future works, we plan to tune other twinSVM hyperparameters to optimize even more our results. Also, we intend to use twinSVM for multinomial classification of different cancer diseases from miRNA RPM expression.

References

Alam, S., Kwon, G.-R., Kim, J.-I., Park, C.-S.: Twin SVM-based classification of Alzheimer's disease using complex dual-tree wavelet principal coefficients and LDA. J. Healthc. Eng. (2017)

Brik, Y., Djerioui, M., Attallah, B.: Efficient heart disease diagnosis based on twin support vector machine. Diagnostyka **22**(3), 1641–6414 (2021)

Buteneers, P., Caluwaerts, K., Dambre, J., Verstraeten, D., Schrauwen, B.: Optimized parameter search for large datasets of the regularization parameter and feature selection for ridge regression. Neural Process. Lett. **38**(3), 403–416 (2013)

Cai, J., Liu, X., Cheng, J., Li, Y., Huang, X., Li, Y., Ma, X., Yu, H., Liu, H., Wei, R.: Microrna-200 is commonly repressed in conjunctival malt lymphoma, and targets cyclin e2. Graefes Arch. Clin. Exp. Ophthalmol. **250**(4), 523–531 (2012)

Calin, G.A., Croce, C.M.: Microrna signatures in human cancers. Nat. Rev. Cancer **6**(11), 857–866 (2006)

Calin, G.A., Dumitru, C.D., Shimizu, M., Bichi, R., Zupo, S., Noch, E., Aldler, H., Rattan, S., Keating, M., Rai, K., et al.: Frequent deletions and down-regulation of micro-RNA genes miR15 and miR16 at 13q14 in chronic lymphocytic leukemia. Proc. Natl. Acad. Sci. **99**(24), 15524–15529 (2002)

Calin, G.A., Sevignani, C., Dumitru, C.D., Hyslop, T., Noch, E., Yendamuri, S., Shimizu, M., Rattan, S., Bullrich, F., Negrini, M., et al.: Human microRNA genes are frequently located at fragile sites and genomic regions involved in cancers. Proc. Natl. Acad. Sci. **101**(9), 2999–3004 (2004)

Chawla, N.V., Bowyer, K.W., Hall, L.O., Kegelmeyer, W.P.: Smote: synthetic minority oversampling technique. J. Artif. Intell. Res. **16**, 321–357 (2002)

Chen, X., Wu, Q.-F., Yan, G.-Y.: Rknnmda: ranking-based KNN for miRNA-disease association prediction. RNA Biol. **14**(7), 952–962 (2017)

Chen, X., Huang, L., Xie, D., Zhao, Q.: Egbmmda: extreme gradient boosting machine for mirna-disease association prediction. Cell Death Dis. **9**(1), 1–16 (2018)

Fonti, V., Belitser, E.: Feature selection using lasso. VU Amsterdam Res. Pap. Bus. Anal. **30**, 1–25 (2017)

Jović, A., Brkić, K., Bogunović, N.: A review of feature selection methods with applications. In: 2015 38th International Convention on Information and Communication Technology, Electronics and Microelectronics (MIPRO), pp. 1200–1205. IEEE (2015)

Khemchandani, R., Chandra, S., et al.: Twin support vector machines for pattern classification. IEEE Trans. Pattern Anal. Mach. Intell. **29**(5), 905–910 (2007)

Kozomara, A., Birgaoanu, M., Griffiths-Jones, S.: Mirbase: from microrna sequences to function. Nucleic Acids Res. **47**(D1), D155–D162 (2019)

Lee, R., Feinbaum, R., Ambros, V., et al.: The C. elegans heterochronic gene lin-4 encodes small RNAs with antisense complementarity to lin-14. Cell **75**, 843–854 (1993)

Lewis, B.P., Burge, C.B., Bartel, D.P.: Conserved seed pairing, often flanked by adenosines, indicates that thousands of human genes are microRNA targets. Cell **120**(1), 15–20 (2005)

Li, A., Deng, Y., Tan, Y., Chen, M.: A novel miRNA-disease association prediction model using dual random walk with restart and space projection federated method. PLoS One **16**(6), e0252971 (2021)

Lv, D., Ma, Z., Yang, S., Li, X., Ma, Z., Jiang, F.: The application of smote algorithm for unbalanced data. In: Proceedings of the 2018 International Conference on Artificial Intelligence and Virtual Reality, pp. 10–13 (2018)

Mir, A.M., Nasiri, J.A.: Lighttwinsvm: a simple and fast implementation of standard twin support vector machine classifier. J. Open Sour. Softw. **4**(35), 1252 (2019)

Mol, C., Vito, E., Rosasco, L.: Elastic-net regularization in learning theory. J. Complex. **25**(2), 201–230 (2009)

Pedregosa, F., Varoquaux, G., Gramfort, A., Michel, V., Thirion, B., Grisel, O., Blondel, M., Prettenhofer, P., Weiss, R., Dubourg, V., et al.: Scikit-learn: machine learning in python. J. Mach. Learn. Res. **12**, 2825–2830 (2011)

Raju, V.G., Lakshmi, K.P., Jain, V.M., Kalidindi, A., Padma, V.: Study the influence of normalization/transformation process on the accuracy of supervised classification. In: 2020 Third International Conference on Smart Systems and Inventive Technology (ICSSIT), pp. 729–735. IEEE (2020)

Sadewo, W., Rustam, Z., Hamidah, H., Chusmarsyah, A.R.: Pancreatic cancer early detection using twin support vector machine based on kernel. Symmetry **12**(4), 667 (2020)

Slimene, I., Messaoudi, I., Elloumi, A., et al.: Microrna expression classification for human disease prediction. In: 2021 18th International Multi-conference on Systems, Signals & Devices (SSD), pp. 1209–1214. IEEE (2021)

Vapnik, V.: The Nature of Statistical Learning Theory. Springer Science & Business Media (1999)

Wagner, G.P., Kin, K., Lynch, V.J.: Measurement of mRNA abundance using RNA-seq data: RPKM measure is inconsistent among samples. Theory Biosci. **131**(4), 281–285 (2012)

Wu, K., Li, L., Li, S.: Circulating microRNA-21 as a biomarker for the detection of various carcinomas: an updated meta-analysis based on 36 studies. Tumor Biol. **36**(3), 1973–1981 (2015)

Chapter 14
Generalized Gaussian Density: An Efficient Tool for Medical Images Texture Analysis

Hela Boulehmi, Hela Mahersia, and Kamel Hamrouni

Abstract Generalized Gaussian Density (GGD) is mainly a wavelet decomposition based technique. As wavelet transform is recognized as a technique for multi-resolution analysis, where image textures are processed by applying both low-pass and high-pass filters to the rows and columns of the image, wavelet coefficients distribution can provide highly discriminative features for images segmentation. In this chapter, we introduce the GGD technique as a tool for medical images texture analysis. First of all, we introduce the GGD analysis technique with theoretical details, and then we describe related works before giving details about the GGD based approach proposed to detect lung nodules, bone sarcoma and breast abnormalities. The results of using GGD analysis on these several tumors detection are illustrated and compared to other existing techniques.

Keywords GGD · Multi-resolution · Medical images · Texture analysis · Breast abnormalities · Lung nodules · Bone lesions · Tumors detection

14.1 Introduction

Computer-aided diagnosis (CAD) systems offer radiologists dependable support in cancer screening. With no cure available for most cancer types, early detection is crucial in lowering the mortality rate by preventing tumors from progressing or spreading to advanced stages.

In this regard, a novel approach for the early detection of various tumors is introduced, utilizing the measurement of generalized Gaussian density (GGD) alongside a supervised classifier, specifically a neural network with Bayesian back-propagation. The following section revisits the mathematical principles behind GGD analysis and

---

H. Boulehmi (✉) · H. Mahersia · K. Hamrouni
SITI Laboratory-ENIT, University of Tunis El Manar, Tunis, Tunisia
e-mail: hela.boulehmi@gmail.com

K. Hamrouni
e-mail: kamel.hamrouni@enit.rnu.tn

N. Derbel and Q. Zhu (eds.), *Advances in Communication, Signal and Image Processing*,
https://doi.org/10.1007/978-981-95-0401-5_14

its historical development. The third section is a description of the proposed medical images texture analysis approach using GGD and applied on three kinds of tumors: breast abnormalities (masses and microcalcifications), lung nodules and bone lesions. In the forth section, we illustrate the different results of these three types of tumors detection using GGD based primitives and a Bayesian back-propagation classifier. The obtained results are finally compared to those of other techniques used for cancer detection from medical images.

14.2 Generalized Gaussian Density and Texture Analysis

The core concept behind Generalized Gaussian Density is wavelet decomposition. While the Fourier transform estimates the frequencies present in a signal, the wavelet transform provides the added advantage of pinpointing the locations where these frequencies emerge or fade, as it operates on the entire signal (Mahersia 2010). Wavelet transform enables multi-resolution analysis, where textures are filtered using two filters (low-pass and high-pass) on both the rows and columns. This process extracts various details, including horizontal, vertical, and diagonal components. Wavelet decomposition has demonstrated its effectiveness, particularly in texture analysis, and has become widely adopted in image processing. Texture analysis through Generalized Gaussian Density was introduced by Do and Vetterli (2002), which involves constructing a histogram to represent the distribution of coefficients derived from the wavelet transform at a specific sub-band (or level). For each sub-band, the goal is to identify a continuous distribution that best fits the histogram. Experimentally, it is observed that the histogram distribution often resembles a Gaussian distribution centered at 0. However, for certain textures, the peak at 0 is less rounded and more akin to a Laplace distribution (Mahersia 2010; Boulehmi 2016).

Do and Vetterli (2002) suggested modeling the behavior of wavelet coefficients at each scale using a generalized Gaussian distribution, which is characterized by three parameters: μ, α and β (Yu et al. 2012):

$$P_{\mu,\alpha,\beta}(x) = \frac{\beta}{2\alpha r(\frac{1}{\beta})} \exp\left(-\frac{lx - \mu l^{\beta}}{\alpha}\right) \tag{14.1}$$

where:

$$r(z) = \int_0^{+\infty} e^{-t}\, t^{z-1}\, dt$$

for $z > 0$ representing the gamma function. μ, α and β denote the mean, scale, and shape parameters, respectively.

The shape of the distribution is influenced by the form factor β, which controls its sharpness. The scale factor α determines the spread of the curve and corresponds

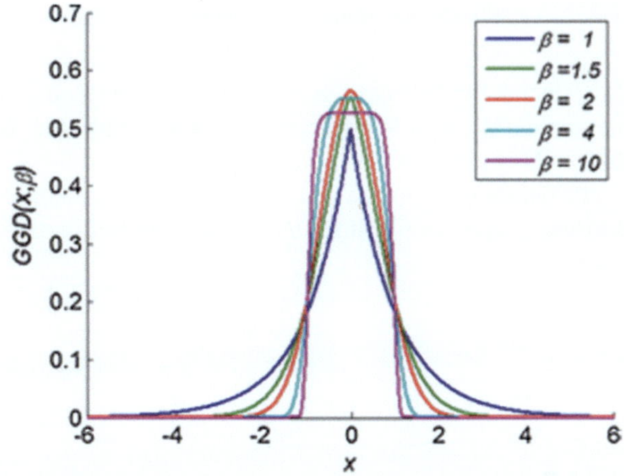

Fig. 14.1 GGD model for $\sigma = 1$ and $\mu = 0$, with various values of β (Fan et al. 2008)

to the standard deviation in the case of a standard Gaussian distribution. It is given as follows (Yu et al. 2012):

$$\alpha = \sqrt[\sigma]{\frac{\Gamma \frac{1}{\beta}}{\Gamma \frac{3}{\beta}}} \tag{14.2}$$

where σ represents the standard deviation of the GGD.

This function appears to be ideally suited for modeling the experimentally observed behavior of texture wavelet coefficients. The modeling process involves an initial phase for evaluating the scale and shape parameters (Fig. 14.1).

In Yu et al. (2012), authors summarized all the evaluation techniques into four categories and compared their performances:

14.2.1 Maximum Likelihood Estimation (MLE)

The maximum likelihood estimator for α is derived from the logarithmic likelihood function, as shown below:

$$L = n\log(\beta) - n\log\left[2\alpha\Gamma\left(\frac{1}{\beta}\right)\right] - \sum_{i=1}^{n}\left(\frac{|x_i|}{\alpha}\right)^{\beta} \tag{14.3}$$

This approach was investigated by Allili and Bouguila (2007) to create a mixture model using the framework of Generalized Gaussian distributions. Its effectiveness was assessed by Novey and Adali (2013).

14.2.2 Moment-Based Method (MM)

Introduced by Varanasi and Aazhang (1989), this method relies on calculating moments. The k-order (where k is a positive integer) absolute central moments of x for GGD is given in the following equation:

Proposed by Varanasi and Aazhang (1989) and based on moment calculating.The k-order (k is a positive integer) absolute central moments of x for GGD is defined as follows:

$$m_k = \alpha^k \frac{\Gamma\left(\frac{k+1}{\beta}\right)}{\Gamma\left(\frac{1}{\beta}\right)} \tag{14.4}$$

14.2.3 Comentropy-Based Method (CM)

Various techniques have been explored in this area, including the generalized entropy matching method (GEM) studied by Kostas and Nandi (2005), the entropy matching method (EM) proposed by Aiazzi et al. (1999), and the negentropy matching method introduced by Prasad et al. (2005). The differential entropy for a GGD signal x is expressed as:

$$H(x) = -\int_{-\infty}^{+\infty} p(x)\, \log[p(x)]\, dx \tag{14.5}$$

14.2.4 Global Convergence Method (GCM)

Building upon the moment-based technique, Song developed the Global Convergence Method (GCM), which offers the advantage of converging more easily than the Maximum Likelihood Estimation (MLE) method (Song 2006). This approach involves the following convex function:

$$Z(\beta) = \frac{E\left(|x|^{2\beta}\right)}{E\left(|x|^{\beta}\right)^2} - (1+\beta) \tag{14.6}$$

The root of $Z(\beta) = 0$ is the used estimator of β.

The estimator for β is obtained by finding the root of $Z(\beta) = 0$. Yu et al. provided a comprehensive theoretical and practical comparison of these four estimation techniques in Yu et al. (2012). For practical purposes, the authors highlighted the following results:

1. For Gaussian signals ($\beta = 2$): MM outperforms MLE, but GCM (with $k = 3$) offers greater precision.

2. For sup-Gaussian signals ($\beta < 2$): MLE is the most effective method, though it is highly sensitive to sample size and the initial estimator.
3. For sub-Gaussian signals ($\beta > 2$): GEM, MLE, or GCM can be used to estimate β, but MLE yields the most accurate results.

GGD was hugely used in literature for different applications, such as speech signals processing (Shin et al. 2005), face recognition (Gonzalez-Jiménez et al. 2007) and texture retrieval (Mahersia 2010).

The Generalized Gaussian Distribution (GGD) has been widely applied in the literature for various purposes, including speech signal processing (Shin et al. 2005), face recognition (Gonzalez-Jiménez et al. 2007), and texture retrieval (Mahersia 2010), among others.

14.3 Related Works

In this section, we provide an overview of the latest research focused on detecting lung, bone, and breast tumors from medical images.

14.3.1 Lung Nodules

Lung cancer affects both men and women and is the leading cause of cancer-related deaths in the United States. In 2015, approximately 158,040 Americans died from lung cancer, accounting for about 27 percent of all cancer deaths (ACS 2015). Lung cancer arises from the abnormal growth of lung cells, which form small, roughly spherical masses of tissue, ranging from 5 to 30 mm in size, known as nodules. According to Lee et al. (2012) and Zhang et al. (2013), lung nodules can be classified into four categories: juxta-vascular, well-circumscribed, pleural tail, and juxta-pleural. Examples of these various types of lung nodules are shown in Fig. 14.2a.

Using CAD systems for lung nodules identification helps improving detection performance and reducing unnecessary biopsies. Several reviews (Mahersia et al. 2015; Parveen and Kavitha 2012) have shown that there are mainly three steps for

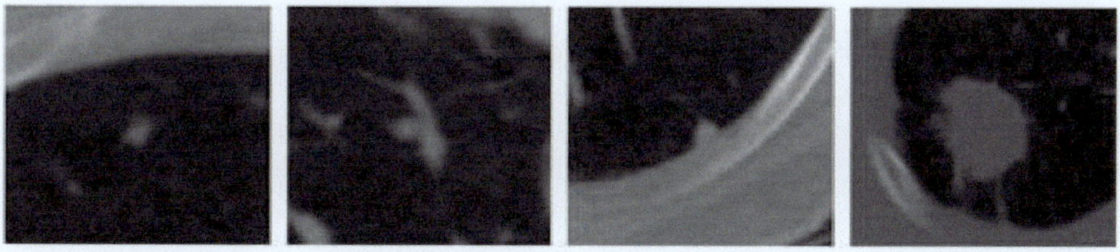

Fig. 14.2 Examples of lung nodules (Mahersia et al. 2015)

extracting lung nodules from CT lung images: preprocessing, nodules segmentation and classification.

The first step consists in denoising and enhancing lung images with several filtering techniques, such as wiener filter (Faizal and Kavitha 2014; Shao and Liu 2012; Sharma and Jindal 2011), bilateral filter (Vijaya and Suhasini 2014), median filter (Ashwin et al. 2012; Blo et al. 1993), Gaussian filter (Pu et al. 2008b; Retico et al. 2008), high-pass filters (Gao et al. 2012), and wavelet-based filters (Gabralla et al. 2015). Various techniques have been employed for both 2D and 3D lung segmentation (Yankelevitz et al. 2000; Kostis et al. 2004). These methods can be categorized into five groups: stochastic methods (ElBaz et al. 2003; Guo et al. 2013), thresholding-based methods (Ashwin et al. 2012; Blo et al. 1993; Choi et al. 2009; Shao and Liu 2012), contour-based methods (Sharma and Jindal 2011), region-based methods (Aggarwal et al. 2013; Taher and Sammouda 2010), and learning-based methods (Jaffar et al. 2009; Antonelli et al. 2011). To the best of our knowledge, only a few studies have investigated wavelet decomposition for lung nodule detection in CT lung images (Talakoub et al. 2007; Vo and Sowmya 2009, 2010; Orozco et al. 2015).

14.3.2 Bone Sarcoma

Bone cancer is an abnormal growth of tissue that can be classified as primary or secondary. Primary bone sarcoma originates from bone cells, while secondary bone sarcoma starts in other parts of the body and spreads to the bone. Although early detection can prevent disabilities and lead to more effective treatments, bone cancer is often misdiagnosed due to the challenges radiologists face when interpreting medical images. Computer-aided diagnosis (CAD) systems can improve the accuracy of bone MRI interpretation and assist radiologists in making more precise diagnoses. There are several types of bone cancer, including osteosarcoma, chondrosarcoma, Ewing sarcoma, and chordoma (Boulehmi et al. 2018) (Fig. 14.3).

To the best of our knowledge, only a few image processing studies have focused on detecting bone sarcoma. Reddy et al. (2016) employed a region growing technique to detect bone cancer and determined the cancer stage by analyzing mean intensity calculations and tumor size measurements. In Mistry and Talati (2016), the authors compared various segmentation techniques for bone tumor detection, including thresholding and morphological operations, K-Means clustering, Fuzzy C-Means clustering, and Rough Fuzzy C-Means clustering. In Nisthula and Yadhu (2018), an edge detection technique based on the Sobel operator was used to identify tumor regions in bone CT images. To reduce noise and smooth MRI images, Sinthia and Sujatha (2016) employed a preprocessing step using average and bilateral filtering, followed by a segmentation step based on mean intensity calculations and tumor size measurements using the K-Means algorithm. In Yao et al. (2014), CT intensity analysis was applied for detecting bone metastasis, while in Ping et al. (2008), graph representation was utilized to identify suspicious bone regions in X-ray images.

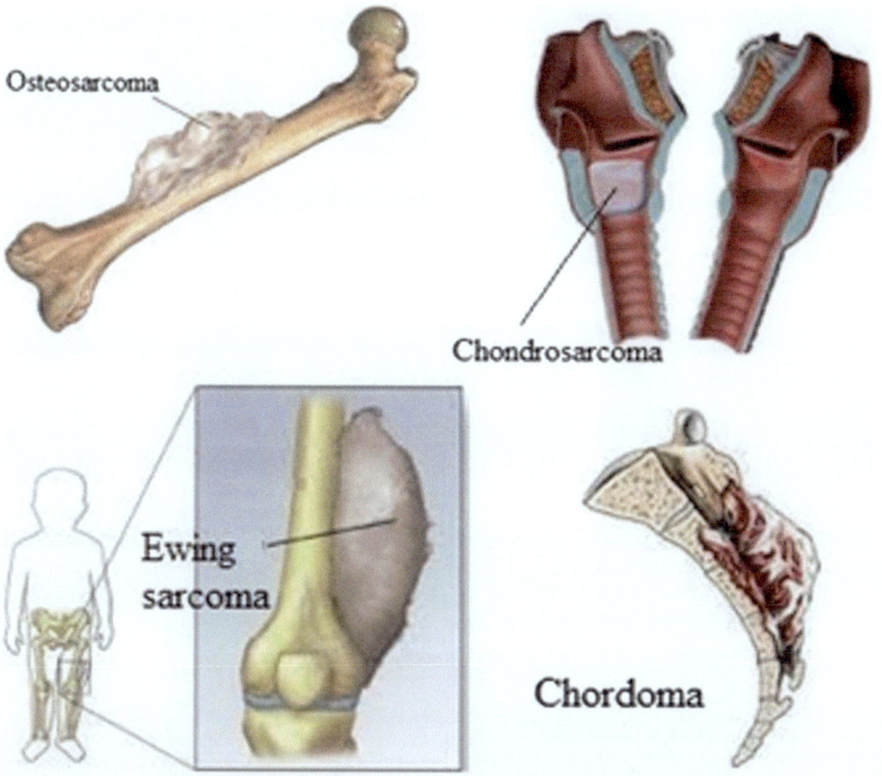

Fig. 14.3 Different forms of bone cancer

14.3.3 Breast Masses and Microcalcifications

Breast abnormalities are typically classified into two main categories: masses and microcalcifications. Breast masses are space-occupying lesions that vary in shape (round, oval, lobulated, or irregular), contour (circumscribed, microlobulated, obscured, indistinct, or speculated), and density (high, medium, or low fat). Masses that consist solely of fat are always benign (Boulehmi et al. 2013) (Figs. 14.4 and 14.5).

Microcalcifications are small calcium deposits, resembling grains of salt, within the soft breast tissue and are often seen as an early sign of a breast tumor. These microcalcifications can appear in various shapes, including annular, round, linear, and vascular, as illustrated in Fig. 14.6a (Boulehmi et al. 2016a).

Several works have been carried on aiming to detect breast masses and/or microcalcifications. In Boulehmi et al. (2013), we gave a review of the different techniques that were exploited for mammograms preprocessing, breast abnormalities segmentation, lesions characterization and classification.

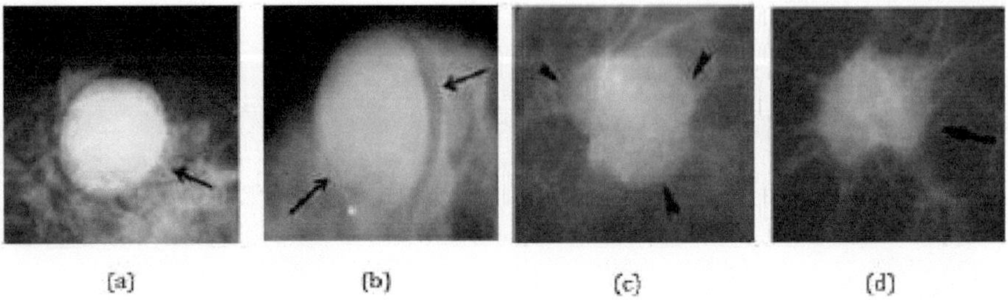

Fig. 14.4 Masses' shapes. **a** Round, **b** oval, **c** lobulated and **d** Irregular

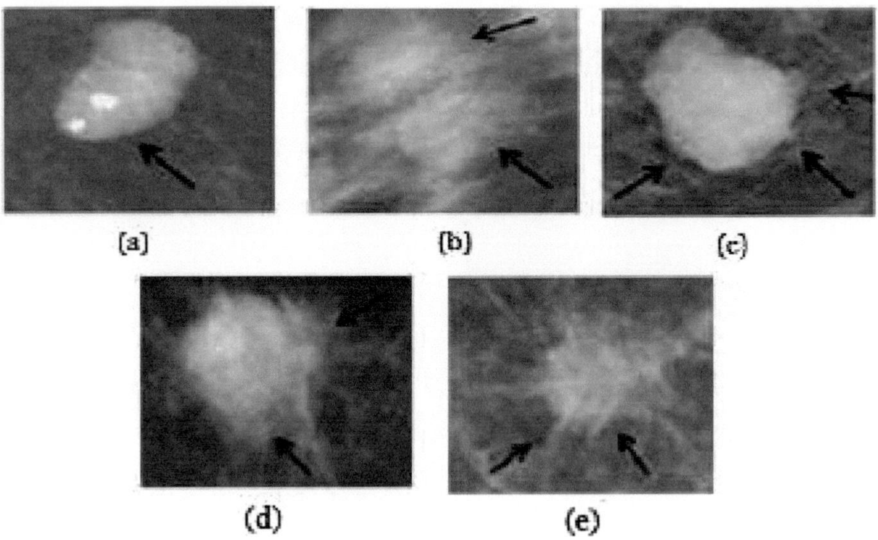

Fig. 14.5 Masses' contours, **a** circumscribed, **b** obscured, **c** microlobulated, **d** indistinct and **e** speculated

Mammogram preprocessing is commonly performed to enhance contrast. In Cheng et al. (2003), various techniques used for contrast enhancement in mammograms were grouped into three main categories: conventional techniques (such as histogram stretching, histogram equalization (Kom et al. 2005), convolution masks (McSweeney et al. 1983), Sobel operators (Laine et al. 1994), and others, region-based techniques (like the region growing algorithm (Morrow et al. 1992)), and feature-based techniques (including morphological operations, wavelet transform, and the fractal approach) (Bouyahia et al. 2009; Laine et al. 1994; Li et al. 1996, 1997).

Breast cancer segmentation involves identifying abnormal lesions within breast tissue. This segmentation can be performed using either a single view of the breast or multiple views:

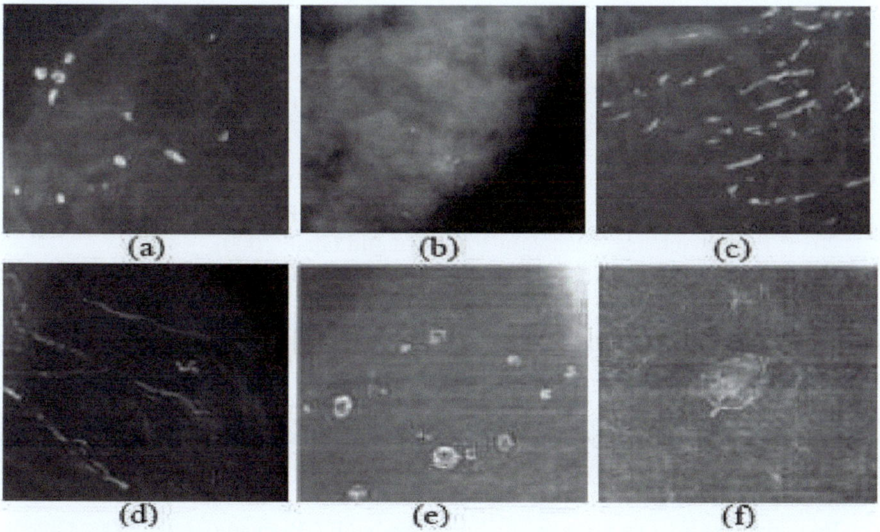

Fig. 14.6 Microcalcifications shapes: **a** annular, **b** round, **c** linear, **d** vascular, **e** with clear center and **f** egg shell

1. Segmentation techniques using a single view: These include region-based approaches such as region growing (Guliato et al. 1998; Hejazi and Ho 2005; Huo et al. 1995; Martý et al. 2003; Woods and Bowyer 1994) watershed methods (Beucher and Lenteuejoul 1979; Gulsrud et al. 2005), and the Split and Merge technique (Chen and Lee 1997). Edge detection-based approaches are also employed (Boucher et al. 2009; Cascio et al. 2006; Davies 1997; Eddaoudi et al. 2011; Fauci et al. 2004; Hong and Brady 2003; Jiang et al. 2008), as well as clustering-based methods (Cao et al. 2004; Catarious et al. 2006; Jain et al. 1999; Li et al. 2002). Additionally, model-based approaches involve comparing current images with previously known healthy or pathological cases (Constantinidis et al. 2001; Freixenet et al. 2008; Lai et al. 1989).
2. Segmentation techniques using multiple views for mass detection: This involves comparing two mammographic images, such as those of the right and left breasts (Christoyianni and Dermatas 2004; Karssemeijer 1998; Lau and Bischof 1991; Méndez et al. 2003; Paquerault et al. 2002), two different views of the same breast (typically mediolateral oblique (MLO) and cranio-caudal (CC) (Pu et al. 2008a)), or two mammograms of the same breast taken at different times to monitor the evolution of a potential anomaly (Marti et al. 2001).

After detecting breast abnormalities, a characterization step is crucial to classify them as benign or malignant. Various primitives have been utilized for this purpose in the literature. In Cheng et al. (2003), the authors summarized the different primitives used to characterize detected breast lesions. These primitives include:

1. Primitives related to the segmented lesions such as perimeter, area, orientation, compactness, thickness, direction, line, elongation, eccentricity, background, foreground, distance and contrast (Bhangale et al. 2000; Bothorel et al. 1997; Chan et al. 1990; Davies and Dance 1990; Jiang et al. 2008; Marias et al. 2002; Yu and Guan 2000).
2. Primitives extracted from the co-occurrence matrix such as energy, entropy and contrast (Blot and Zwiggelaar 2001; Bovis and Singh 2002; Eddaoudi et al. 2011; Enderwich and Tzanakou 1997; Yu and Guan 2000).
3. Primitives extracted from the surround region dependence matrix—SRDM (Kim and Park 1999; Lee et al. 1998).
4. Primitives extracted from the gray level run length—GLRL (Kim and Park 1999).
5. Primitives extracted from the gray level difference—GLD (Geronimo et al. 1994; Kim and Park 1999).
6. Primitives extracted from the wavelet decomposition (Clarke et al. 1994; Dheeba and Jiji 2010; Eltoukhy et al. 2010; Gagnon et al. 1995; Gurcan et al. 1997; Kestener 2003; Yu and Guan 2000).
7. Primitives extracted from the Gabor filter bank (Bhangale et al. 2000; Dheeba and Jiji 2010; Nakayama et al. 2006).
8. Primitives extracted from the space scale (Netsch and Peitgen 1999).
9. Primitives extracted from the fractal dimension (Enderwich and Tzanakou 1997; Oliver et al. 2006).
10. Primitives extracted from the characteristics of clusters (Bothorel et al. 1997; Cordella and Vento 2000; DaPonte and Sherman 1991; Davies and Dance 1990; Petroudi et al. 2003).

Characterizing techniques performance should not be evaluated separately from classification approach as they are tightly related.

Several classifiers were used to identify malignant breast lesions from those benign (Cheng et al. 2003; Oliver et al. 2006). The most commonly explored techniques are Neural networks (Chen and Lee 1997; Dheeba and Jiji 2010; Jasmine et al. 2010; Kavitha and Kumaravel 2007), Bayesian classifier (Kavitha and Kumaravel 2007; Nakayama et al. 2006; Oliver et al. 2008), K-nearest neighbors (Bhangale et al. 2000; Karssemeijer 1998; Kramer and Aghdasi 1999), Quadratic classifier (Marti et al. 1999), Linear classifier (Davies and Dance 1990; Ema et al. 1995), Genetic algorithms (Castella 2009), Fuzzy decision tree (Bothorel et al. 1997), Expert system (Cordella and Vento 2000), Binary decision tree (Woods and Bowyer 1994), SVM (Eddaoudi et al. 2011; Nanni et al. 2011) and Adaptive thresholding (Sharma and Jindal 2011).

The performance of these classification techniques is highly sensitive to the primitives selected during the characterization step.

14.4 Proposed Approach

This section is dedicated to explain the main principle of the proposed cancer early detection approach. The process starts with a preprocessing phase depending on the medical images nature (mammogram, MRI or CT). This step aims to denoise and enhance the contrast of the processed image.

The same segmentation procedure is used for all the three considered cancer forms (lung, bone and breast): The process starts by dividing the image into blocks of a specified size, followed by applying wavelet decomposition to each block to estimate the GGD parameters, which then form the feature vector of the block.

A Bayesian back-propogation classifier wil then explore these features to classify the different blocks into "Tumor" and "Not Tumor" categories.

Once detected, the images showing abnormalities will later be classified into "Benign" and "malignant" cases using textural primitives and an ANFIS system (Boulehmi et al. 2016a, b, 2018; Mahersia et al. 2020).

Figure 14.7 illustrates the flowchart of the medical image processing method based on GGD measurement.

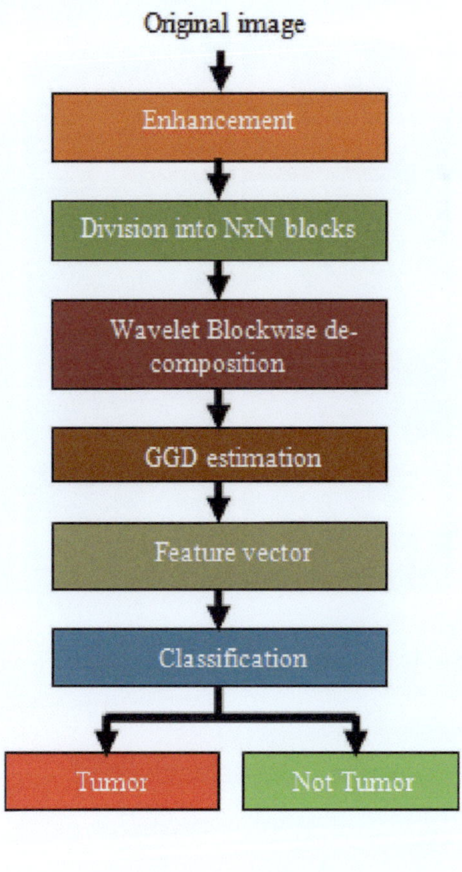

Fig. 14.7 Flowchart of the tumor detection approach using GGD estimation

14.5 Testing Results

This section illustrates the different results obtained for lung, bone and breast cancer detection using the proposed GGD based approaches.

14.5.1 Lung Nodules Segmentation

The primary contribution of the proposed method is the use of Gaussian wavelet-based models as a spatio-frequency textural descriptor to identify lung nodules. In Mahersia et al. (Mahersia et al. 2020), we gave further details about our GGD based approach. Figure 14.8 gives some examples of lung nodules detected using our algorithm on images from LIDC database.

The confusion matrix for the tests conducted with the Bayesian Back-propagation Classifier is presented in Table 14.1.

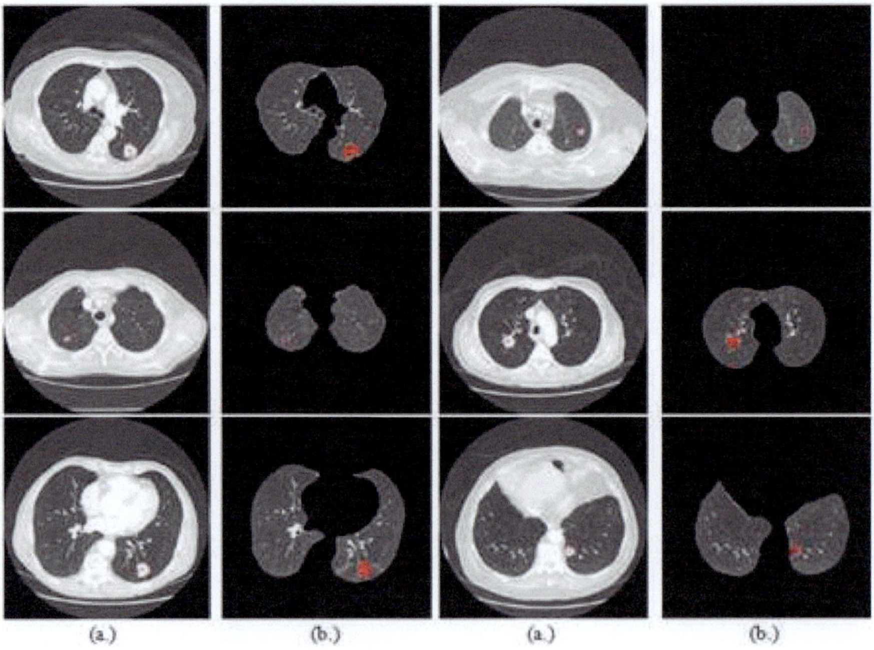

Fig. 14.8 Lung nodules detection examples: **a** original ground-truth images and **b** results of our algorithm

Table 14.1 Confusion matrix for ten tests conducted using the Bayesian Back-propagation classifier

	Average recognition rate (%)	
	Nodule	Not nodule
Nodule	**94.19**	0.82
Not nodule	05.81	**99.18**

14.5.2 Sarcoma Detection from MRI

The GGD based algorithm was tested on several bone MRI. Figure 14.9 illustrates some of the obtained bone tumors segmentation results (Boulehmi et al. 2018).

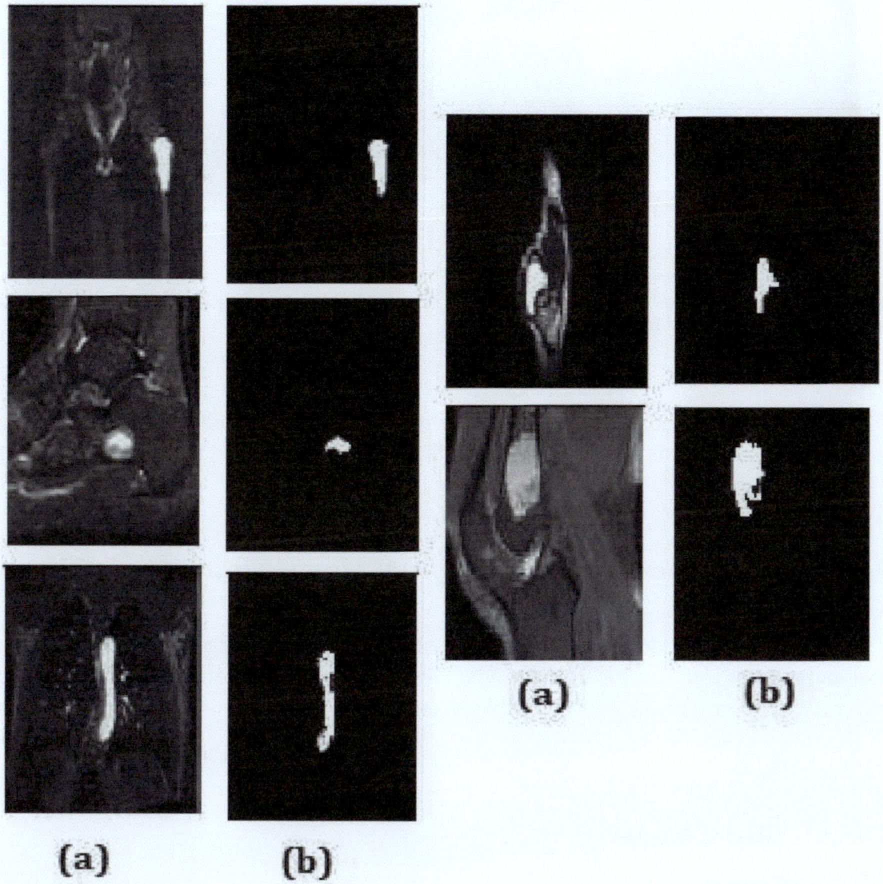

Fig. 14.9 Examples of bone sarcoma detection: **a** Original ground-truth images and **b** results from the proposed algorithm

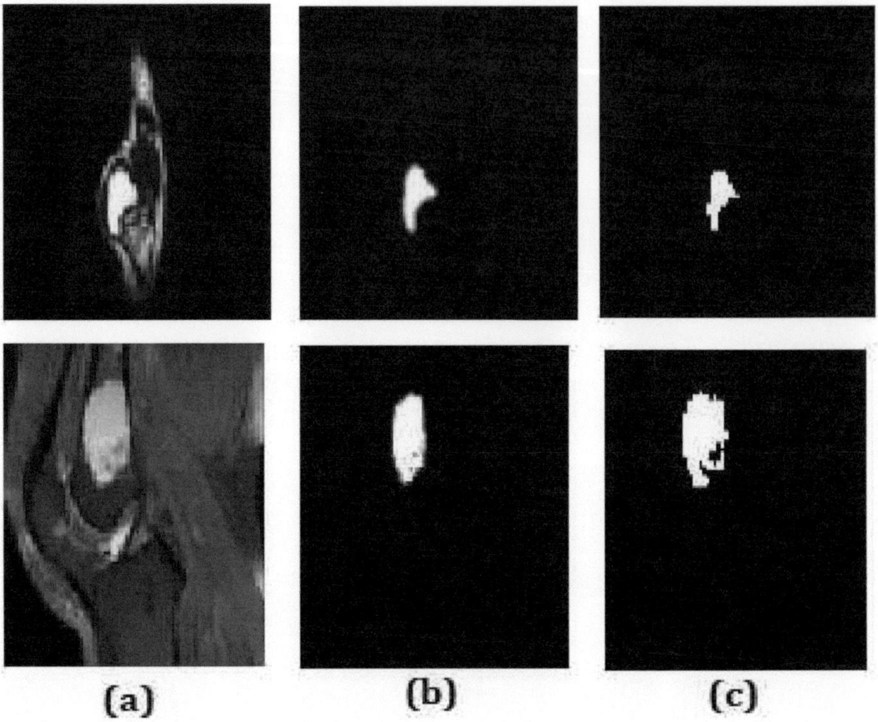

Fig. 14.10 Evaluation of bone cancer detection results. **a** Original MRI; **b** rough fuzzy c-means result (Mistry and Talati 2016); **c** obtained results

These results are comparable to those obtained by Mistry and Talati (2016) using a rough fuzzy c-means algorithm (Fig. 14.10).

14.5.3 Breast Lesions Detection from Mammograms

The proposed cancer detection algorithm was tested on two different forms of breast lesions: masses and microcalcifications. Despite the huge difference between these two breast abnormalities, our GGD based technique gave promising results for both of them.

14.5.3.1 Breast Masses

We evaluated our algorithm using the MIAS database, which is one of the most commonly used mammography databases due to its accessibility and ease of use. It

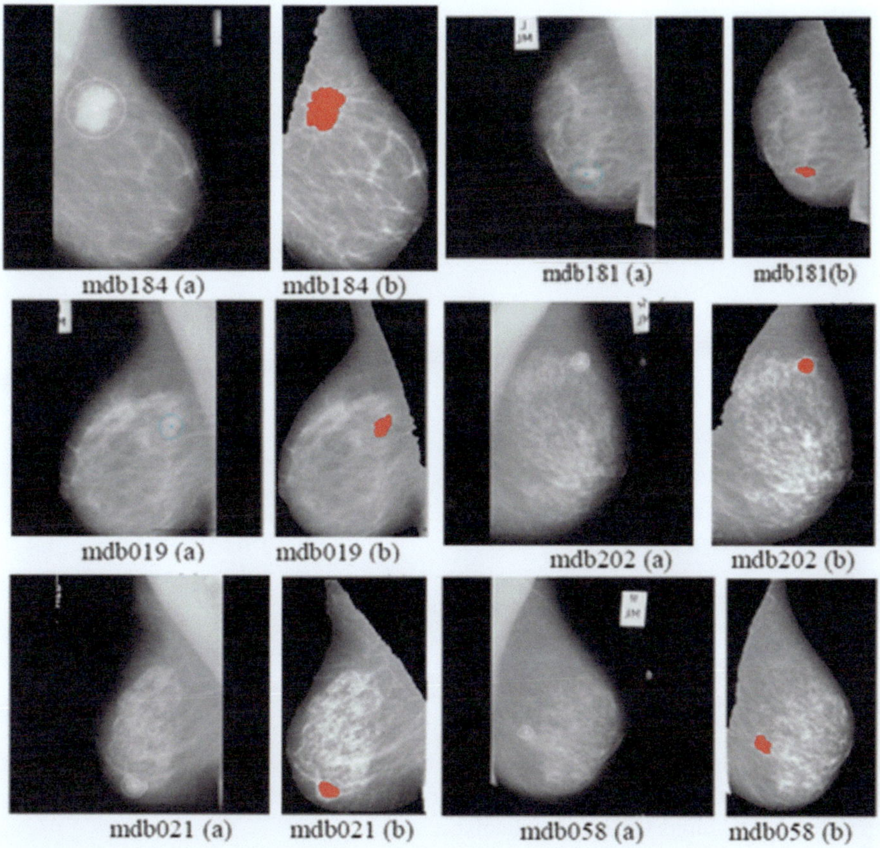

Fig. 14.11 Examples of breast masses detection results obtained on MIAS database

includes 322 medio-lateral oblique (MLO) mammograms: even-numbered images correspond to left MLO, while odd-numbered images correspond to right MLO. The database contains a comprehensive range of diagnoses: 208 normal images, 56 with masses, 25 with microcalcifications, 18 with architectural distortions, and 15 with imbalances. Figure 14.11 gives examples of detected breast masses.

14.5.3.2 Breast Microcalcifications

The same process is applied on mammograms from MIAS database showing microcalcifications. An example of microcalcifications cluster detection in given in Fig. 14.12.

The average rate for microcalcifications detection is 94.44%.

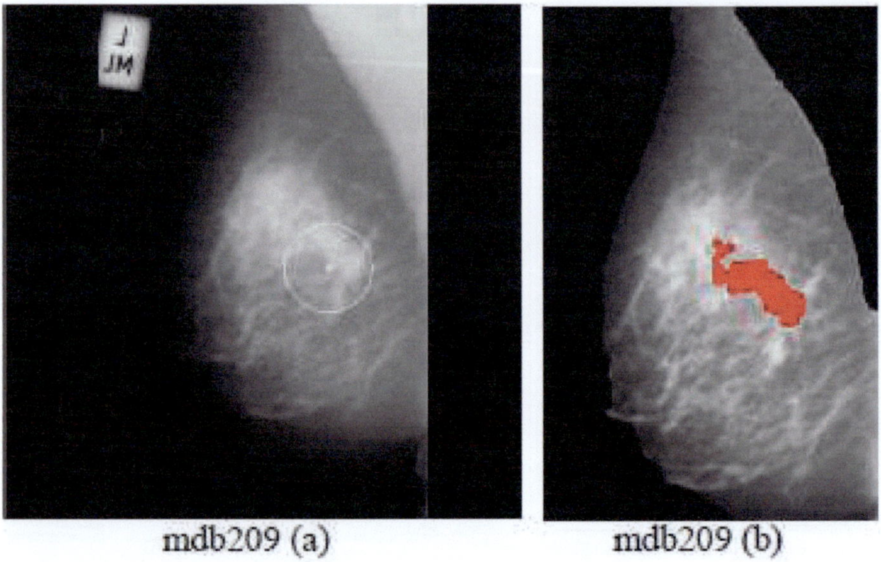

Fig. 14.12 Example of detected microcalcifications clusters

14.5.3.3 Discussion

The proposed GGD based segmentation technique has proved its efficiency on several medical images. Table 14.2 shows comparison of this method with previous works tested on LIDC lungs CT images database.

For bone cancer, our GGD-based method proves to be faster, as it requires less computation time compared to the Rough Fuzzy C-Means clustering technique used by Mistry and Talati (2016) (Table 14.3).

Results obtained for breast masses are competitive. Table 14.4 is a comparison of the proposed algorithm with other works carried on to detect breast masses from digital mammograms.

Table 14.2 Comparison between proposed and previous works using LIDC

Related work	Proposed method	Average recognition rate %
Vo and Sowmya (2009)	Wavelet + GGD + SVM	92.25
Vo and Sowmya (2010)	Wavelet + GGD + KNN	90.12
Netto et al. (2012)	Growing neural gaz + SVM	90.65
Santos et al. (2014)	Region growing + GMMs + SVM	88.4
Proposed algorithm	**Wavelet + GGD + NN**	**97.88**

Table 14.3 Computation time comparison

Related work	Proposed method	Average computation time (s)
Mistry and Talati (2016)	Rough fuzzy C-means clustering technique	11
Our work	Wavelet + GGD + NN	6

Table 14.4 Comparison of breast masses detection results

Related work	Proposed method	Average recognition rate (%)
Abdalla et al. (2011)	Statistical features + ANN	84.17
Wei et al. (2005)	Textural features + ANN	86
Tourassi et al. (2006)	Morphological and textural features + SVM	91
Proposed algorithm	**GDD + ANN**	**97.08**
Delogu et al. (2007)	Textural features + ANN	97.8

Table 14.5 Comparison of breast microcalcifications detection results

Related work	Proposed method	Average recognition rate (%)
Arai et al. (2012)	textural features + ANN	70.8
Anuradha and Preeti (2013)	Wavelet + SVM	79.58
Papadopoulos et al. (2004)	Morphological + ANN	84
Proposed algorithm	**GDD + ANN**	**94.44**
Krishnaveni et al. (2014)	Textural features + ANN	96.25

Finally, Table 14.5 compares GGD based proposed technique to other breast microcalcifications detection approaches.

14.6 Conclusion

In this chapter, a GGD-based method for detecting various forms of cancer has been described. The proposed algorithm was tested on CT lung images, bone MRI, and digital mammograms. The recognition rates reached 97.88% for lung nodules, 97.08% for breast masses, and 94.44% for breast microcalcifications. All tests were conducted using MATLAB R2014 on a system with an Intel Core i5 CPU at 2.53 GHz and 4GB of RAM.

References

Abdalla, M., Dress, S., Zaki, N.: Detection of masses in digital mammogram using second order statistics and artificial neural network. Int. J. Comput. Sci. Inf. Technol. (IJCSIT) **3**, 176–186 (2011)

Aggarwal, P., Vig, R., Sardana, H.K.: Semantic and content-based medical image retrieval for lung cancer diagnosis with the inclusion of expert knowledge and proven pathology. In: IEEE Second International Conference on Image Information Processing ICIIP'2013 (2013)

Aiazzi, B., Alparone, L., Baronti, S.: Estimation based on entropy matching for generalized Gaussian PDF modeling. IEEE Signal Process. Lett. (1999)

Allili, M., Bouguila, N.: Finite generalized Gaussian mixture modeling and applications to image and video foreground segmentation. In: Fourth Canadian Conference on Computer and Robot Vision (2007)

Antonelli, M., Cococcioni, M., Lazzerini, B., Marcelloni, F.: Computer-aided detection of lung nodules based on decision fusion techniques. Pattern Anal. Appl. **14**, 295–310 (2011)

Anuradha, C., Preeti, P.: Detection and classification of microcalcifica-tions using discrete wavelet transform. Int. J. Emerg. Trends Technol. Comput. Sci. **2** (2013)

Arai, K., Abdullah, I., Okumura, H.: Automated detection method for clustered microcalcification in mammogram image based on statistical textural features. Int. J. Adv. Comput. Sci. Appl. **3**, 12–16 (2012)

Ashwin, S., Kuma, S., Ramesh, J., Gunavathi, K.: Efficient and Reliable Lung Nodule Detection using a neural network based computer aided diagnosis system. In: Proceedings of the ICE-TEEEM'2012 (2012)

Beucher, S., Lenteuejoul, C.: Use of watersheds in contour detection. In: Proceedings of the International Workshop on Image Processing: real-time Edge and Motion Detection/Estimation, vol. 2 (1979)

Bhangale, T., Desai, U., Sharma, U.: An unsupervised scheme for detection of microcalcifications on mammograms (2000)

Blo, S., Freedman, M., Lin, J., Mun, S.: Automatic lung nodule detection using profile matching and back-propagation neural network techniques. J. Digit. Imaging **6**, 48–54 (1993)

Blot, L., Zwiggelaar, R.: Background texture extraction for the classifi-cation of mammographic parenchymal patterns. In: Medical Image Understanding and Analysis-MIUA (2001)

Bothorel, S., Meunier, B., Muller, S.: A fuzzy logic based approach for semiological analysis of microcalcifications in mammographic images. Int. J. Intell. Syst. **12**, 819–848 (1997)

Boucher, A., Jouve, P., Cloppet, F., Vincent, N.: Segmentation du muscle pectoral sur une mammographie. In: Congrès des jeunes chercheurs en vision parordinateur-ORASIS'09 (2009)

Boulehmi, H., Mahersia, H., Hamrouni, K., Boussetta, S., Mnif, N.: Breast cancer detection: a review on mammograms analysis techniques. In: Proceedings of the International IEEE Multi-conference on Systems, Signals and Devices-SSD (2013)

Boulehmi, H.: Analyse d'images mammographiques par analyse de texture: application à la détection précoce du cancer. Ph.D. thesis. National Engineering School of Tunis, Campus Tunis Elmanar (2016)

Boulehmi, H., Mahersia, H., Hamrouni, K.: Breast masses diagnosis using supervised approaches. In: Proceedings of The Second IEEE International Image Processing, Applications and Systems Conference-IPAS (2016a)

Boulehmi, H., Mahersia, H., Hamrouni, K.: A new cad system for breast microcalcifications diagnosis. Int. J. Adv. Comput. Sci. Appl. IJACSA **7** (2016b)

Boulehmi, H., Mahersia, H., Hamrouni, K.: Bone cancer diagnosis using GGD analysis. In: Proceedings of the International IEEE Multi-conference on Systems, Signals and Devices-SSD (2018)

Bouyahia, S., Mbainaibeye, J., Ellouze, N.: Wavelet based microcalcifications detection in digitized mammograms. ICGST-GVIP J. **8**. ISSN 1687-398X (2009)

Bovis, K., Singh, S.: Classification of mammographic breast density using a combined classifier paradigm. In: 4th International Workshop on Digital Mammography, pp. 177–180 (2002)

Cao, A., Song, Q., Yang, X.L., Liu, S.: Breast mass segmentation on digital mammograms by a combined deterministic annealing method. In: IEEE International Symposium on Biomedical Imaging, vol. 2, pp. 1303–1306 (2004)

Cascio, D., Fauci, F., Magro, R., Raso, G., Bellotti, R., Tangaro, F.D.C.S., Nunzio, G.D., Quarta, M., Forni, G., Lauria, A., Fantacci, M., Retico, A., Masala, G., Oliva, P., Bagnasco, S., Cheran, S., Lopez-Torres, E.: Mammogram segmentation by contour searching and mass lesions classification with neural network. IEEE Trans. Nucl. Sci. **53**(5) (2006)

Castella, C.: Breast texture synthesis and estimation of the role of the anatomy and tumor shape in the radiological detection process: from digital mammography to breast tomosynthesis. Ph.D. thesis, Ecole Polytechnique Fédérale de Lausanne, France (2009)

Catarious, D.M., Baydush, A., Floyd, C.: Characterization of difference of gaussian filters in the detection of mammographic regions. Med. Phys. **33**(11), 4104–4114 (2006)

Chan, H., Doi, K., Vyborny, C., Schmidt, R., Metz, C., Lam, K., Ogura, T., Wu, Y., Maxmahon, H.: Improvement in radiologists' detection of clustered microcalcifications on mammogram: the potential of computer-aided diagnosis. Investigative Radiol. **25**, 1102–1110 (1990)

Chen, C., Lee, G.: Image segmentation using multiresolution wavelet analysis and expectation-maximization (EM) algorithm for digital mammography. Digit. Imag. **8**, 491–504 (1997)

Cheng, H., Cai, X., Chen, X., Hu, L., Lou, X.: Computer-aided detection and classification of microcalcifications in mammograms: a survey. Pattern Recogn. **36**, 2967–2991 (2003)

Choi, W., Majid, A., Choi, T.: Computerized detection of pulmonary nodule based on two-dimensional PCA. Comput. Sci. Appl. ICCSA **2009**(5593), 693–702 (2009)

Christoyianni, I., Dermatas, E.C.E.: Computerized detection of pulmonary nodule based on two-dimensional PCA automatic detection of abnormal tissue in bilateral mammograms using neural networks. In: Methods and Applications of Artificial Intelligence, pp. 267–275 (2004)

Clarke, L., Kallergi, M., Qian, W., Li, H., Clark, R., Silbiger, M.: Three-structured nonlinear filter and wavelet transform for microcalcification segmentation in digital mammography. Cancer Lett. **77**, 173–181 (1994)

Constantinidis, A., Fairhust, M., Rahman, A.: A new multiexpert decision combination algorithm and its application to the detection of circum-scribed masses in digital mammograms. Pattern Recogn. **34**, 1527–1537 (2001)

Cordella, L., Vento, F.T.M.: Combing experts with different features for classifying clustered microcalcifications in mammograms. In: Proceedings of 15th International Conference on Patten Recognition, pp. 324–327 (2000)

DaPonte, J.S., Sherman, P.: Classification of ultrasonic image texture by statistical discriminant analysis and neural networks. Comput. Med. Imag. Graph. **15**, 3–9 (1991)

Davies, D., Dance, D.: Automatic computer detection of clustered calcifications in digital mammograms. Phys. Med. Biol. **35**, 1111–1118 (1990)

Davies, E.: Machine Vision, 2nd edn. Academic Press (1997)

Delogu, P., Fantacci, M., Kasae, P., Retico, A.: Characterization of mammographic masses using a gradient based segmentation algorithm and a neural classifier. Comut. Biol. Med. **37**(10), 1479–1491 (2007)

Dheeba, J., Jiji, G.W.: Detection of microcalcification clusters in mammograms using neural network. Int. J. Adv. Sci. Technol. **19** (2010)

Do, M., Vetterli, M.: Wavelet-based texture retrieval using generalized gaussian density and Kullback-Leibler distance. IEEE Trans. Image Process. **11**, 146–158 (2002)

Eddaoudi, F., Regragui, F., Mahmoudi, A., Lamouri, N.: Masses detection using SVM classifier based on textures analysis. Appl. Math. Sci. **5**, 367–379 (2011)

ElBaz, A., Farag, A., Falk, R., LaRocca, R.: Automatic identification of lung abnormalities in chest spiral CT scans. In: International Conference on Acoustics, Speech, and Signal Processing (ICASSP '03), vol. 2, pp. 261–264 (2003)

Eltoukhy, M., Faye, I., Samir, B.: A comparison of wavelet and curvelet for breast cancer diagnosis in digital mammogram. Comput. Biol. Med. **40**, 384–391 (2010)

Ema, T., Doi, K., Jiang, R.N.Y., Papaioannou, J.: Image feature analysis and computer-aided diagnosis in mammography: reduction of false-positive clustered microcalcifications using local edge-gradient analysis. Med. Phys. **22**, 161–169 (1995)

Enderwich, C., Tzanakou, E.: Classification of mammographic tissue using shape and texture features. In: Proceedings of the 19th International Conference-IEEE/EMBS, pp. 810–813 (1997)

Faizal, K., Kavitha, V.: An effective segmentation approach for lung CT images using histogram thresholding with EMD refinement. In: International Conference on Internet Computing and Information Communications, Advances in Intelligent Systems and Computing, vol. 216, pp. 483–489 (2014)

Fan, S., Lin, Y., Wu, C.: Image thresholding using a novel estimation method in generalized gaussian distribution mixture modeling. Neurocomputing **72**, 500–512 (2008)

Fauci, F., Bagnasco, S., Bellotti, R., Cascio, D., Cheran, S., Carlo, F.D., Nunzio, G.D., Fantàcci, M., Forni, G., Lauria, A., Torrez, E.L., Magro, R., Masala, G., Oliva, P., Quarta, M., Raso, G., Retico, A., Tangaro, S.: Mammogram segmentation by contour searching and massive lesion classification with neural network. In: IEEE Nuclear Science Symposium Conference Record, vol. 5, pp. 2695–2699 (2004)

Freixenet, J., Oliver, A., Lado, X., Marti, R., Pont, J., Perez, E., Zwig-gelaar, E.D.R.: Eigendetection of masses considering false positive reduction and breast density information. Med. Phys. **35**, 1840–1853 (2008)

Gabralla, L., Mahersia, H., Zaroug, M.: Denoising CT images using wavelet transform. Int. J. Adv. Comput. Sci. Appl. (IJACSA) **6**, 125–129 (2015)

Gagnon, L., Lina, J., Goulard, B.: Sharpening enhancement of digitized mammograms with complex symmetric daubechies wavelets. In: Annual International Conference of the IEEE Engineering in Medicine and Biology, vol. 17, pp. 543–544 (1995)

Gao, T., Sun, X., Wang, Y., Nie, S.: A pulmonary nodules detection method using 3d template matching. Found. Intell. Syst. **122**, 625–633 (2012)

Geronimo, J., Hardin, D., Massopust, P.: Fractal functions and wavelet expansions based on several scaling functions. J. Approx. Theory **78**, 373–401 (1994)

Gonzalez-Jiménez, D., Pérez-Gonzalez, F., Comesana-Alfaro, P., Alba-Castro, L.P.-F.J.: Modeling Gabor coefficients via generalized Gaussian distributions for face recognition. In: Image Processing-ICIP, vol. 4 (2007)

Guliato, D., Rangayyan, R., Carnielli, W., Zuffo, J., Desautels, J.: Segmentation of breast tumors in mammograms by fuzzy region growing. In: IEEE Conference on Engineering in Medicine and Biology Society, vol. 20, pp. 1002–1005 (1998)

Gulsrud, T., Engan, K., Hanstveit, T.: Watershed segmentation of detected masses in digital mammograms. In: IEEE Conference on Engineering in Medicine and Biology Society, pp. 3304–3307 (2005)

Guo, Y., Zhou, C., Chughtai, H.C.A., Wei, J., Hadjiiski, L., Kazerooni, E.: Automated iterative lung segmentation for image analysis in thoracic computed tomography. Med. Phys. **40**, 081912/1–081912/11 (2013)

Gurcan, M., Yardimci, Y., Cetin, A., Ansari, R.: Automated detection and enhancement of microcalcification on digital mammograms using wavelet transform techniques. Department of Radiology. University of Chicago

Hejazi, M., Ho, Y.: Automated detection of tumors in mammograms using two segments for classification. In: Lecturer Notes in Computer Science, vol. 3767, pp. 910–921 (2005)

Hong, B., Brady, M.: A topographic representation for mammogram segmentation. In: Lecturer Notes in Computer Science, vol. 2879, pp. 730–737 (2003)

Huo, Z., Giger, M., Vyborny, C., Bick, U., Lu, P., Wolverton, D., Schmidt, R.: Analysis of spiculation in the computerized classification of mammographic masses. Med. Phys. **22**, 1569–1579 (1995)

Jaffar, M., Hussain, A., Nazir, F.J.M., Mirza, A.: Ga-svm based lungs nodule detection and classification. Commun. Comput. Inf. Sci. **61**, 133–140 (2009)

Jain, A., Murty, M., Flynn, P.: Data clustering: a review. ACM: Comput. Surv. **31**, 264–323 (1999)

Jasmine, J., Govardhan, A., Baskaran, S.: Classification of microcalcification in mammograms using nonsubsampled contourlet transform and neural network. Euro. J. Sci. Res. **46**, 531–539 (2010). ISSN 1450-216X

Jiang, L., Song, E., Xu, X., Ma, G., Zheng, B.: Automated detection of breast mass spiculation levels and evaluation of scheme performance. Acad. Radiol. **15**, 1534–1544 (2008)

Karssemeijer, N.: Automated classification of parenchymal patterns in mammograms. Phys. Med. Biol. **43**, 365–378 (1998)

Kavitha, K., Kumaravel, N.: A comparative study of various microcalcification cluster detection methods in digitized mammograms. In: Systems, Signals and Image Processing, pp. 405–409 (2007)

Kestener, P.: Analyse multifractale 2d et 3d à l'aide de la transformation en ondelettes : application en mammographie et en turbulence développée. Ph.D. thesis, Ecole doctorale de sciences physiques et de l'ingénieur. University of Bordeaux I (2003)

Kim, J., Park, H.: Statistical textural features for detection of microcalcifications in digitized mammograms. IEEE Trans. Med. Imag. **3**, 231–238 (1999)

Kom, G., Tiedeu, A., Kom, M., Gonsu, C.N.J.: Détection automatique des opacités dans les mammographies par la méthode de minimisation de la somme de l'inertie. Elsevier ITBM-RBM **26**, 347–356 (2005)

Kostas, K., Nandi, A.: Exponent parameter estimation for generalized gaussian probability density functions with application to speech modeling. Signal Process. **85**, 1852–1858 (2005)

Kostis, W., Yankelevitz, D., Reeves, A., Fluture, S., Henschke, C.: Small pulmonary nodules, reproducibility of three dimensional volumetric measurement and estimation of time to followup ct. Radiology **231**, 446–452 (2004)

Kramer, D., Aghdasi, F.: Texture analysis techniques for the classification of microcalcifications in digitized mammograms. In: Fifth IEEE AFRICON Conference Electrotechnical Service for Africa, pp. 395–400 (1999)

Krishnaveni, S., Bhanumathi, R., Pugazharasan, T.: Study of mammogram microcalcification to aid tumour detection using naïve bayes classifier. Int. J. Adv. Res. Electr. Electron. Instrum. Eng. **3** (2014)

Lai, S., Li, X., Bischof, W.: On techniques for detecting circumscribed masses in mammograms. IEEE Trans. Med. Imag. **8**, 377–386 (1989)

Laine, A., Schuler, S., Fan, J., Huda, W.: Mammographic feature enhancement by multiscale analysis. IEEE Trans. Med. Imaging **13**, 725–740 (1994)

Lau, T.K., Bischof, W.F.: Automated detection of breast tumors using the asymmetry approach. Comput. Biomed. Res. **24**, 273–295 (1991)

Lee, C., Kim, J., Park, H.: Computer-aided diagnostic system for breast cancer by detecting microcalcification. SPIE **3335**, 615–626 (1998)

Lee, S., Kouzani, A.Z., hu, E.: Automated detection of lung nodules in computed tomography images: a review. Mach. Vis. Appl. **23**, 151–163 (2012)

Li, H., Liu, K., Lo, S.: Fractal modeling of mammogram and enhancement of microcalcifications. In: IEEE Nuclear Science Symposium and Medical Imaging Conference, vol. 3, pp. 1850–1854 (1996)

Li, H., Liu, K., Lo, S.: Fractal modeling and segmentation for the enhancement of microcalcifications in digital mammograms. IEEE Trans. Med. Imag. **16**, 785–798 (1997)

Li, L., Clark, R.A., Thomas, J.: Computer-aided diagnosis of masses with fullfield digital mammography. Acad. Radiol. **9**, 4–12 (2002)

Mahersia, H., Boulehmi, H., Hamrouni, K., Boussetta, S., Mnif, N.: Development of intelligent systems based on Bayesian regularization network and neuro-fuzzy models for mass detection in mammograms: a comparative analysis. Comput. Methods Programs Biomed. **126**, 46–62 (2015)

Mahersia, H.: Contribution à l'analyse de texture par des méthodes spa-tio-fréquentielles. Ph.D. thesis, National Engineering School of Tunis, Campus Tunis Elmanar (2010)

Mahersia, H., Boulehmi, H., Hamrouni, K.: CAD system for lung nodules detection using wavelet-based approach and intelligent classifiers. In: Proceedings of the International IEEE Multi-Conference on Systems, Signals and Devices-SSD (2020)

Marias, K., Petroudi, S., English, R., Adams, R., Brady, M.: Subjective and computer-based characterisation of mammographic patterns. In: International Workshop on Digital Mammography, pp. 552–557 (2002)

Marti, J., Batle, J., Cufi, X., and Espanol, J. (1999). Microcalcification evaluation in computer assisted diagnosis for digital mammography. In: Colloquium on Digital Mammography, pp. 1–6

Marti, R., Zwiggelaar, R., Rubin, C.: Tracking mammographic structures over time. In: British Machine Vision Conference, page 143–152 (2001)

Martý, J., Freixenet, J., Muñoz, X., Oliver, A.: Active region segmentation of mammographic masses based on texture, contour, and shape features. In: Lecture Notes in Computer Science, vol. 2652, pp. 478–485 (2003)

McSweeney, M., Sprawls, P., Egan, R.: Enhanced image mammography. AJR **140**, 9–14 (1983)

Méndez, A., Tahoces, M.S.P., Vidal, J.: Computer aided diagnosis for breast masses detection on a telemammography system. Comp. Med. Imag. Graph. **27**, 497–502 (2003)

Mistry, D.K., Talati, J.B.: An approach to detect bone tumor using comparative analysis of segmentation technique. Int. J. Innov. Res. Comput. Commun. Eng. **4**, 8176–8184 (2016)

Morrow, W., Paranjape, R., Rangayyan, R., Desautels, J.: Region-based contrast enhancement of mammograms. IEEE Trans. Med. Imag. **11**, 392–406 (1992)

Nakayama, R., Uchiyama, Y., Namba, K.: Cad scheme using filter bank for detection of mcc in mammograms. IEEE Trans. Bio Med. Eng. **53**, 273–283 (2006)

Nanni, L., Brahnam, S., Lumini, A.: Combining different local binary pattern variants to boost performance. Expert Syst. Appl. **38**, 6209–6216 (2011)

Netsch, T., Peitgen, H.: Scale-space signatures for the detection of clustered microcalcifications in digital mammograms. IEEE Trans. Med. Imag. **18**, 774–786 (1999)

Netto, S., Silva, A., Nunes, A., Gattass, M.: Automatic segmentation of lung nodules with growing neural gas and support vector machine. Comput. Biol. Med. **42**, 1110–1121 (2012)

Nisthula, P., Yadhu, R.: A novel method to detect bone cancer using image fusion and edge detection. Int. J. Eng. Comput. Sci. **2**, 2012–2018 (2018)

Novey, M., Adali, T.: A complex generalized gaussian distribution characterization generation and estimation. IEEE Trans. Signal Process. **58** (2013)

Oliver, A., Freixenet, J., Marti, R., Zwiggelaar, R.: A comparison of breast tissue classification techniques. MICCAI **2006**(4191), 872–879 (2006)

Oliver, A., Freixenet, J., Marti, R., Pont, J., Perez, E., Denton, E., Zwiggelaar, R.: A novel breast tissue density classification methodology. IEEE Trans. Inf. Technol. Biomed. **12** (2008)

Orozco, H., Villegas, O., Snchez, V., Domnguez, H., Alfaro, M.: Automated system for lung nodules classification based on wavelet feature descriptor and support vector machine. Biomed. Eng. **14**, 1–20 (2015)

Papadopoulos, A., Fotiadisb, I., Likas, A.: Characterization of clustered microcalcifications in digitized mammograms using neural networks and support vector machines. Artif. Intell. Med. (2004)

Paquerault, S., Petrick, N., Chan, H., Sahiner, B., Helvie, M.: Improvement of computerized mass detection on mammograms: fusion of two-view information. Med. Phys. **29**, 238–247 (2002)

Parveen, S., Kavitha, C.: A review on computer aided detection and diagnosis of lung cancer nodules. Int. J. Comput. Technol. **3**, 393–400 (2012)

Petroudi, S., Kadir, T., Brady, M.: Automatic classification of mammographic parenchymal patterns: a statistical approach. Eng. Med. Biol. Soc. **1**, 798 – 801 (2003). ISBN: 0-7803-7789-3

Ping, Y.Y., Yin, C.W., Kok, L.: Computer aided bone tumor detection and classification using x-ray images. In: International Conference on Biomedical Engineering, pp. 544–547 (2008)

Prasad, R., Saruwatari, H., Shikano, K.: Estimation of shape parameter of GGD function by negentropy matching. Neural Process. Lett. **22**, 377–389 (2005)

Pu, J., Leader, J.K., Zheng, B., Gur, D.: An ellipse-fitting based method for efficient registration of breast masses on two mammographic views. Med. Phys. **35**, 487–494 (2008a)

Pu, J., Yi, C., Roos, J.: Adaptive border marching algorithm: automatic lung segmentation on chest CT images. Comput. Med. Imaging Graph. **32**, 452–462 (2008b)
Reddy, C.K.K., Anisha, P., Prasad, L.N.: A novel approach for detecting the bone cancer and its stage based on mean intensity and tumor size. In: Recent Researches in Applied Computer Science, pp. 162–171 (2016)
Retico, A., Delogu, P., Fantacci, M.: Lung nodule detection in low-dose and thin-slice computed tomography. Comput. Biol. Med. **38**, 525–534 (2008)
Santos, A., Filho, A., Silva, A., dePaiva, A., Nunes, R., Gattass, M.: Automatic detection of small lung nodules in 3d CT data using gaussian mixture models, Tsallis entropy and SVM. Eng. Appl. Artif. Intell. **36**, 27–39 (2014)
Shao, H., Liu, L.C.Y.: A detection approach for solitary pulmonary nodules based on CT images. In: 2nd International Conference on Computer Science and Network Technology, pp. 1253–1257 (2012)
Sharma, D., Jindal, G.: Identifying lung cancer using image processing techniques. In: International Conference on Computational Techniques and Artificial Intelligence (ICCTAI'2011), pp. 115–120 (2011)
Shin, J., Chang, J., Kim, N.: Statistical modeling of speech signals based on generalized gamma distribution. IEEE Signal Process. Lett. **12** (2005)
Sinthia, P., Sujatha, K.: A novel approach to detect bone cancer using k-means clustering algorithm and edge detection method. Asian Research Publishing Network. ARPN J. Eng. Appl. Sci. **11**, 8002–8007 (2016)
Song, K.: A globally convergent and consistent method for estimating the shape parameter of a generalized gaussian distribution. IEEE Trans. Inf. Theory **52**, 510–527 (2006)
Taher, F., Sammouda, R.: Identification of lung cancer based on shape and color. In: 4th International Conference, ICISP'2010 (2010)
Talakoub, O., Alirezaie, J., Babyn, P.: Lung segmentation in pulmonary CT images using wavelet transform. In: IEEE International Conference on Acoustics, Speech and Signal Processing ICASSP, vol. 2007, no. 1, pp. 453–456 (2007)
Tourassi, N.E.G., Fadeev, A., Elmaghraby, A.: Significance of mpeg-7 textural features for improved mass detection in mammography. In: IEEE EMBS Annual International Conference, pp. 4779–4782 (2006)
Varanasi, M., Aazhang, B.: Parametric generalized gaussian density estimation. J. Acoust. Soc. Am. **86**, 1404–1415 (1989)
Vijaya, G., Suhasini, A.: An adaptive pre-processing of lung CT images with various filters for better enhancement. Acad. J. Can. Res. **7**, 179–184 (2014)
Vo, K., Sowmya, A.: Diffuse lung disease classification in HRCT lung images using generalized gaussian density modeling of wavelets coefficients. In: 16th IEEE International Conference on Image Processing (ICIP), pp. 2645–2648 (2009)
Vo, K, Sowmya, A.: Multiple kernel learning for classification of diffuse lung disease using HRCT lung images. In: 32nd Annual International Conference of the IEEE EMBS, pp. 3085–3088 (2010)
Wei, L., Yangi, Y., Nishikawa, R., Jiang, Y.: A study on several machine-learning methods for classification of malignant and benign clustered microcalcifications. IEEE Trans. Med. Imaging **24**(3), 371–380 (2005)
Woods, K., Bowyer, K.: Computer detection of stellate lesions. In: International Workshop on Digital Mammography, pp. 221–229 (1994)
Yankelevitz, D., Reeves, A., Kostis, W., Zhao, B., Henschke, C.: Small pulmonary nodules: volumetrically determined growth rates based on CT evaluation. Radiology **217**, 251–256 (2000)
Yao, J., Burns, J., Summers, R.: Computer aided detection of bone metastases in the thoracolumbar spine. In: Lecture Notes in Computational Vision and Biomechanics, vol. 18, pp. 97–130 (2014)
Yu, S., Guan, L.: A cad system for the automatic detection of clustered microcalcifications in digitized mammogram films. IEEE Trans. Med. Imag. **2**, 115–126 (2000)

Yu, S., Zhang, A., Li, H.: A review of estimating the shape parameter of generalized gaussian distribution. J. Comput. Inf. Syst. **8**, 9055–9064 (2012)

Zhang, F., Cai, W., Song, Y., Lee, M., Shan, S., Feng, D.: Overlapping node discovery for improving classification of lung nodules. In: 35th Annual International Conference of the IEEE EMBS, pp. 3–7 (2013)

Chapter 15
Pectoral Muscle Removal Techniques from Digital Mammograms: A Survey

Hela Boulehmi, Hela Mahersia, and Kamel Hamrouni

Abstract Breast cancer remains one of the leading causes of cancer-related deaths among women worldwide. Early detection of breast lesions signifi-cantly improves the chances of successful treatment and survival. However, identifying small or subtle abnormalities can be particularly challenging for radiologists, especially in women with dense breast tissue. This highlights the urgent need for computer-aided diagnosis (CAD) systems to support radiologists in detecting breast lesions at an earlier, more treatable stage. Many challenges are encountered when using CAD systems on digital mammograms. The first obstacle is the existence of elements that have a density similar to that of breast lesions such as artifacts and pectoral muscle. A pre-processing step is therefore essential in order to remove these components and thus optimize the results of breast anomalies segmentation. This chapter provides an overview of various image processing methods developed since 2000 for the detection and removal of the pectoral muscle in digital mammograms, specifically using the mini-MIAS database.

Keywords Breast cancer early detection · CAD systems · Pectoral muscle removal · Digital mammograms · MIAS database

15.1 Introduction

According to global statistics released by the World Health Organization (WHO) in March 2021, approximately 2.3 million women worldwide were diagnosed with breast cancer in 2020. During that same year, the disease claimed the lives of around 685,000 individuals. Since no direct cause has been confirmed for this disease, the only way to increase the chance of survival for women with breast cancer is early

H. Boulehmi (✉) · H. Mahersia · K. Hamrouni
SITI Laboratory-ENIT, University of Tunis El Manar, Tunis, Tunisia
e-mail: hela.boulehmi@gmail.com

K. Hamrouni
e-mail: kamel.hamrouni@enit.rnu.tn

N. Derbel and Q. Zhu (eds.), *Advances in Communication, Signal and Image Processing*,
https://doi.org/10.1007/978-981-95-0401-5_15

detection. As post-menopausal women and those with a family history are more likely to develop breast cancer, a mammogram is imposed on them once a year. A Computer-Aided Diagnosis (CAD) system supports radiologists in analyzing mammograms, aiming to detect abnormalities at an early stage.

One of the key challenges in segmenting breast anomalies on medio-lateral oblique (MLO) mammograms is the presence of the pectoral muscle. This muscle often appears as a bright region on the image, closely resembling masses or clusters of microcalcifications, making accurate segmentation more difficult.

As a result, removing the pectoral muscle region becomes a crucial part of the preprocessing phase. This step enhances the segmentation of breast tissue abnormalities and contributes to improving the overall accuracy and effectiveness of the CAD system.

The second section of this chapter reviews a range of image processing techniques developed since 2000 for detecting and removing the pectoral muscle in mammographic images. These methods have been evaluated using mammograms from the MIAS database.

Illustrative examples of the results produced by the discussed techniques are presented in the third section. Finally, the chapter concludes with a comparative analysis of the outcomes achieved by the various approaches.

15.2 Related Works

As shown in Fig. 15.1, the pectoral muscles are anatomical structures located in the front of the chest, linking the upper arm and shoulder bones to the thoracic region. In digital MLO mammograms, the pectoral muscle typically appears in one of the upper corners of the image, as illustrated in Fig. 15.2.

While the problem of pectoral muscle removal in digital mammograms has been addressed for many years, the scope of this chapter is confined to research efforts conducted since the year 2000, presented in chronological order. Early approaches for detecting the pectoral muscle were based on simple, intuitive methods that aimed to identify a straight boundary line separating the muscle from the breast region. These methods relied on the assumption that the pectoral muscle appears as a roughly triangular region in mammographic images-a hypothesis that does not consistently hold true, as shown in Fig. 15.3.

Building on the work of Karssemeijer (1998), Ferrari and his colleagues (2000) proposed a segmentation method that utilizes the Hough transform to detect the pectoral muscle. While this approach proved effective in many cases, it encountered limitations when dealing with muscles exhibiting curved boundaries.

In 2001, Kwok et al. introduced an enhanced algorithm that combined the Hough transform with an iterative thresholding technique to estimate the straight boundary of the pectoral muscle region (Kwok et al. 2001). This initial step was followed by gradient analysis to identify the muscle's contour. Finally, a cliff detection algorithm was applied to achieve a smoother and more accurate line fitting.

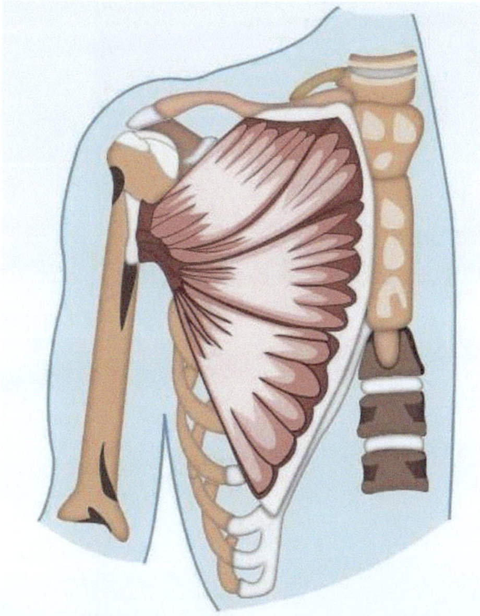

Fig. 15.1 Abdominal head of pectoralis major muscle

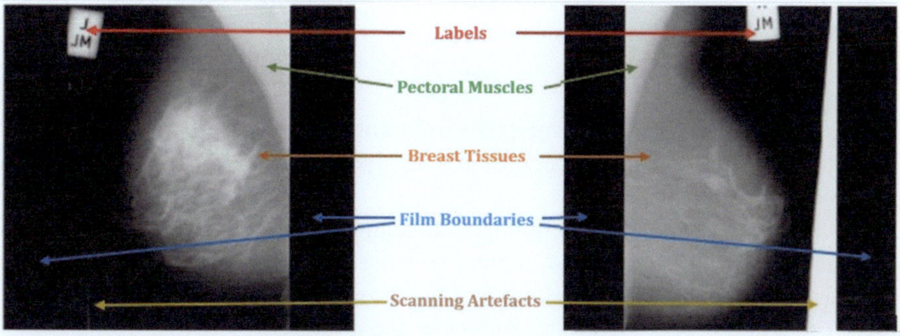

Fig. 15.2 Mammogram components

In 2004, Ferrari et al. introduced a method utilizing Gabor wavelets to detect curved pectoral muscle contours (Ferrari et al. 2004). This approach addressed the limitations of their earlier work, which relied on straight-line detection using the Hough transform. The proposed technique begins by applying specially designed Gabor filters to the region of interest containing the pectoral muscle. A vector summation combined with a magnitude propagation process generates an image that highlights potential edges. The final contour of the pectoral muscle is then extracted through a dedicated post-processing step.

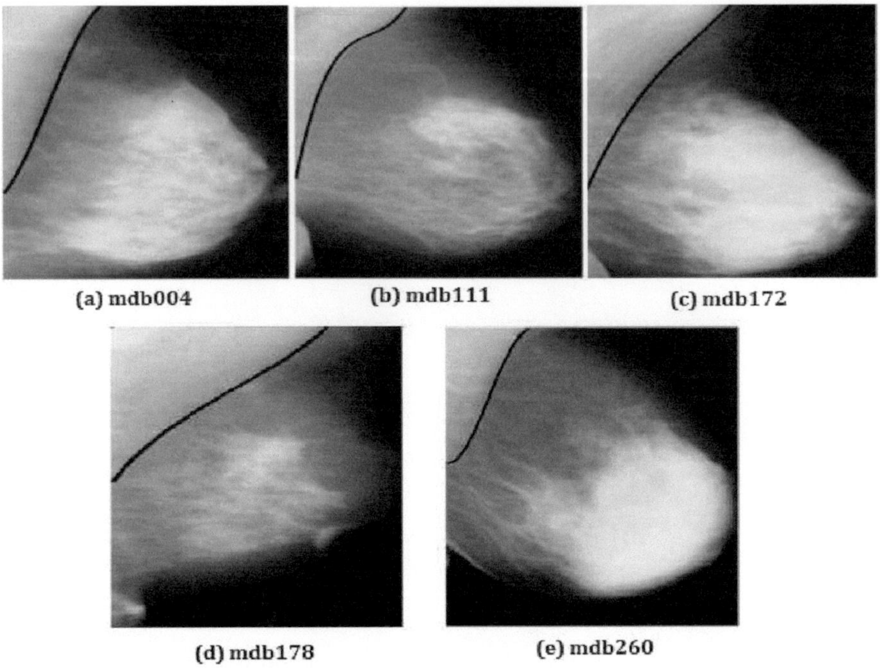

Fig. 15.3 Examples of nonlinear pectoral muscle boundaries (Vikhe and Thool 2017)

Also in 2004, Kwok et al. developed a different strategy based on edge detection and surface smoothing, aimed at approximating the muscle boundary using a straight line (Kwok et al. 2004).

In 2005, (Raba et al. 2005) employed an adaptive histogram equalization technique combined with a selective region growing algorithm to detect and remove pectoral muscles from digital mammograms. In the same year, (Bajger et al. 2005) introduced a graph-based approach known as the minimum spanning tree (MST) to segment the pectoral muscle. Additionally, Ma et al. (2005) applied the MST method in conjunction with the Adaptive Pyramid (AP) algorithm to segment the pectoral muscle in screening mammograms, with further developments in 2007 (Ma et al. 2007).

In 2008, Kinoshita et al. proposed a technique based on the Radon transform, which identifies the longest straight line in the mammogram to detect the boundaries of the pectoral muscle (Kinoshita et al. 2008).

In 2009, Camilus et al. introduced a novel approach for pectoral muscle removal, combining a graph cut and merging method with a Bezier curve algorithm (Camilus et al. 2009).

Several techniques were explored in 2010 for detecting and removing the pectoral muscle. For instance, (Subashini et al. 2010) employed connected component labeling (CCL) for the task. In a similar year, (Domingues et al. 2010) proposed a shortest-path-based approach to construct the pectoral muscle contour, first determining the end points and then applying a supervised learning regression algorithm. Additionally, Chen and Zwiggelaar (2010) utilized a region growing algorithm to remove the pectoral muscle from digital mammograms.

In Camilus et al. (2011), the authors developed a pectoral muscle removal method based on the watershed transform. The watershed transformation of the mammogram reveals a distinct watershed line that corresponds to the pectoral muscle boundary. A merging algorithm is then applied to combine the relevant catchment basins, effectively segmenting the pectoral muscle region.

In 2012, Mario Mustra et al. combined contrast enhancement techniques with adaptive histogram equalization to improve the contrast of mammograms. They then used a polynomial curvature estimation method to identify the region of interest. Following this, they applied the Hough transform, followed by k-means thresholding, to remove any artifacts (Mustra and Grgic 2012). In 2013, Chandra Boss et al. utilized a connected component labeling technique with an 8-neighborhood histogram-based approach to remove the pectoral muscle (Boss et al. 2013). The same year, Yanfeng Li et al. leveraged anatomical features, homogeneous texture, and high-intensity deviations to segment the pectoral region, applying a Kalman filter to refine the detected muscle edges (Li et al. 2013).

In 2014, Alhsnony et al. introduced a novel method based on bit depth and edge processing algorithms to segment and remove the pectoral muscle from digital mammograms (Alhsnony et al. 2014). That same year, Boulehmi et al. proposed a preprocessing technique for mammograms, specifically targeting the removal of film boundaries, artifacts, and the pectoral muscle from images in the MIAS database (Boulehmi et al. 2014). After eliminating the film boundaries and artifacts, a contrast enhancement technique was applied, which involved extracting the low-frequency component $L(x, y)$ from the processed image $I(x, y)$, followed by denoising using a Gaussian low-pass filter. Subsequently, both white and black top-hat transformations (denoted ToHb and ToHn) were applied to $L(x, y)$ (Boulehmi et al. 2015, 2016a, b; Boulehmi 2016; Mahersia et al. 2016). A soft denoising algorithm based on 2-level Daubechies wavelet decomposition was then applied, followed by dynamic thresholding of the detail coefficients. Finally, wavelet reconstruction was performed to enhance the mammogram contrast, with the pectoral muscle removed using an adaptive thresholding algorithm.

In Alam and Islam (2014), the authors removed the pectoral muscle using a K-means clustering approach to eliminate the triangular region. They also employed morphological operators and a seeded region growing (SRG) technique for noise reduction and artifact removal.

Djaroudib et al. used textural descriptors in 2015 to detect pectoral muscle from a limited number of mammograms (Djaroudib et al. 2015) and Sreedevi et al. exploited global thresholding to identify pectoral muscle (Sreedevi and Sherly 2015).

In 2016, Vikhe et al. introduced an intensity-based algorithm for detecting the pectoral region contour. A filtering technique was then applied to enhance the boundary points, which were identified using a thresholding method. These points were subsequently connected to define the contour between the pectoral muscle and the breast tissue region (Vikhe and Thool 2016). The same year, Yoon et al. developed a pectoral muscle removal technique that combined morphological operators with the random sample consensus (RANSAC) algorithm (Yoon et al. 2016).

In 2017, Andrik Rampun et al. successfully segmented the pectoral region using a thresholding algorithm combined with an active contour model (Rampun et al. 2017), while Saied Asgari et al. merged geometric rules with a region-growing algorithm to detect the pectoral muscle (Aghanaki et al. 2017).

A genetic algorithm based on wavelet transform was used by Shen et al. (2018) to determine multilevel thresholds for pectoral muscle segmentation. Morphological operators were then applied to select the optimal contour. In 2019, (Ma et al. 2019) introduced a deep neural network for pectoral muscle segmentation, comparing its results to those obtained from conventional methods.

In 2020, Suganthi et al. used an active contour algorithm in conjunction with an intensity-based thresholding technique to detect the boundary between the pectoral muscle and breast tissue (Suganthi et al. 2020).

In Saeed and Saleh (2020), the authors presented the HBBRG algorithm, combining the BB and RG algorithms to refine pectoral muscle segmentation. That same year, Pascal Vagssa et al. applied global thresholding and the Hough transform to detect the pectoral muscle (Vagssa et al. 2020).

Finally, in 2021, Rahimeto et al. used a global thresholding algorithm combined with morphological operations to successfully detect the pectoral muscle (Rahimeto et al. 2021).

Recently, in 2022, (Yu et al. 2022) introduced a new deep learning framework called PeMNet for segmenting the breast pectoral muscle in digital mammograms. The PeMNet approach incorporated a novel attention module, the Global Channel Attention Module (GCAM), which contributed to improved segmentation performance.

15.3 Experimentations and Results

All the previously discussed methods were applied to mammograms from the MIAS database. The MIAS database is one of the most widely used resources by researchers worldwide, as it provides ground truth annotations for each image and is easily accessible for download. It contains a total of 322 medio-lateral oblique (MLO) mammograms, with left MLO images indexed by even numbers and right MLO images by odd numbers, as shown in Fig. 15.4.

These 322 mammograms are labeled by expert radiologists and encompass a range of diagnostic cases: 208 images of normal breast tissue, 56 images with masses, 25 images with microcalcifications, 18 images with architectural distortions, and 15 images with imbalances.

Each of the techniques described earlier was evaluated using a specific number of mammograms from the MIAS database.

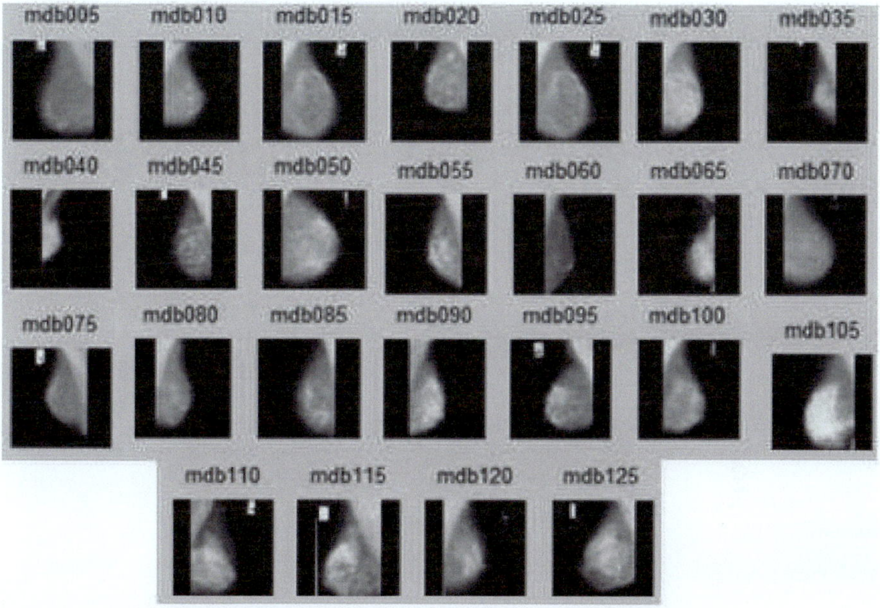

Fig. 15.4 Amples of mammographic images sourced from MIAS database (Boulehmi et al. 2016b)

Subashini et al. tested their proposed connected component labeling (CCL) algorithm on MIAS database mammograms (Subashini et al. 2010). Figure 15.5 shows two examples of results obtained by Subashini.

Figure 15.6 displays results from (Alam and Islam 2014).

Figure 15.7 presents results of pectoral muscle removal using global thresholding, as proposed by Sreedevi and Sherly (2015).

Figure 15.8 provides examples of results achieved by Suganthi et al., who used an active contour and intensity-based thresholding approach to identify the boundary between the pectoral muscle region and the rest of the breast tissue (Suganthi et al. 2020).

In Vagssa et al. (2020), the authors employed global thresholding and the Hough transform to remove the pectoral muscle. Figure 15.9 illustrates some results obtained using this technique.

Figure 15.10 shows results obtained by Boulehmi et al. (2014, 2015, 2016a, b) and Boulehmi (2016).

Figure 15.10 displays results obtained by Boulehmi et al. (2014, 2015, 2016a, b) and Boulehmi (2016).

Finally, Fig. 15.11 showcases the results reported by Yu et al. (2022).

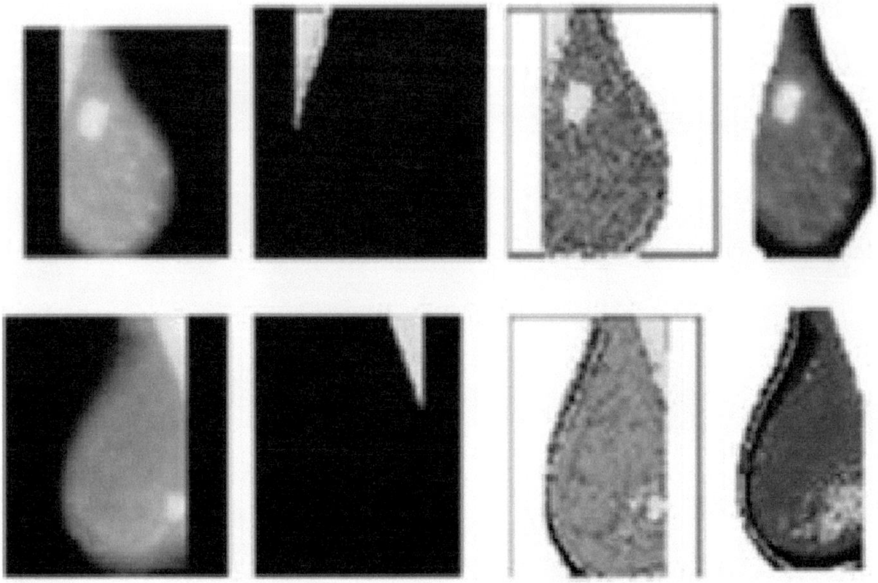

Fig. 15.5 Illustrative outcomes from the method proposed by Subashini et al. (2010)

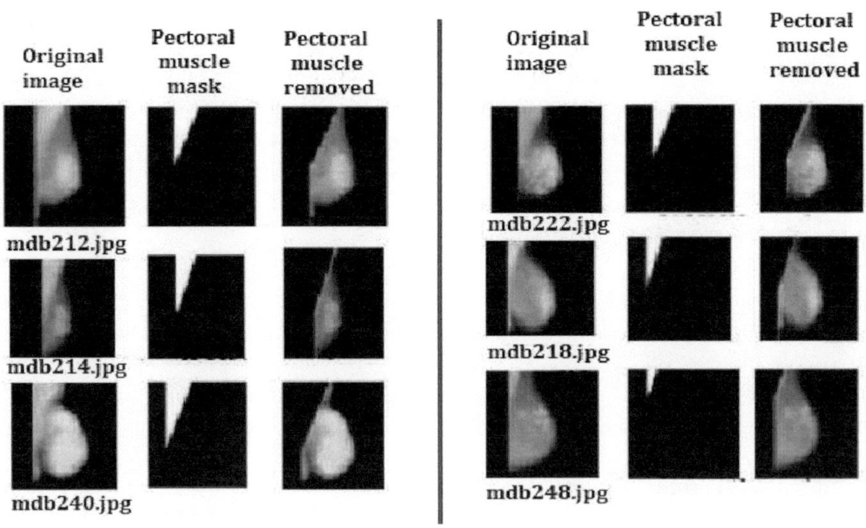

Fig. 15.6 Illustrative outcomes from the method proposed by Alam and Islam (2014)

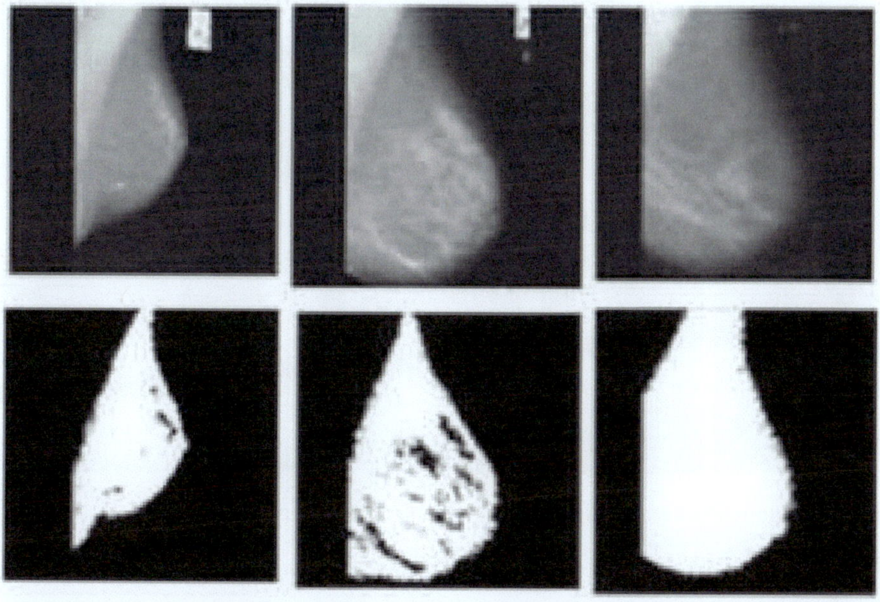

Fig. 15.7 Illustrative outcomes from the method proposed by Sreedevi and Sherly (2015)

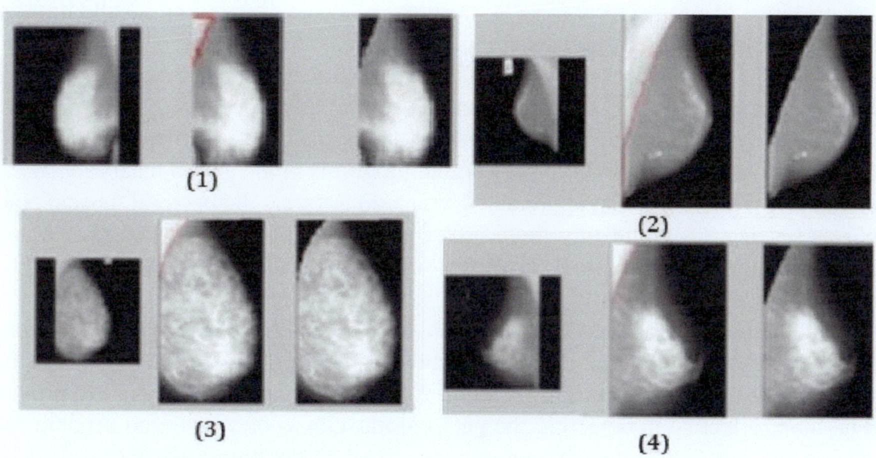

Fig. 15.8 Illustrative outcomes from the method proposed by Suganthi et al. (2020)

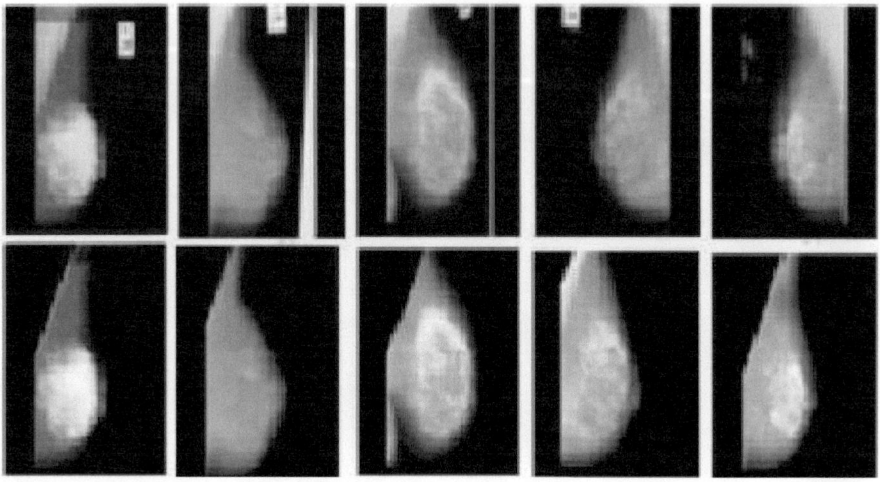

Fig. 15.9 Illustrative outcomes from the method proposed by Vagssa et al. (2020)

15.4 Discussion

The presence of the pectoral muscle in digital mammograms often results in false positive detections when attempting to identify breast lesions.

Removing the pectoral muscle during the preprocessing stage is a critical and delicate task, as its density closely resembles that of breast lesions.

Table 15.1 provides a comparison of several previously described techniques used for pectoral muscle segmentation from mammograms in the MIAS database.

Table 15.1 indicates that approaches which do not treat the pectoral muscle contour as linear are significantly more effective. This is because the boundaries and appearance of the pectoral muscle vary from one mammogram to another, as illustrated in Fig. 15.10.

Some of the pectoral muscle region could remain in the resulting image when considering a linear pectoral muscle contour while segmenting it, which would reduce the performance of breast lesions segmentation results. Adaptive methods have proven to be more effective in detecting the pectoral muscle, as they take into account the specific characteristics of each mammogram, particularly breast density. Table 15.1 demonstrates that with adaptive techniques, the detection rate can exceed 97%, which is a promising result.

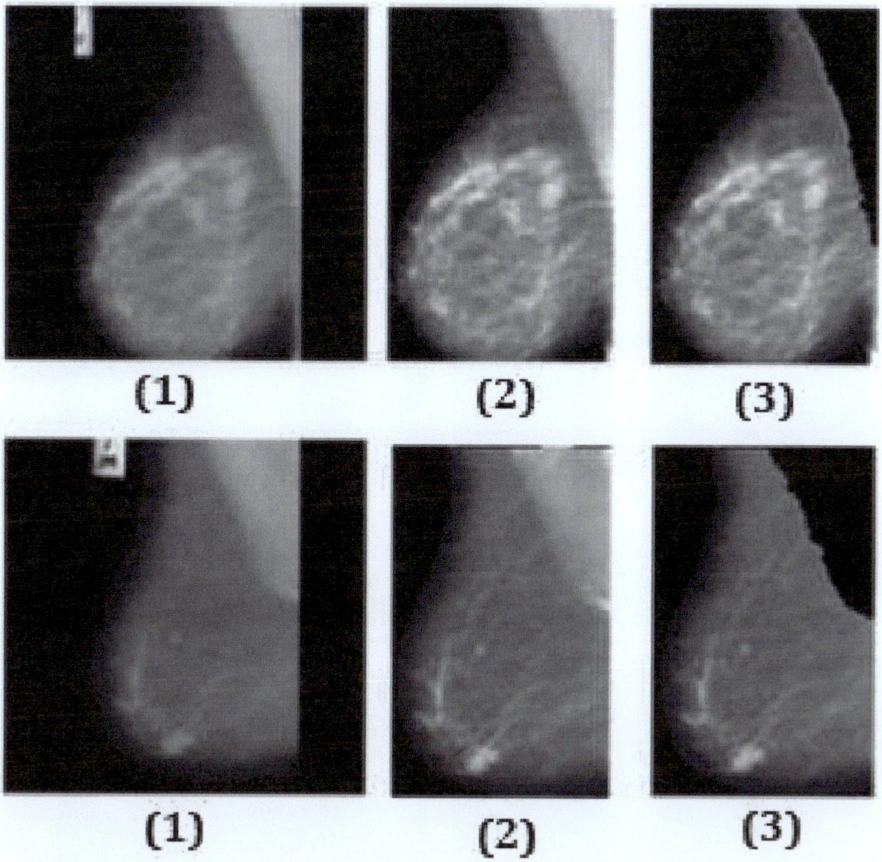

Fig. 15.10 Pectoral muscle removal outcomes from (Boulehmi et al. 2014): **a** Original images, **b** preprocessed and enhanced images, and **c** Post-removal images

15.5 Conclusion

This chapter provides an overview of techniques for pectoral muscle removal from digital mammograms, emphasizing its importance as a critical preprocessing step in CAD systems.

All the methods discussed were evaluated on mammograms from the MIAS database and achieved excellent pectoral muscle segmentation results.

While techniques that treat the pectoral muscle boundary as linear showed promising results, their efficiency decreases with higher breast density.

Adaptive approaches, on the other hand, demonstrated greater effectiveness in pectoral muscle segmentation and removal from digital mammograms.

However, further work is needed to minimize false positives caused by poor pectoral muscle removal and to reduce unnecessary biopsies.

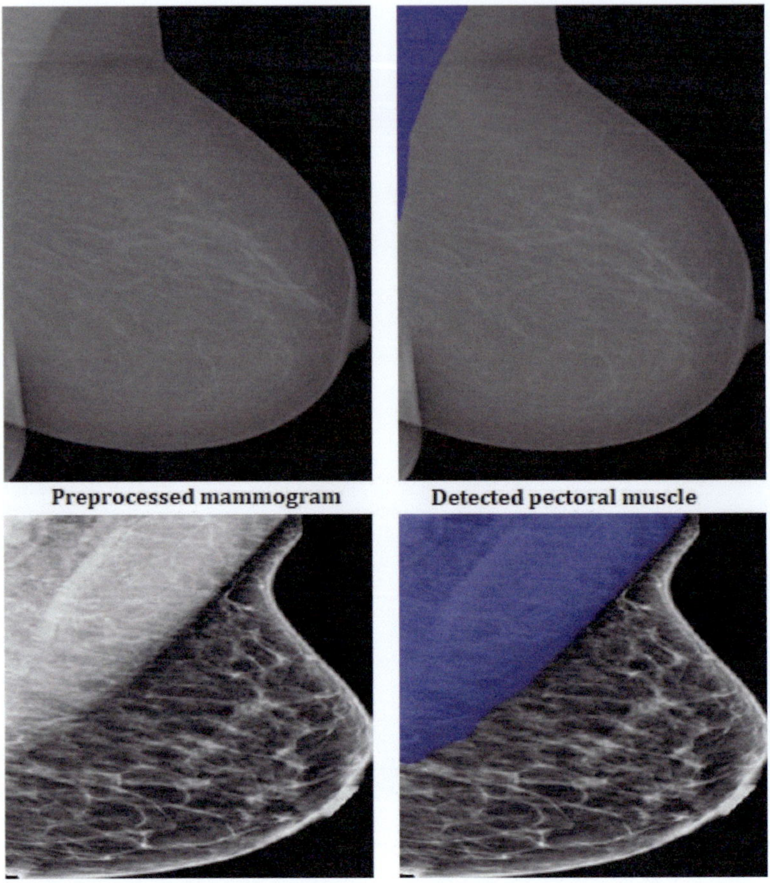

Fig. 15.11 Pectoral muscle removal results from (Yu et al. 2022)

Table 15.1 Results comparison

Related works	Database size	Detection rate (%)
Mustra et al. (2009)	40 images	85
Boulehmi et al. (2014)	282 images	97.58
Yoon et al. (2016)	322 images	92.2
Vagssa et al. (2020)	322 images	93.8
Kinoshita et al. (2008)	322 images	94
Chen et al. (2015)	322 images	97.2

References

Aghanaki, S., B.M., Y. L., Hamarneh, G.: Geometry based pectoral muscle segmentation from MLO mammogram views. IEEE Trans. Biomed. Eng. **64**(11), 2662–2671 (2017)

Alam, N., Islam, M.J.: Pectoral muscle elimination on mammogram using k-means clustering approach. Int. J. Comput. Vis. Signal Process. **4**(1), 11–21 (2014)

Alhsnony, F., Abdolrasol, M., Abadelrsool, S.: Auto-identification of pectoral muscle region in digital mammogram images. Int. J. e-Educ. e-Bus. e-Manag. eLearn. **4**, 345–367 (2014)

Bajger, M., Ma, F., Bottema, M.: Minimum spanning trees and active contours for identification of the pectoral muscle in screening mammograms. In: Proceedings in Digital Image Computing, Techniques and Applications (2005)

Boss, R.C., Thangavel, K., Daniel, D.A.P.: Automatic mammogram image breast region extraction and removal of pectoral muscle. Int. J. Sci. Eng. Res. **4**(5), 229–235 (2013)

Boulehmi, H., Mahersia, H., Hamrouni, K.: Rehaussement d'images mammographiques pour la segmentation des masses. In: Proceedings of the Conférence Internationale en Traitement et Analyse de l'Information, Methodes et Application-TAIMA (2015)

Boulehmi, H.: Analyse d'images mammographiques par analyse de texture: application à la détection précoce du cancer. Ph.D. thesis, National Engineering School of Tunis, Campus Tunis Elmanar (2016)

Boulehmi, H., Mahersia, H., Hamrouni, K.: Breast masses diagnosis using supervised approaches. In: Proceedings of The Second IEEE International Image Processing, Applications and Systems Conference-IPAS (2016a)

Boulehmi, H., Mahersia, H., Hamrouni, K.: A new cad system for breast microcalcifications diagnosis. Int. J. Adv. Comput. Sci. Appl. IJACSA **7** (2016b)

Boulehmi, H., Mahersia, H., Hamrouni, K., Boussetta, S., Mnif, N.: Unsupervised masses segmentation technique using generalized gaussian density. Int. J. Image Process. Graph. (IJIPG) **1**, 15–24 (2014)

Camilus, K., Govindan, V., Sathidevi, P.: Computer-aided identification of the pectoral muscle in digitized mammograms. J. Digit. Imaging **23**(5), 562–580 (2009)

Camilus, K., Govindan, V., Sathidevi, P.: Pectoral muscle identification in mammograms. J. Appl. Clin. Med. Phys. **12**(3), 123–134 (2011)

Chen, Z., Zwiggelaar, R.: Segmentation of the breast region with pectoral muscle removal in mammograms. Int. J. Comput. Appl. **1**(6), 1–5 (2010)

Chen, C., Liu, G., Wang, J., Sudlow, G.: Shape-based automatic detection of pectoral muscle boundary in mammograms. J. Med. Biol. Eng. **35**(3), 315–322 (2015)

Djaroudib, K., Lorenz, P., Ahmed, A., Zidani, A.: Pectoral and breast segmentation technique based on texture information, pp. 219–228. Springer International Publishing (2015)

Domingues, I., Cardoso, J.S., Amaral, I., Moreira, I., Passarinho, P., Comba, J.S., Correia, R., Cardoso, M.: Pectoral muscle detection in mammograms based on the shortest path with endpoints learnt by SVMs (2010)

Ferrari, R., Rangayyan, R., Desautels, J., Borges, R., Frère, A.: Segmentation of mammograms: identification of the skin-air boundary, pectoral muscle, and fibroglandular disc. In: Proceedings of the 5th International Workshop on Digital Mammography (2000)

Ferrari, R., Rangayyan, R., Desautels, J., Borges, R., Frère, A.: Automatic identification of the pectoral muscle in mammograms. In: Proceedings of the IEEE Transactions on Medical Imaging (2004)

Karssemeijer, N.: Automated classification of parenchymal patterns in mammograms. Phys. Med. Biol. **43** (1998)

Kinoshita, S., Azevedo-Marques, P., Pereira, R., Rodrigues, J., Rangayyan, R.: Radon-domain detection of the nipple and the pectoral muscle in mammograms. J. Digit. Imaging **21**(1), 37–49 (2008)

Kwok, S., Chandrasekhar, R., Attikiouzel, Y.: Automatic pectoral muscle segmentation on mammograms by straight line estimation and cliff detection. In: Proceedings of the Seventh Australian and New Zealand Intelligent Information Systems (2001)

Kwok, S., Chandrasekhar, R., Attikiouzel, Y.: Automatic pectoral muscle segmentation on mammograms by straight line estimation and cliff detection. In: Proceedings of the IEEE Transactions on Medical Imaging (2004)

Li, Y., Chen, H., Yang, Y., Yang, N.: Pectoral muscle segmentation in mammograms based on homogenous texture and intensity deviation. Pattern Recogn. **46**(3), 681–691 (2013)

Ma, F., Bajger, M., Bottema, M.: Extracting the pectoral muscle in screening mammograms using a graph pyramid. In: Proceedings in APRS Workshop on Digital Image Computing. University of Queensland, Brisbane, Australia (2005)

Ma, F., Slavotinek, M.B.J., Bottema, M.: Two graph theory based methods for identifying the pectoral muscle in mammograms. Pattern Recogn. **40**(9), 2592–2602 (2007)

Ma, X., Wei, J., Zhou, C., Helvie, M.A., Chan, H., Hadjiiski, L., Lu, Y.: Automated pectoral muscle identification on MLO-view mammograms: comparison of deep neural network to conventional computer vision. Med. Phys. **46**(5), 2103–2114 (2019)

Mahersia, H., Boulehmi, H., Hamrouni, K.: Development of intelligent systems based on Bayesian regularization network and neuro-fuzzy models for mass detection in mammograms: a comparative analysis. Comput. Methods Progr. Biomed. **126**, 46–62 (2016)

Mustra, M., Bozek, J., Grgic, M.: Breast border extraction and pectoral muscle detection using wavelet decomposition. In: Proceedings of the IEEE Eurocon (2009)

Mustra, M., Grgic, M.: Robust automatic breast and pectoral muscle segmentation from scanned mammograms. Signal Process. (Elsevier) **93**(10), 2817–2827 (2012)

Raba, D., Oliver, A., Mart, J., Peracaula, M., Espunya, J.: Breast segmentation with pectoral muscle suppression on digital mammograms. In: Proceedings of the Springer-Verlag (2005)

Rahimeto, S., Debelee, T., Yohannes, D.: Automatic pectoral muscle removal in mammograms. Evol. Syst. **12**, 519–526 (2021)

Rampun, A., Morrow, P., Scotney, B.W., Winder, J.: Fully automated breast boundary and pectoral muscle segmentation in mammograms. Artif. Intell. Med. **79**, 28–41 (2017)

Saeed, E. Saleh, H.: Pectoral muscles removal in mammogram image by hybrid bounding box and region growing algorithm. In: International Conference on Computer Science and Software Engineering (CSASE) (2020)

Shen, R., Yan, K., Xiao, F., Chang, J., Jiang, C., Zhou, K.: Automatic pectoral muscle region segmentation in mammograms using genetic algorithm and morphological selection. J. Digit. Imaging **31**(5), 680–691 (2018)

Sreedevi, S., Sherly, E.: A novel approach for removal of pectoral muscles in digital mammogram. In: International Conference on Information and Communication Technologies (ICICT) (2015)

Subashini, T., Ramalingam, V., Palanivel, S.: Pectoral muscle removal and detection of masses in digital mammogram using CCl. Int. J. Comput. Appl. **1**(6) (2010)

Suganthi, G., Sutha, J., Parvathy, M., Devi, C.D.: Pectoral muscle segmentation in mammograms. Biomed. Pharmacol. J. **13**(3), 1357–1365 (2020)

Vagssa, P., Doudou, N., Jolivo, T., Videme, O., Kolyang, D.: Pectoral muscle deletion on a mammogram to aid in the early diagnosis of breast cancer. Int. J. Eng. Sci. Technol. **12**(3), 57–65 (2020)

Vikhe, P., Thool, V.: Intensity based automatic boundary identification of pectoral muscle in mammograms. In: Proceedings of 7th International Conference on Communication, Computing and Virtualization (2016)

Vikhe, P.S., Thool, V.R.: Detection and segmentation of pectoral muscle on MLO-view mammogram using enhancement filter. J. Med. Syst. **41** (2017)

Yoon, W., Oh, J., Chae, E., Kim, H., Lee, S., Kim, K.: Automatic detection of pectoral muscle region for computer-aided diagnosis using MIAS mammograms. In: Hindawi Publishing Corporation BioMed Research International (2016)
Yu, X., Wang, S., Górriz, J., Jiang, X., Guttery, D., Zhang, Y.: Pemnet for pectoral muscle segmentation. Biology **2022**, 11 (2022)

MIX
Papier aus verantwortungsvollen Quellen
Paper from responsible sources
FSC® C105338

If you have any concerns about our products,
you can contact us on
ProductSafety@springernature.com

In case Publisher is established outside the EU,
the EU authorized representative is:
Springer Nature Customer Service Center GmbH
Europaplatz 3, 69115 Heidelberg, Germany

Printed by Libri Plureos GmbH
in Hamburg, Germany